The world's largest collection of visual travel guides

East African Wildlife

Edited and Produced by Geoffrey Eu
Editor in Nairobi: Deborah Appleton
Principal photography by Karl Ammann

Editorial Director: Brian Bell

APA PUBLICATIONS
Part of the Langenscheidt Publishing Group

Involvement with ecological issues has been a characteristic of Insight Guides ever since **Hans Höfer** founded Apa Publications in 1970. Following the success of country and city guides, which set out to give insights into peoples and cultures all over the world, he instituted the Discover Nature series, which seeks to give equally perceptive and entertaining insights into the private lives of animals and plants. Apart from the present book, the series includes *Asian Wildlife*, *Indian Wildlife*, *Amazon Wildlife*, *The Great Barrier Reef* and two books on *US National Parks*.

Eu and friend

Ammann

Project editor **Geoffrey Eu** arrived in Nairobi armed with a book outline, names of people to contact and, as a concession to technology, a fax machine. Although some unsympathetic customs officials relieved him of the latter item, Eu was eventually rescued by hotelier-turned-photographer **Karl Ammann**, a long-time Kenya resident and principal photographer for this guide. Together, they set their sights on East Africa's pool of talented writers and wildlife experts.

Insight Guide: East African Wildlife is aimed at the safari-seeking tourist, but it also discusses some of the major environmental and conservation issues and the search for a Project Editor who would help turn a basic outline into a fully-fledged Insight Guide was a tough one. It ended at the door of **Deborah Appleton**. Since her arrival in Nairobi several years earlier, she had written numerous travel articles on Kenya. She was a consultant for the United Nations Environment Programme (UNEP) in Nairobi. She liaised with and briefed all the writers involved with this book.

Appleton

Harvey Croze

Dr Harvey Croze acted as advisor on this project. His animal conservation and environmental expertise and his extensive knowledge of Kenyan game parks added immeasurably to *East African Wildlife*. The Kenya-based Croze had been involved in many wildlife projects, and was acting as coordinator for the Global Resource Information Database at UNEP when commissioned.

Stiles

Another UN staffer, **Dr Daniel Stiles**, wrote the cultural and anthropological chapters for this guide. Stiles previously contributed several chapters to *Insight Guide: Kenya*.

Peter Davey, an avid ornithologist from the illustrious Ker and Downey Safaris, wrote the section on birdlife and also supplied numerous photos. **James Ashe**, a Mombasa-based expert on all that slithers, contributed the informative piece on reptiles. **Sir Michael Blundell**, who wrote the piece on flora, is a renowned expert on the subject and the author of an extensive work on East African flowers. **Mary Anne Fitzgerald**, a journalist and Apa contributor with a passion for the safari, is a former Nairobi correspondent for London's *Financial Times*. Having spent a good deal of time out in the bush, she wrote the chapters on early hunting safaris, customised safaris and camel safaris.

Davey

Fitzgerald

Other important contributors to the safari section include angling expert **Peter Usher**, who gets us hooked on freshwater fishing, and **Dudley Chignall**, who extols the pleasures of hot-air ballooning. Usher is an environmental meteorologist and long-time member of the Kenya Fly-Fishers'

Club, while Chignall is a professional balloonist who first lifted off from Kenyan game parks in 1976. **Leslie Duckworth** wrote the short feature on Rusinga Island and **Jackie MaConnell** penned the Tana Delta piece.

One of Kenya's finest climbers, **Iain Allan** from Tropical Ice Safaris in Nairobi, wrote about East Africa's amazing snow-capped peaks. Allan also provided some sweeping shots from the summits.

Many years of living in Kenya laid the groundwork for **Cristina Boelcke** and **Anselm Croze** to write a book about the country's game parks. Boelcke was working as an agronomist at UNEP while Croze is a Nairobi-based artist/writer. **Jeanette Hanby** and **David Bygott** have written a great deal on Tanzania and were able to impart valuable information on both the popular game parks and some of the lesser-known regions. **Fred de Vries** journeyed through much of Uganda to give us the most up-to-date information on the country's game parks.

Dr J. Chris Hillman, an advisor to the Ethiopia Wildlife Conservation Organisation, and a long-term resident of Ethiopia, readily admitted to being bullish on what the country has to offer, despite continuing economic and political setbacks. Hillman provided the original written and pictorial account of the little-known game parks of Ethiopia.

Jones

David Keith Jones, a wildlife expert with many years of experience in East Africa, contributed valuable text and pictures. Jones, a former editor of *SWARA*, the East African Wildlife Society magazine, wrote the section on game parks in Rwanda and successfully tackled the box story on mountain gorillas. His splendid photography graces many pages of this edition of the book.

SWARA editor **Shereen Karmali** provided the original Travel Tips information on wildlife and conservation organisations in East Africa.

Many of the superb photographs in *East African Wildlife* were taken by Karl Ammann. An award-winning photographer, Ammann has travelled throughout Africa in search of his sometimes elusive but always fascinating wildlife subjects. His work in this book and in his photo books *Cheetah* and *The Hunters and the Hunted* are fine examples of his patient, sensitive and rewarding style.

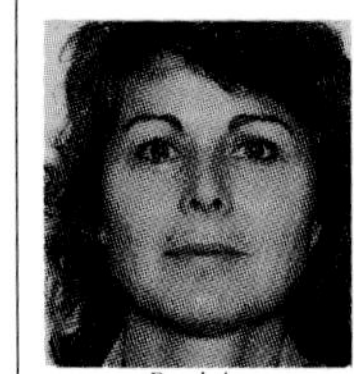
Boelcke

The film stills for **Eva Ndavu's** entertaining article on Hollywood in Africa were taken from **The Kobal Collection**. Additional images came from **Topham Picturepoint, Tony Church, David Coulson, Alan Binks**, and **Nicky Martin.**

Insight Guide: East African Wildlife was completed with the assistance of many other people. **Colin Church**, of Church Orr & Associates in Nairobi, provided valuable help during the initial stages of the book and also contributed a lively piece on his grand passion: deep sea fishing. **Marti Colley**, a relative newcomer to East Africa, put her editorial skills to work and assisted throughout the project. **Cathy Beech** and **Cassie McIlvaine** provided technical assistance and devoted valuable time to fact checking.

Colley

The Travel Tips section for this latest edition has been thoroughly updated by **Annie Moller** and **David Light** of Africa Expeditions, a safari operator based in Nairobi, and by **Wendy Stone**, a Nairobi-based photographer.

CONTENTS

Introduction

History

Wildlife

Features

Places

Uganda

Ethiopia

Rwanda

Maps

TRAVEL TIPS

Kenya

Tanzania

Uganda

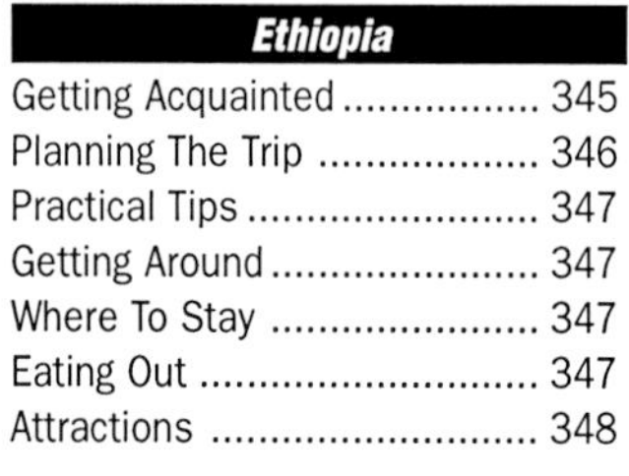

Ethiopia

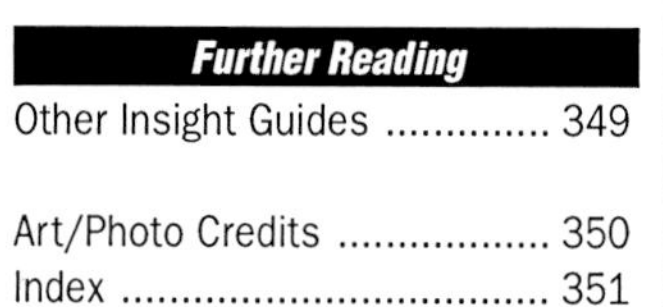

Further Reading

WELCOME

Visitors to East Africa are never short of superlatives when describing a trip there. It is a region of endless variety and stunning beauty. Africa's image of mystery and romance has long been fueled by tales of early explorers. Well-known adventurers from presidents and kings to writers and renegades have also been similarly awed.

In recent decades, mere mortals like ourselves have had the opportunity to sample some of East Africa's multifarious attractions. Chief among its delights is its wildlife, which still roam the open grasslands and mountain habitats in large numbers. Kenya, a long-established holiday destination, is a traditional starting point for people going on safari. Together with neighbouring Tanzania, Kenya shares one of the most spectacular backdrops in Africa, majestic Mount Kilimanjaro.The scenery and the large quantities of wildlife in these two countries, helped by a well-developed tourist infrastructure, make them the most popular choices for travellers to East Africa.

Though lesser known, other countries in the region can boast of a number of impressive game parks and reserves. Uganda's enigmatic Mountains of the Moon, the mountain gorillas of Rwanda and eastern Zaire and the hot springs of Ethiopia's Awash National Park are just some of the attractions awaiting the initiated.

Where once the African safari was a romantic sojourn for the rich and famous, tours now feature experts who conduct safaris on geography, history, flora, birds, archaeology and other topics of special interest. Adventure safaris can take you up ice-clad peaks or down crocodile-infested rivers. A flight in a hot air balloon features fabulous views of migrating wildebeest while an overland trip across a barren landscape leads to a sea-green lake covered with thousands of pink flamingoes. *Insight Guide: East African Wildlife* provides information on safaris and the wildlife you are likely to encounter on the way. It is both a field guide and an armchair companion. Animals are categorised, game parks are discussed and there's even a look at Hollywood's place in Africa. With *East African Wildlife*, it's always likely to be a case of safari, so good.

Preceding pages: wildebeest make their annual journey across the Mara River; silverback mountain gorilla is a study in concentration; two of Africa's tallest—Mount Kilimanjaro and Maasai giraffe; elephants bathed in warm evening light, Tsavo West; rainy day in the Rift Valley. Left, shy young cheetah hides behind the high grass.

Origins Of Early Man

East Africa has provided more evidence of the physical and cultural evolution of Early Man than any other region on earth. But the story of the search for the "missing link" and the evolution of humankind is as much a question of differing opinions and personalities as it is of science.

The Great Rift Valley stretching from southern Turkey, through Israel and the Red Sea, and down the length of Ethiopia, Kenya and Tanzania into Mozambique contains a wealth of fossil remains. This huge, uneven trough in the earth has acted as a geological museum to collect, preserve and display the remains of animals and plants which lived, evolved or became extinct over the past several million years.

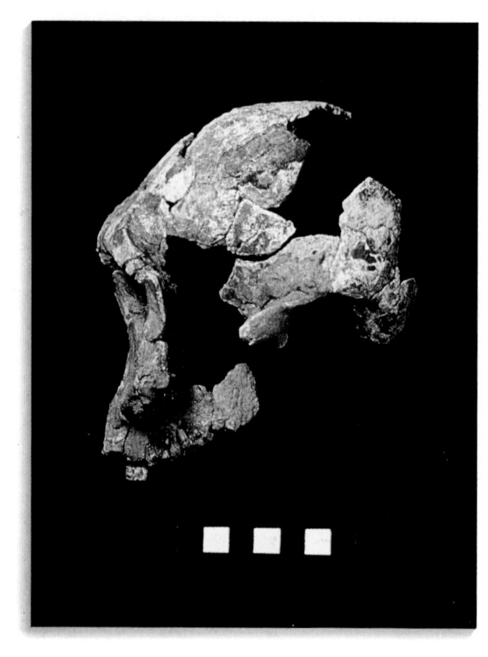

A long series of lakes exists on the floor of the rift. These lakes fluctuate in size over time, sometimes disappearing completely. Bones of animals which died nearby were covered first by water, then by silts, a process that preserved them as fossils. After a lake dries up, or its shoreline recedes, soil erosion may expose the fossilized bones which can then be discovered by curious *Homo sapiens* —modern man.

The oldest history: The first curious *Homo sapiens* to look for human remains in East Africa was L.S.B. Leakey, sometimes called the father of East African archaeology. Louis Leakey was born in Kenya to an English missionary family. He intended to follow in his father's footsteps as a missionary, but after a rugby accident in 1926 he was sent to "rest" on a dinosaur fossil expedition to Tanganyika with the British Museum.

Over the next 45 years he conducted archaeological and palaeontological research into virtually every period between the Miocene, some 25 million years ago, up to the Later Stone Age only a few centuries past. Leakey's ideas about the evolution of early mankind influenced theories up to the present day. Since Leakey's death in 1972, his wife Mary, and their son Richard have continued to support his most controversial belief. He was convinced that the genus *Homo* had very ancient origins, contemporary with or even preceding the genus most experts accepted as the rightful ancestor, *Australopithecus.* Most debates during the 1960s and 1970s about early human evolution centred on this belief.

The most important site worked by the Leakeys was Olduvai Gorge in northern Tanzania. It was discovered in 1913 by a German, Hans Reck, who was chasing a butterfly across the Serengeti Plains. The Leakeys explored Olduvai from 1931 to 1959 without discovering a human fossil of any importance.

Then in 1959, Mary found the skull and jaws of a primitive hominid which was eventually *Australopithecus boisei.* This species became extinct about one million years ago.

At the time of discovery this large-toothed "Nutcracker Man" created a great sensation which led the National Geographic Society in the U.S. to fund the Leakeys' research. In subsequent years many other hominid fossils were discovered at Olduvai, ranging in age from 1.85 million years to only a few thousand years old. But rather than solving the evolutionary puzzle each new fossil created more confusion as researchers interpreted the size, shape and markings of the skull parts, teeth and limb bones in different ways.

The most controversial find was one identified as *Homo habilis* by Louis Leakey. He believed it to be the maker of the oldest stone tools. It lived some 1.8 to 1.6 million years ago at the same time as another small form called *Australopithecus africanus*, first discovered in limestone caves in South Africa. If his interpretation were correct, it meant there were two forms of *Australopithecus*, one large and one small, living alongside a genus *Homo*. This was difficult for many to accept. Leakey defended his *Homo* by claiming that it had a larger brain than *A. africanus*.

Controversial bones: Most scientists agree that a type of forest ape of the *dryopithecine* family was the ancestor of the first hominid. The first definite hominids begin to appear about four million years ago. A very early hominid can be recognised by bones which indicate that it stood erect and walked habitually on two legs. Further identifying traits are dental features such as small canines and a parabolic shaped jaw, and a bigger brain in relation to the face and body than shown by other apes.

The oldest evidence of upright walking, dating back almost four million years, was discovered at Maka in the Awash River Valley of Ethiopia, by J. Desmond Clark and Tim White of the University of California at Berkeley (UCB) in 1981.

The most exciting finds were made in 1974 and 1975 at Hadar in the Afar Triangle of Ethiopia by a team led by Donald Johanson, now director of the Institute for Human Origins in Berkeley. In 1974 the team found a three-million-year old fossil skeleton which was 40 percent complete: it was named "Lucy" after a *Beatles* song popular in the camp. In 1975 the remains of 13 individuals dating back three and a half million years were recovered. Johanson sparked a heated controversy with the Leakey camp when he changed his mind about what these and other fossils from Hadar represented.

He had originally agreed with Mary and Richard Leakey that some of the fossil hominds were of the genus *Homo*. Later discoveries of hominds dating back 3.7 million years found by Mary Leakey at Laetoli near Olduvai showed similarities to Johanson's Hadar fossils and seemed to vindicate Louis Leakey's theory. Then in 1979 Johanson published a paper with Tim White which announced an entirely new species of homind, *Australopithecus afarensis*, named after the Afar Triangle.

The race to find earliest man: In 1968 Richard Leakey began a research project on the east side of Lake Turkana, then Lake Rudolph, in collaboration with the late Glynn Isaac of UCB. Many significant fossils and stone tool sites were found around Koobi Fora, but in 1972 the most spectacular skull of all was discovered and named KNM-ER 1470 (Kenya National Museum-East Rudolf). A photograph of this skull made the cover pages of *Time, National Geographic* and the front pages of newspa-

Left, *Homo Sapiens* ancestor? Above, petrified wood in Kenya's Sibiloi National Park dates back over 20 million years.

pers around the world. It caused a sensation because it was then thought to be 2.9 million years old and with its large brain it was undisputably a *Homo*.

Richard assigned KNM-ER 1470 to the species *habilis* created by his father. This seemed to settle the debate about an early *Homo,* and some distinguished opponents of the idea even conceded defeat.

But then questions began to appear about the accuracy of KNM-ER 1470's age. Some scientists condemned as faulty the dating methodology, called potassium/argon, practised by Richard Leakey's team.

The potassium/argon method was used to date fossils of ancient pigs and elephants found in different layers of rock at East Turkana. Similar specimens had already been discovered and securely dated by a team working in the nearby Omo valley in southern Ethiopia. Although stages of animal evolution were contemporary in each sample, neither team could agree on the age of the rock deposits: the East Turkana dates were consistently found to be older than those at Omo.

This uncertainty over the accuracy of potassium/argon dating meant Leakey's theory of an early *Homo* was again in dispute. The argument was still raging when Johanson withdrew his support of the existence of an early *Homo* at Hadar and Laetoli. To settle the matter, further rock samples from East Turkana were sent to be dated at UCB using a different technique from the potassium/argon method. The UCB dates were a million years younger.

In 1980, after fighting the evidence for more than five years, Richard Leakey finally conceded that KNM-ER 1470 was not 2.9 million, but rather closer to 1.8 or 1.9 million years old. This fit well with the dates of *Homo habilis* at Olduvai, but now there were only the disputed Hadar and Laetoli fossils to support an early *Homo*.

The question remains: Experts quibbled endlessly about whether certain skull or teeth features should be called *Australopithecus* or *Homo*. The remarkable thing about this long controversy is the absence of any real science being applied. The Leakey belief in the existence of an ancient *Homo* was really nothing more than a hunch, or intuition. First Louis, then Mary and Richard, devoted their efforts to trying to find fossils to support their predetermined belief.

The Leakey family spent their whole lives trying to solve the mysteries of evolution. Mary Leakey considers her greatest discovery to be a trail of hominid footprints found at Laetoli in 1978.

Their discovery was only overshadowed by Mary Leakey's find in the same area of preserved footprints of a man, woman and child who, 3.7 million years ago, walked across a field of soft, volcanic ash deposited by a recent eruption.

The hunt for fossils has overshadowed the study of humankind's cultural past which in real terms is more important for gaining an understanding of how we came to be as we are. Archaeological evidence is made up of cultural artifacts—stone, bone and wooden tools, structures, camp fires, the remains of meals, and the spatial distribution of it all. These ancient garbage heaps provide clues to our ancestors' behaviour and society.

Popular excavation sites: East Africa has an impressive wealth of archaeological sites from the oldest in the world, at Hadar, Omo, and west Lake Turkana in Kenya, all dating to more than two million years ago, up to Swahili coastal ruins of towns only a few centuries old.

Tourists can visit some of the more interesting sites, including Melka Konture in Ethiopia where there are several sites of Oldowan (pebble tools) and Acheulean (hand axe culture) dating to several hundred thousand years ago. This site is located in the Awash valley a few kilometres south of Addis Ababa. It is not practicable to try to visit Hadar or Omo.

In Kenya at Koobi Fora there is an interesting museum and simple *bandas* (huts) and a campsite where one can stay on the edge of Lake Turkana.

There are several ruins of Swahili stone towns, mosques and tombs between Mombasa and the border with Somalia in the north. Some of the more notable are the abandoned 14th-century towns of Gedi, on the turn-off to Watamu 14 kilometres (eight miles) south of Malindi, and Takwa, on Manda island across from Lamu.

The National Museum in Nairobi has a very interesting display on human evolution and archaeology.

In Tanzania visitors can explore a small museum and several of the sites at Olduvai Gorge excavated by Mary and Louis Leakey. Unfortunately, the site at Laetoli is not open to the public.

There are no excavation site museums in Uganda, Rwanda, Burundi or eastern Zaire, but archaeological research is going on in these countries. Interested travellers can ask for information at the local museums.

Left, prehistoric turtle fossil at Koobi Fora. Above, the Leakeys inspect another dig.

THE SAFARI EXPERIENCE

"There are no words that can tell the hidden spirit of the wilderness, that can reveal its mystery, its melancholy and its charm," penned a ponderous but enchanted Theodore Roosevelt in his book *African Game Trails*. That nostalgic passage refers to one of the first commercial hunting safaris.

In 1909 Mr Roosevelt came to Kenya on an expedition to collect natural history specimens for museums in the United States. Accompanied by two legendary professional hunters, Frederick Selous and Philip Percival, as well as some 600 porters, the former U.S. president spent several months in the bush satisfying scientific curiosity and outraging some of the earliest conservation consciences by bagging over 500 animals and sending their skins back home.

Celebrity safaris: Roosevelt's caravan set off from the Norfolk Hotel, which today stands in downtown Nairobi and still is probably the finest lodging the city has to offer. Otherwise, safaris and Kenyan tourism have changed a lot since those turn-of-the-century years when setting off on foot across the African plains meant taking your life into your hands.

Edward, Prince of Wales, went on a shooting safari about the same time as Roosevelt and helped to popularise the concept of paying to go hunting in Africa by fascinating his friends with tales of shooting rhinos, elephants and lions at close range.

Farmers and ranchers were quick to see the potential and many did seasonal work, as some still do today, taking wealthy sportsmen and women from the United States and Britain big game hunting. In the 1950s, the dilettante aspect of the sport faded and hunting blossomed into a commercial enterprise, the forerunner of today's tourist industry.

Then in 1977, hunting was banned by the late President Jomo Kenyatta, who stated that a ban would help to stop poaching. Some hunters went to Tanzania where they still hunt today; others tried their hand, with less success, in Sudan and the Central African Republic. Many more reluctantly packed their guns away, sent their trackers home to their farms in the bush and looked for other jobs.

The late 1970s marked the end of a golden era of hunting that spanned well over a century. The men who took clients out in search of trophies were risking their lives as much for pleasure as for profit.

An art: Hunting symbolised the gentleman's code by which European explorers,

farmers and settlers lived. For them, the safari experience embodied all the attractions that had brought them to Africa in the first place.

The sport was conducted along lines as carefully marked out as the rules of a gentleman's club off Picadilly. Apart from the exhilaration of testing vitality under dangerous conditions, there was the mystical communion with nature that culminated in the chase. Often an intimacy developed between hunter and hunted that, in the hunter's mind at least, was endowed with unspoken feelings of respect and friendship.

On the safaris that predated the elephant

Left, turn-of-the-century drawing satirizes "The Great White Hunter." Above, hunting plains game.

crisis, when out of work hunters turned a profit on their elephant licence by selling the ivory at a pound a pound in Mombasa or Dar es Salaam, the bulls were felled in a buisnesslike manner. But some clients who hunted regularly would walk for weeks and hundreds of miles without firing a single shot. For them the sport was in the chase and the trophy told all. When the wily old bull whose tusks weighed 45 kilograms (100 pounds) each evaded them, they chose to walk away empty handed.

Similar respect was sometimes shown for the rest of the "Big Five", as the dangerous game were called. The stealth and beauty of the leopard inspired visiting sportsmen as they crawled through the bushes in a hopelessly clumsy fashion in the predawn dark. Leopards are so evasive that they are customarily shot from a hide overlooking the bait—usually a zebra strung in a tree.

For many hunters, buffalos were the most awesome animals. The thick boss of horn that curls over their head, a three-inch thick hide and overlapping rib cage makes them the Sherman tank of the animal world. Every professional hunter knows that buffalos are possibly the most difficult animals to down with one clean shot. If wounded, the tables are turned, for buffalos ambush their prey, lying up in the thick grass, ears twitching to catch the snap of a broken twig as the hunter approaches. Then they erupt from the shadows in an explosion of fury, bent head and very little else exposed to the poised gun.

It is hardly surprising that many hunters and, on rare occasions, clients too, have been gored and mauled not only by elephant, leopard and buffalo but also by lion and rhino. Most survived their misadventures but there were those who did not.

Vivienne de Watteville, then 24, and her father Bernard set out from Nairobi in 1923 on a prolonged expedition to collect specimens for the Berne museum. While shooting his 19th lion, Bernard de Watteville was mauled. He managed to thrust the muzzle of his rifle under its jaw and kill it. But the claws contracted in his body as *rigor mortis* set in. He had to tear them painfully out one by one before he could push the huge body off him. The elder de Watteville walked two hours back to camp, bleeding profusely from his wounds. His daughter washed and dressed the deep cuts, but his stoic performance was in vain. He died later that night.

The redoubtable Vivienne de Watteville continued the safari, using the bushlore she had learned from her father to complete the collection for the Berne museum. Like her peers, she had fallen under the sway of Africa's breathtaking beauty. The sight of un-

trammelled plains rimmed with volcanoes and carpetted with thousands of antelope and wildebeest moved the hearts of rogues as well as romantics.

Freedom: But above all, the attraction of the safari life was the unparallelled freedom it allowed. "England is too small. Much too small. I shall go to Africa. I need space," declared Denys Finch Hatton before sailing to Mombasa in 1911.

Until World War II, the writ of the government did not reliably extend very far beyond the towns. Once in the bush, a professional hunter could dispense justice, advice and medicine by virtue of the fact that he simply happened to be there. Neither was it likely, as hunters roamed from country to country, that anyone would question their own actions—as long as they adhered to an unwritten gentleman's code born in cold schoolrooms and embodied in an ill-defined notion of honour.

In the early days hunters travelled with their clients, as everyone did, on foot. The *Swahili* word *safari* is rooted in the Arabic *safariya*, which means a "voyage". Colonial settlers went on safari when they stocked up with supplies from the nearest market town just as today a Kenyan marketing manager making his rounds of the company's branches will say that he is going on safari.

There was little to differentiate a commercial safari from any other except for the extra larding of gunbearers, trackers and skinners. Clients slept in tents just as they do today and wallowed in hot water poured into collapsible canvas baths. Their trips sometimes took months and necessitated a long, snaking line of porters.

Everything that was needed was carried on a man's head. The baggage would be considered excessive by today's standards. On a hunting trip in 1891 to what is now Zimbabwe, Lord Randolph Churchill took with him a piano as well as a red and gold wheelchair for the gout-afflicted Lobengula, King of Mashonaland.

Left, loading tusks at Mombasa, circa 1920. Above, 1930s safari scene.

The commercialisation of the sport was sired partially by the British imperative to empire build and partially by the Kenyan settlers' desire to pad out their bank accounts. By the time Teddy Roosevelt arrived in Nairobi in 1909, it was a libidinous frontier town that squatted on a swamp. Spear-carrying *Maasai* warriors mingled in the dusty streets with the restless younger generation of the British aristocacy. Ten years earlier, Nairobi had been the railhead for the Uganda

Railway while workers rested before tackling the inhospitable Kikuyu Escarpment.

The hunters of these years had either won their spurs hunting for ivory, as Frederick Selous had done, or they were farmers who moonlighted when the crops failed, or they simply wanted to earn more money and have some fun while doing it. There were the Cole brothers, Galbraith and Berkeley, brothers-in-law to Lord Delamare; Karen Blixen's husband Baron Bror von Blixen, and her lover, Denys Finch Hatton.

The tortured relationship between the two lovers was aggravated further by Denys Finch Hatton's habit of disappearing into the bush for months at a time. Karen Blixen was

particularly peeved when both her husband and her lover took the Prince of Wales shooting, leaving her behind to mind the coffee farm.

Lord Cranworth's entrepreneurial spirit extended even further. He was a founding figure in Newland & Tarlton, the first safari outfitter. By 1914 it was the largest employer in Kenya, a point which angered colonial bureaucrats. They blamed Lord Cranworth and his partners for the labour shortage. The porters were paid the equivalent of US$4.50 a month.

Poor communications meant that publicity hardly existed except by word of mouth. It took eight weeks for a letter to reach England so safaris were arranged at least a year in advance.

Hunting safaris gained their élitist reputation as a pastime for the rich and famous thanks to clients such as the Duke and Duchess of Connaught. He was attended by an equerry; she was looked after by a lady-in-waiting. Seven-course dinners ended with savouries such as giraffe marrow on toast.

"White hunters": The term "white hunter" was coined by Lord Delamare. He employed two men to control the wildlife on his ranch, an Ethiopian and a professional hunter called Alan Black. To avoid confusion, he referred to Black as "the white hunter". The term remained in use until well after Kenya and Tanzania gained independence from Britain.

The introduction of vehicles in the 1920s changed both the pace and the ethics of hunting. Clients travelled to campsites in wooden box body *Fords* that rolled through mud and sand on spoked wheels. There were no windows but canvas blinds could be rolled down when it rained.

Motorised transport encouraged some to shoot from the car, a practise that outraged the hunters. Denys Finch Hatton and others lobbied against instances such as the slaughter of 323 lion by two Americans. Eventually, the game department in Kenya ruled that no one could shoot within 200 yards of a vehicle.

Philip Percival, considered the finest sportsman of that period, traced the rise and decline of hunting with his own career. He came to Kenya in 1905 to join his brother Blainey, who later headed a fledgling game department. He took Teddy Roosevelt on safari and after the World War I went into partnership with Bror von Blixen.

Old photos show him in the hunting outfit that was *de rigueur* for both sexes—a terai hat with a double thickness of felt against sunstroke, breeches, puttees and boots. He chased lion on horseback over the stony ground of his ranch near Machakos and charged £10 to shoot a lion while on the train going to Nairobi. Wanting a proper education for his children, he chose a Bristol boarding school because it was so near a zoo the lions could be heard roaring at night.

Philip Percival taught Ernest Hemingway his bushlore over two decades that spanned World War II. The author was not always as fine a shot as he would have liked to be but he made up for this in the lyrical passages of his novels set in Africa and at the mess table in the evenings. Philip Percival kept his ego under control by appealing to his wife Pauline to "throw a drink into the beast and he'll quiet down."

After World War II, Hollywood finally discovered Africa. America exported its fiery emotions and inflatable budgets into the bush to dramatise the literature of Hemingway, Haggard and others. In the vanguard was *The Macomber Affair*, filmed in Kenya in 1946.

The film's production company became the first clients of Donald Ker and Syd Downey, who celebrated the defeat of the Germans by creating Ker & Downey. It was to become the largest and most successful safari company in East Africa. Their theatrical tradition lives on. In 1985 Ker & Downey mounted a gigantic field operation that put 340 people under canvas for the making of *Out of Africa*.

While hunting continues in Tanzania, its heyday has long since vanished. The ban in Kenya marked the end of an era that represented an expansive lifestyle lived to the full. Those who were part of it mourn its passing and the rising incidence of poaching. They loved not only the life they led but also the animals.

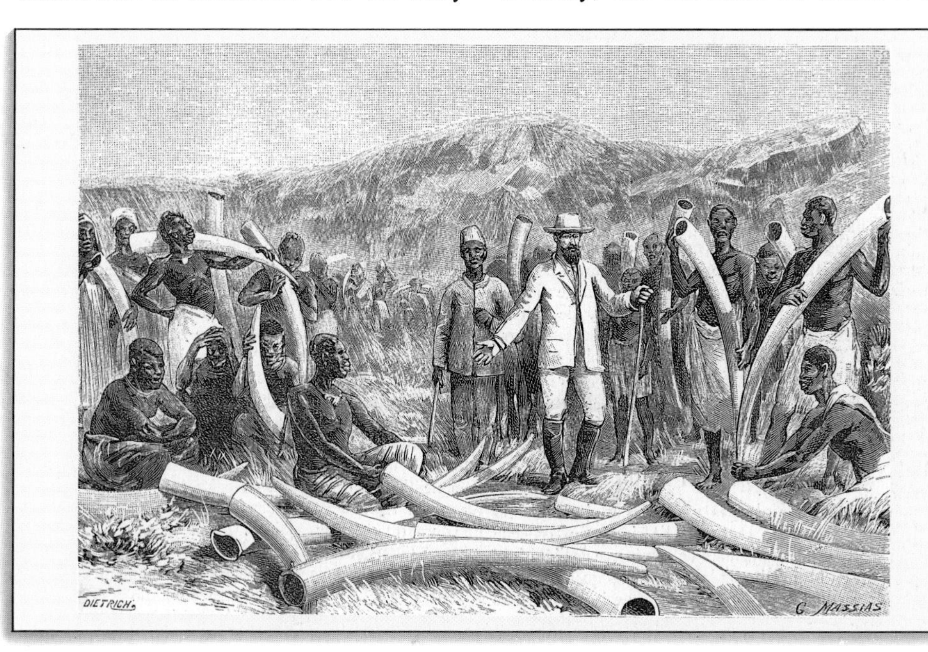

Left, "I gave it the second bullet in the neck." Above, this type of scene was all too familiar in the late 1800s.

BECHUANA H

NG THE LION.

Wildlife and African Culture

Wildlife and the peoples of East Africa have had an intimate relationship since time immemorial. Wild animals and birds of the plains and mountains have always played an integral part in African culture and, more recently, in African economies. There is no other place on earth where humans still live in such close proximity to large numbers of free-roaming animals, which is both a curse and blessing for the governments and people concerned.

African attitudes towards wildlife vary depending on tribal history and current local situations. These cultural dispositions are important as they will ultimately determine the future survival of wildlife. But vast areas of East Africa encompass a great number of different cultural groups which results in many predilections.

Hunter-gatherer: To simplify things, present day African attitudes can be classified according to profession. The oldest profession is hunter-gatherer, followed by agriculturalist and livestock pastoralist and finally, modern urban man. Between two million and 5,000 years ago all East Africans were hunter-gatherers. Wild plants and animals, in that order, were the basis of all life. Humankind was one small cog in the ecological nature machine, operating in small bands and graduating from the technology of wood and stone to metal only a few centuries ago. Controlled fire was the most influential factor affecting wildlife and environment, though the extent of its effect is still being debated. Some say that the great grass plains of Kenya and Tanzania, which today provide the necessary habitat for masses of wildlife, are a result of burning, first by hunters and then by pastoralists.

Not surprisingly, hunter-gatherers have a very positive attitude towards wildlife since their lives depend upon it. No solely hunter-gathering peoples survive in East Africa, though groups still exist who, until recently, were hunter-gatherers and who still retain many of the cultural aspects, including occasional illegal hunts. Wildlife is not, and never was, threatened by these people. Animals were never killed indiscriminately en masse as they are by poachers today. Traditional hunters killed for a reason, whether it was for food, ritual, or something to sell. Each species represented a unique value in a cosmology interlinking natural and spiritual worlds. To kill an animal without reason was to violate all that was held sacred, and would surely result in later retribution for the killer from the spirits.

Hunter-gatherer society was originally classless and life was simple and mobile. Then 5,000 years ago domestic livestock and cultivated crops were brought south from Ethiopia and the Nile Valley by new groups of people. Experts speculate that hunter-gatherers at this time spoke *Khoisan*, the click language still used by southern African bushmen. The first immigrants spoke Southern *Cushite* languages from Ethiopia and Southern *Nilotic* languages from the Nile. Over the centuries many different groups came from the north speaking Eastern *Cushitic* and Southern and Eastern *Nilotic* languages and about 1,500 years ago *Bantu* speakers appeared from the west. The result today is a complex mixture of many languages with ancient Semitic language-speakers still widespread in Ethiopia.

The true hunters: Hunter-gatherers now became subservient to the better organised immigrants. Over time a caste relationship evolved in many parts of East Africa, with hunters occupying the lowest rung along with potters, iron smiths and tanners. In Ethiopia today the hunting caste is called *Wata* and similar peoples in the south are the *Manjo* and *Funa*. With the changing political situation in Ethiopia and diminishing wildlife, these people are being slowly absorbed by the dominant groups with whom they live.

Wata also exist in north and east Kenya where they traditionally lived in symbiosis with the *Oromo* and *Somali* pastoralist groups. They traded wildlife products and labour for protection, milk and use of land. In Kenya they are called variously *Sanye* or *Arianguio*, often with the *Bantu* prefix *Wa*, which refers to a people. Two former hunting groups live in the Lamu District of Kenya, the *Aweer (Boni)* and the *Dhalo*. These groups, especially the *Wata*, were, until the 1960s, infamous long-bow elephant

Preceding pages: hunters and the hunted. Left, framed with ostrich feathers. Above, headress of bird carcasses, worn after circumcision ceremony.

hunters. Some of the *Wata* aces had bows which exceeded the famous English longbows in drawing power.

Other hunting groups in Kenya are the *Okiek* and *Dorobo*, associated with *Kalenjin* and *Maasai*. They are culturally a very mixed bag: the same people switch cultural identities to suit the situation. They are generally highland people who trap and spear animals rather than use bows and arrows. Honey gathering is also important. The *Dorobo* and *Maasai* extend into Tanzania, though Kenya is their main home. In Tanzania there are also three small remnant hunting groups who today have taken up farming. These are Southern *Cushitic Iraqw*, and the *Sandawe* and *Hazda*, whose language contains elements of the bushmen clicks. The *Hazda* live around Lake Eyasi and some of them still hunt and gather, against government wishes.

Uganda, Rwanda and Burundi still have low caste hunting groups called *Twa* or *Batwa*, the Bantu equivalent of the *Cushitic Wata* and *Nilotic Okiek/Dorobo*. Hunting is no longer practiced in Rwanda and Burundi, due to the high density population which is eliminating wildlife. A notable exception is in Rwanda's volcanic mountains, home to the mountain gorilla. Gorillas and monkeys were traditionally an important source of meat for the *Batwa*, and for some of the surrounding *Bantu* agriculturalists. But in the mid 1900s gorillas became more important for the money paid by foreign zoos and stuffed animal collectors for specimens, dead or alive. In spite of laws passed to protect mountain gorillas they were in danger of extinction by the early 1980s. Conservation organizations, local government support and the pioneering work of the late Dian Fossey have all contributed to reduce gorilla poaching dramatically in Rwanda, eastern Zaire and Uganda. The main strategy is to familiarize gorillas with humans: when these animals can be used as a tourist attraction the resulting income earned and daily human presence helps to discourage poachers.

The last hunting group in East Africa is the famous *Mbuti* pygmies of the Ituri Forest in eastern Zaire. The *Mbuti* hunt with both nets and bows and arrows and are proficient trappers. Nowadays, they still hunt for forest elephant as much for its meat as for the small tusks, but their infrequent kills are not a threat to elephant survival.

All these hunting groups have, or had until recently, strong cultural beliefs and practices involving wildlife. Wild animals and birds were important in ceremonies such as initiation, marriage and prayer. They were used in divination and prophecy, for medicine, clothing and, of course, food. Without wildlife these cultures would have no meaning and could not exist.

Herders: The herding peoples of East Africa also tend to have a benign attitude towards most wildlife species. Unlike their European or American counterparts, African livestock herders accept the right of animals other than cattle, sheep and goats to share land and water resources. Many pastoralist peoples even recognise the cape buffalo, eland and some antelope and gazelle species as honorary cattle. Most other animals, and especially birds and fish, are not regarded as fit for human consumption.

Just as many pastoralist groups created a caste system for people, so they have also established a hierarchy for animals. The *Borana* of southern Ethiopia and Kenya, for example, have their own "Big Four"—the lion, elephant, rhinoceros and buffalo. To kill one animal single-handedly for the first time is cause for celebration in the community and the person becomes a "man". Each subsequent kill increases the man's reputation. Similarly, *Maasai moran* (warriors) organise formal lion hunts in which one *morani* tries to spear to death a cornered lion. Success means great prestige for the killer. Other wild, hoofed animals occupy a middle caste position: they are not fit to eat, but neither are they despised.

In normal times pastoralists did little hunting, since livestock were expected to satisfy all needs. It was a loss of prestige to have to resort to wild animals for subsistence. Associated hunter-gatherers were supposed to be the only ones to defile themselves in hunting activities, which were regarded in most cases as ritually impure. In abnormal times however, following drought, animal epidemics or raids, which resulted in greatly reduced numbers of livestock, pastoralists had to revert to hunting and gathering for survival. Some speculate that elephant and rhino were practically wiped out in Ethiopia and northern Kenya following a series of natural disasters in the late 1880s, as pastoralists sold tusks and rhino horn to traders coming from the coast. Herds were rebuilt with the resulting income, and even today *Boran* and *Gabra* can point to cattle which are descendants of those bought with ivory.

Wildlife is important to pastoralists both culturally and economically. Birds and

Pastoral scenes: Left, tending camels in Turkana. Above, crossing the dusty plains of Amboseli.

feathers are particularly significant in ritual and dress. *Maasai* and *Samburu* boys, for example, make stuffed bird crowns to wear after circumsion, and ostrich feathers are worn in ceremonial headdresses by many different peoples.

The camel pastoralists: Three-quarters of Kenya is inhabited by pastoralists. The *Somali* are in the north-east and east, as are the *Borana*, who also extend down the Kenya coast and hinterland as the *Orma* group. In the north, to the east of Lake Turkana, one finds the *Gabra* and *Rendille* camel pastoralists, and to the south are the *Maasai*-related *Samburu*. West of Lake Turkana are the *Turkana* people and south of them live the *Pokot*; the semi-sedentary *Njemps* live south of Lake Baringo. Raiding and cattle rustling still occur, and the introduction of modern firearms at some raids has created serious problems for governments trying to control poaching.

The major pastoralist groups in Uganda are the *Karamajong* in the north-east who are related to the *Turkana*, and the *Bantu Ankole* in the south. The latter are famous for their long horn cattle. Burundi and Rwanda are occupied by the *Rundi* and the *Banyarwanda* respectively. In Burundi, the *Tutsi* (or *Watutsi*) and *Hima* are upper caste groups associated with livestock. The *Tutsi* also are present as a minority in Rwanda. The *Hutu* are a majority caste in Rwanda and a minority in Burundi. They are higher caste than the *Batwa* but fall below the *Tutsi* and *Hima*. This has caused considerable social strife since independence as the *Hutu* try to throw off *Tutsi* domination.

There are two main theories to explain the social and physical differences between the three main caste groups of Rwanda, Burundi and southern Uganda. The first suggests that *Batwa* hunters, generally short and dark, are descendants of the original inhabitants; *Hutu* are descendants of *Bantu* agriculturalist immigrants; and the tall *Tutsi* and *Hima* are the descendants of later immigrant pastoralists of *Cushitic* or *Nilotic* stock who, because of their warlike nature, came to dominate the hunters and farmers.

The second theory hypothesizes that castes developed long ago among a homogenous people and that selective marriage resulted in the breeding of three different physical types.

In Tanzania the main pastoral peoples are the *Maasai* in the north, with most other groups practising mixed agro-pastoralism and living sedentary lives.

The farmers: If hunter-gatherering and pastoralism were the only economies practised in East Africa, wildlife would be sure of a secure future. But farmers have very different attitudes to the sympathetic depiction of animals and birds in folk tales. An all out war has been raging for centuries over control of the land. The farmer is winning. Animals and birds which attack crops are his greatest enemy.

The first agricultural peoples to reduce wildlife populations to controllable limits were the *Amharic* and *Tigrinya* speakers living in highly structured states in highland Ethiopia. Over a period of 2,000 years, high density population, deforestation and use of weapons by formal armies from the 19th century onwards, resulted in the annihilation of wildlife. Internal and overseas trade stimulated localized extinction of desired species. The rise of state systems in Uganda among the *Baganda*, *Banyore*, *Batoro* and other *Bantu* peoples had a similar effect, though with lower population densities wildlife survived in large numbers in various parts of the country. But in the 1970s, social strife and the breakdown of effective government resulted in the slaughter of thousands of elephants and other animals.

Without habitats, wildlife cannot survive. Only a few small mountain areas still containing wildlife exist in Rwanda and Burundi, which are the most densely settled areas in Africa. Burundi's animal population, in particular, is sadly non-existent.

Fortunately, there are now encouraging attempts being made by indigenous wildlife clubs and conservation organizations to create awareness and respect for wildlife. But without traditional cultural values and with the increase in population and poverty, the future for wildlife in East Africa is in a precarious state. Tourism is one of the strongest forces working towards conserving the magnificent natural heritage of this part of the world.

Right, beaded and braided member of the Samburu tribe. Note his traditional ivory earplugs.

A Vast and Varied Land

Visitors to East Africa should realise at the outset that the environment is not fragile, an adjective frequently used to beguile people into supporting environment and conservation related activities. Better adjectives would be harsh, unpredictable, fundamentally impoverished, or resilient. In fact, most evolution and species development in the region has occurred under harsh conditions which have been the rule since the Pleistocene era, some two million years ago. If the environment were truly fragile, as in the

delicate chemical balance of a coastal marine ecosystem or a conservative climax forest, then it would hardly have survived in more or less the same form for such a long time. Most African plants and animals are remarkably hardy, adaptable and able to survive despite what we persist in doing to them.

The ecosystem: The African environment is governed by a handful of ecological laws which appear at first glance almost simple in their concept. There is a finite amount of minerals and elements on the earth, just as there is a more or less finite, but of course smaller amount in an ecosystem. An ecosystem is the collection of plants and animals, together with the soil storehouse of materials which create the building blocks of plants and animals, which occur in a particular region. The energy which brings such otherwise inert materials into life is provided by the sun, which is essential for the first food production—*photosynthesis*. Plants mobilize the nutrients in the soil together with the constituents of water to produce simple sugars which form the beginning of all terrestrial food chains. Herbivores eat plants, carnivores eat herbivores, and at one stage or another, decomposers, from bacteria to vultures, eat them all. So the basic materials are returned to the soil to be picked up by plants and sent through the ecosystem again in this sun-driven, water-lubricated carousel of life.

Limited jungles: Current popular attention to the Africa described by authors such as Karen Blixen, Elspeth Huxley, Beryl Markham or Doris Lessing has finally dispelled the Edgar Rice Burroughs image of Tarzan's equatorial jungle stretching from the Atlantic to the Indian Ocean. In fact, evergreen forest is the minority of vegetation cover in the continent, confined in East Africa to belts around major mountains such as the Virunga volcanoes in Rwanda, the Ruwenzoris in Uganda, Mount Kilimanjaro in Tanzania and Mount Kenya. Ground water forests occur where there is enough water seeping out of springs, such as at the base of the escarpment above Lake Manyara; riverine forests exist along major perennial water courses; along with remnant patches of low-lying coastal forest. Although tree species differ, the physiognomy is more or less the same: dense, layered canopies, trees up to 70 metres (230 feet) and a relatively thin undergrowth due to the shade of the dominant trees. To enter a forest from the surrounding grassland is like escaping inside on a hot summer day. The light dims, the temperature drops, the wind drops. One can almost feel the additional moisture in the soil and the air which the forest, by its very presence, retains.

The sweeping plains: The vast majority of East Africa is covered by vegetation which is essentially variations on the grassland theme. They are most commonly described by ecologists in terms of their physiognomy. Open grassland is the backdrop, such as the long grass plains of the Serengeti covered with red oat grass (*Setaria and Themeda*). Add scattered shrubs, such as the patches of "toothbrush" bush (*Salvadora*) in Queen Elizabeth National Park in Uganda, and one has bushed-grassland. Replace the bushes with widely-spaced, flat-topped acacias and spiny desert dates (*Balanites*), and the scene is wooded grassland, most typical of *Out of Africa*.

If trees and shrubs are about equal in number, one has, not surprisingly, wooded and bushed grassland, like the *Acacia-Commiphora* region of Tsavo East in Kenya, which stretches to the north and east in what cynical old African hands call "MMBA"—miles and miles of bloody Africa! If trees are more numerous, with their canopies almost touching, it is woodland, such as the vast *Brachystegia* woodlands in southern Tanzania and beyond. Dwarf shrub grasslands are found on desert fringes, hill thickets, coastal thickets, swamps, and the like.

The "seasons": The climate, comprised of average and actual precipitation and temperature, is a critical factor limiting the form and abundance of life in any ecosystem. The difference between an annual average rainfall and the actual pattern of delivery is important to consider. The Serengeti in Tanzania, and Ireland have roughly the same amount of water falling on their territories each year. In Ireland, it rains almost constantly. In the Serengeti it rains in two peaks—April and November—when powerful storms bucket down more rain than the soil can possibly absorb at a time. Water runs down the slopes, boils down the watercourses, swells the great rivers and plunges eventually into one of the Rift Valley lakes or the Indian Ocean.

In the long rains, usually April to May, there may be some constantly rainy days. But more characteristically, rainfall is discontinuous. Both plants and animals have to

Left, forest stream in Kilimanjaro. Above, the Virunga volcanoes mark the border between Rwanda and Zaire.

be adaptive, mobile or both. "Adaptive" means, roughly, the ability to breed quickly when there is sufficient moisture. "Mobile" means having the ability to be carried, such as a seed clinging to a paw, or to migrate, demonstrated by the wildebeest in their 1,000-kilometre (600-mile) trek around the Serengeti, to places where the chance of reproduction and survival are better. Only trees and tortoises in East Africa stay in one place, and both do their best to make sure their progeny have the chance to get out of the neighbourhood if necessary.

The other important atmosphere-related characteristic, generally only the concern of those who visit East Africa's splendid beaches, is the intensity of the sunlight. Incoming solar radiation is particularly strong at the equator and when combined with the relatively high altitudes of much of East Africa (Nairobi is 1,500 metres/4,920 feet) above sea level) the light is intense indeed. In fact, it has led to the evolution of a chlorophyll type slightly different to that of temperate latitudes. The sun, which provides the power to generate the ecosystems, takes its toll: nearly 80 percent of rainfall is evaporated back into the atmosphere.

East Africa's geology: Much of East Africa looks unfinished: the gash of the Great Rift Valley which cuts through the country from the Red Sea to Zimbabwe has rugged step faults which drop, for example, from the Ngong Hills 2,100 metres (6,888 feet) outside Nairobi, to the steaming soda of Lake Magadi at 610 metres (2,000 feet). There are active or recently extinct volcanoes—Virunga in Rwanda, Ol-Doinyo Lengai in Tanzania, Shetani and Teleki's in Kenya. Whole hill ranges are remnants of volcanic activities over the most recent geological eras, such as the Ruwenzoris in Uganda and the Chyulu Hills in Kenya.

All of these are associated with the restless state of the earth's crust at the edge of the two tectonic plates which make up either edge of the Rift. In East Africa the two plates are slowly pulling apart, unlike southern California where two others are pushing together. The relatively recent activity means that many of the soils of East Africa have a high content of ash delivered from nearby volcanoes: for example, the southern Serengeti from the Ngorongoro highlands; the Nairobi National Park and adjacent Athi-Kapiti plains from the Aberdares and Mount Kenya, and others.

Recent lava flows are common and look like black treacle poured over the landscape. Older lava flows boulders are exposed as escarpments and great granitic boulders as picturesque *kopyes* (or *Inselbergs*) from the age-old and continual erosion of the land surface.

Much of the variety of plant and animal life in East Africa is due to the variation in elevation, from sea-level to 6,000 metres (19,680 feet). This produces a range of temperature and humidity with extremes similar to both southern Florida and the Alps. Altitude produces impenetrable barriers to some species: baobabs are never found above 1,000 metres (3,280 feet), rock hyraxes rarely below 500 metres (1,640 feet).

Periodic drought: For thousands of years Africa has been subjected to short term fluctuations in the rains. Annual failures of the rains happen at least once every 10 years and some decades are drier than others. The effect is to change temporarily the relative numbers of species and the reproductive success of those less tolerant to dry periods. For example, in certain areas of Kenya, during the dry mid-1970s, perennial grasses all but disappeared in favour of the more opportunistic annuals that re-seed each year. When the rains improved at the end of the decade, the balance shifted back in favour of the perennials.

Animal populations, too, respond to such fluctuations. In dry periods, reproductive success is suppressed by a failure to ovulate or for the foetus to implant. This is related to the mother's nutritional state which is lowered because of the lower plane of nutrition offered by plants in drier periods. The impact is relatively greater on herbivores than carnivores since lions can live quite well off skinny wildebeest, but wildebeest cannot survive on dust. When the rains come back and the vegetation is verdant, then herbivore populations, even elephants, are able to rebuild their numbers remarkably quickly.

Bush fires: Long before modern man, who loves lighting fires, African ecosystems were subjected to sporadic burning from natural causes such as lightning strikes. This is evidenced by the fact that many trees and shrubs are "fire-adapted", that is, not only able to withstand burning, but actually flourish or use the fire signal to put out flowers and seed. Fire keeps many of the grasslands of East Africa in an early stage of succession: if they were not burned periodically, by whatever means, they would start to develop towards bushland.

Modern herders and park wardens take advantage of this fact and deliberately set the grass alight each year. This both clears the

Left, scenic lake in Uganda. Above, spectacular soil erosion in Ethiopia.

regenerating shrubs and ensures that new grass can catch the next available rainfall unencumbered by stands of rank hay. Burning is best done early, before the standing crop of dry grass accumulates into a fuel dump that will destroy trees and soils alike when burnt.

In the annual burning, some animals win and some lose. Tortoises are often caught short and roasted in their shells; snakes and small mammals usually cannot outrun a fire fanned by a strong wind. Birds of prey and egrets have a feast at the leading edge of the fire.and later, the local herbivores, both wild and domestic, will have an unencumbered sward of fresh green grass.

Wildlife makes its mark: Animal signs are everywhere. Some are subtle and fleeting, like the nests of birds and insects; others are etched into the very face of the ecosystem, such as the animal trails which criss-cross major wildlife areas. Some elephant trails have been followed by human engineers when executing a road cut up a difficult escarpment, and others have been recorded by cartographers who mistook the trails for all-terrain vehicle roads on aerial photographs.

In general, the major herbivores make their mark on the vegetation, and therefore the look of an area, more than the carnivores. Grass along large rivers is kept to croquet lawn height by hippos grazing at night. Feeding giraffes hedge the smaller acacia trees and create a browsing table which is more dense than if the shrub were not browsed. By their feeding actions giraffes stimulate the plants to produce more foliage.

Most large herbivores, such as buffalos and elephants, are fond of taking mud and dust baths, to cool themselves and help rid the skin of parasites. This habit creates pits of activity, "animal furniture", around the ecosystem, spots which are temporarily clear of vegetation, but rich in the nutrients of the animals which visit them. When the wallow falls out of favour, the vegetation which recolonizes the spot is particularly lush.

Man's role: In modern times the greatest modifier of natural systems is undeniably man. His activities have effects at varying levels of intensity: He can remove materials and put them more or less back, as in subsistence agriculture; he can remove materials and export them to other ecosystems, as in commercial agriculture or mining; or he can just make use of the space for dwelling or commerce and manufacturing, as in villages, urban complexes or industrial parks. Each of these have different levels of energy requirements and differing degrees of physical impact on the look of the land.

There is little point in placing value judgements on these varied levels of use—the

Noble Savage versus the Land Baron versus the Urban Slave—they are all here to stay.

Man continues to cut down forests, plough up the soil and kill or displace wild animals, just as he has always done over the last few millennia. Some modification is beneficial and non-destructive, such as early burning of grasslands, replacing some—not all—of the forest with tea or coffee plantations, or shifting cultivation with a cycle long enough to allow the cleared land to recover in an undisturbed fallow. Other forms of modification leave a lasting and denatured mark: forest clearing on steep slopes, the dumping of industrial wastes into the nearest waterway or the overexploitation of a vulnerable species, as in the poaching of elephants for their ivory tusks.

Feeling the heat: The current portent of change is, as they say in B-grade movies, likely to change life on earth as we know it today. There is now little doubt that the effect of increasing, largely man-produced gases in the upper atmosphere, the so-called "green house gases" (carbon dioxide, methane, ozone and others), will over the next three or four decades alter all major temperature and moisture regimes around the world.

First indications are that the centre of continents in the region of the equator will get hotter and drier. East Africa, however, may become on average wetter and warmer. Most of the models suggest that the sea level will almost certainly rise a metre or two, although history and better data may well prove these models wrong. Such changes will alter the shape of natural ecosystems both in terms of plant and animal species as well as physiognomy.

The loss of soil from the land under the indigenous farmer's plough or the hooves of the pastoralist's cattle is a favourite bugbear of well-meaning conservationists. The solution most usually put forward is to teach the natives how to farm properly or to demand that the herds be reduced in number. But a layer of silt at the bottom of the Indian Ocean off the East African coast proves that soil erosion has been occurring at varying rates over the last 25 millennia. The real reason for soil loss today is that over this last, relatively dry, geological period in the region, soil has not been made.

Erosion tends to be cyclical. Vegetation thins out over a couple of very dry years; soils are sluiced away in the following excessively "good" rains; vegetation re-establishes itself during the rains, and the rate of soil loss slows down.

Conservation consciousness: Land management strategies must be practicable in the African setting, useful to African people and linked to their society and economy. Wise conservation schemes must involve local people, not just by informing them what is going on, but by making them a direct part of the conservation process. At the simplest level this means getting some of the wildlife-generated revenue, such as game park takings, back to the landowners who bear the cost of having wildlife graze on their land.

Left, lion lounging on a *kopje*. Above, volcanic rock formations in northern Kenya.

We are only just beginning to learn that conservation attempts work best when they are woven into the desires and aspirations of those they purport to help. But now that we accept that the environment, in all its manifestations, is not the exclusive trophy of the West, perhaps developing countries will become more involved in helping to conserve the world which is, after all, the home of all mankind.

WILDLIFE: A WONDERFUL WORLD

Few experiences can replace the visceral excitement of standing amidst a herd of resting elephants, floating in a hot-air balloon over thousands of migrating wildebeest, being in on 'the kill' with a pride of hungry lions, or making eye contact with an imposing mountain gorilla.

Wildlife lovers will admire everything from the awesome majesty of the Big Five—elephant, rhino, lion, buffalo and leopard—to the astonishing variety of beautiful birdlife. They will also marvel in the knowledge that they are following in the footsteps of their *Homo habilis* ancestors who lived here over three million years ago.

Today, about 30 species of herbivores, two dozen carnivores, a dozen primates, over 1,000 species of birds, numerous reptiles, hundreds of coral reef fish and many more invertebrates offer an opportunity to discover the wonders of the East African ecosystem. This section of the guide discusses the larger mammals and some of the more common smaller species, including their normally observed behaviour and habitats. Separate chapters are also devoted to birds and flora.

Preceding pages: taking a real licking—big cats in a joint cleaning exercise; happiness is a wet wallow; migration spearhead: crossing a lake in Serengeti. Left, mother and cubs survey the Mara plains.

THE AFRICAN ELEPHANT

The African elephant (*Loxodonta africana*) is the largest living land animal. Most of its special characteristics are a consequence of unique body size of three- plus metres (10 feet) at the shoulder and up to five tonnes in weight. They have a disproportionately large head; toe nails instead of hooves; two breasts between the forelegs; testes carried in the body cavity, next to the kidneys; specialised grinding teeth; tusks nearly worth their weight in gold; a unique trunk; enormous ears and a 60-year life span.

Elephants once roamed throughout Africa. Now they occupy only one-fifth of the continent. Although they now number between 700,000 and one million, their range and numbers are dwindling rapidly. Poaching for ivory has over the past 20 years reduced most East African populations by 90 percent. Despite international concern for their threatened status, even conservative figures suggest that elephants will be virtually eliminated within the next 20 years.

Elephants occupy all African habitats from near desert to closed canopy forest. The inherent mobility of the animal allows it to select foods from a variety of habitats over a home range which may be thousands of square kilometres. The availability of grass for a good part of the year is important; the presence of perennial water within their range is essential. Elephants move daily and seasonally between different parts of the habitat: from woodland to grassland, from bushland to swamp, and back again.

Elephants are active both night and day, since their 16- to 20-hour waking period must necessarily spill into the dark hours. The bulk of time is spent feeding. A midday siesta in shade is common, and a period of deep sleep, with most of the group even lying down, will occur at night if the group feels secure.

Mud baths: Elephants bathe, wallow or dust whenever possible, at least once a day, for cooling to help get rid of parasites. Trunkfuls of liquid or dust are blown be-

A dust-covered herd heads for some liquid refreshment.

tween the legs, on top of the head or along the flanks. The habit of frequent mud bathing and dusting leaves elephants more the colour of local soil types than the natural grey of their skins. Adopted colours range from light grey, red, to dark brown, allowing the observer to guess from where a group of elephants has recently come.

The only serious threat to an adult elephant is humankind, who will eventually either displace all wild elephant populations as he spreads into marginal lands, or kill them directly in poaching for ivory. Elephants do, rarely, get their own back. An unarmed human has little chance at close range against an angry elephant. The frequently

observed head shake, accompanied often with an audible ear snap, is a warning to keep away. A serious charge is strangely quiet: the elephant runs at some 35 kph (22 mph), ears out, head lowered, trunk curled under.

Elephants are as sociable as primates. Greeting elephants will either put their trunks in each others' mouths, or touch and therefore smell each others' temporal glands. Such greeting gestures probably reveal subtle states of mood as well as identity. Like many other African herbivores, elephants may urinate or defecate to mark territory and for recognition. When two family units reunite after several days or weeks of separation, there is much squealing and trumpeting, pirouetting, backing up, greeting and excreting.

The family: The basic social group is the family unit, comprising several related adult females and their immature offspring. The family unit is led by the eldest cow, the matriarch. The herd bull does not exist, except in the preconceptions of observers. Adult males tag along with family units for short periods to inspect females for their readiness to mate. Females are "tasted" by putting the trunk to their vulvas, collecting a residue of urine, and inserting the sample into the mouth. There is a special organ in the palate sensitive to the hormonal content of the urine, and hence the female's reproductive state. These visiting bulls have no leadership role in the group, although they will assist in defence.

Bulls drift from family unit to family unit, and from time to time into loosely knit bull groups: two to 20 males who move together for a day, a week or a season. Bull group composition changes continually. Larger bulls go off to test females and mate: young bulls, newly ousted from their family unit, quickly find companionship, teachers and safety in numbers in bull groups.

A family unit of 15 is large. Beyond that size the group is likely to split into two, each new group going its own way, one led by the old matriarch, the other by one of her sisters or cousins. Family units may show strong associations with certain other units in a particular area. Such associations are the likely result of a blood relation between the matriarchs: they were probably sisters or cousins in a family unit which grew and split in the past.

With the rains and abundant grass growth, however, family groups are likely to join together in larger assemblages which, in places like the Serengeti, Selous, Luangwa and Tsavo, have numbered more than 1,000 elephants in sight from the same hilltop. Bull groups satellite such assemblages, and many bulls are within them testing females. The frequency of mating is high. Seasonal gathering of the clans may facilitate breeding in a wide-ranging and mobile beast.

Old elephants provide the family unit with a historical memory of watering holes, the location of seasonally available fruits, and other dispersed features of the elephant's world. Even a menopausal cow can retain her role as matriarch, suggesting that wisdom, and not just sex appeal, is a predominant quality for an elephant decision maker.

Recent research has shown that elephants are highly vocal animals. Low frequency sounds, well below or at the very edge of human perception, allow contact to be maintained for up to 10 kilometres (six miles). The deep rumble heard from time to time is a contact vocalisation ("here I am; where are you?") which just enters the range of human hearing. Low frequency sound is efficient for transmission where there is interference from vegetation. Elephants also roar and scream audibly through the trunk to produce classical trumpeting, either in anger or exultation, depending on the situation.

Check mate: Females may come into season every two or three months, if not pregnant or lactating. This, and the generally wide dispersal of an elephant population makes it necessary for males to move among family units, constantly testing for female readiness to mate. A testing male walks past a female's rear end, surreptitiously sniffs her vulva and then puts the trunk tip in his mouth to confirm the test.

It was once thought that only Asian elephant males came into sexual season, a period of ill temper called musth. Recent studies show that African elephant males also show seasonal fluctuations in their temper and sexual motivation and display conspicuous physiological indicators, known to elephant watchers as the "green penis syndrome". The annual two to three months of irascibility is accompanied by the penis dangling, dribbling, and taking on a characteristic greenish hue and strong smell. The odour is detectable to human observers, so it must be nearly overwhelming to other elephants. The signs of musth, usually attributed only to irritable males, are now known to be inherent in both sexes. A liquid oozes from the temporal gland, a modified tear gland halfway between the ear and the eye, leaving a conspicuous dark stain along the side of the face. The secretion, called temporin, accompanies states of excitement, such as when there is a frightening disturbance or if close relatives reunite after a period of separation. A bull in musth is more likely to displace his peers when it comes to winning the favours of a female.

Occasionally, a female in oestrous and a large courting bull will consort for a time:

Left, this baby gets its back scrubbed during a mud bath. Above, a gathering at Samburu.

they stay close together, pay attention to one another, and exclude others from any intimacy. Consorting may be very subtle, or actually take the pair some distance away from the family unit and other bulls. This may account for the old tale of elephant "marriages".

Consorting may not always be so tender, and the male may have to catch her first. If the consorting female breaks into a run, the male takes after her. Females in general can outrun males if they wish, and getting caught is probably the female's way of choosing who catches her. Females mate most with bulls in musth.

The female usually stops when the male touches her, particularly if he is able to lay his trunk along her back or across her shoulders. He then rests his head and tusks on her rump and heaves himself up on his hind legs. By squatting slightly and hooking upward his highly mobile erect penis, he is able to enter her. Copulation lasts less than a minute. The rest of the family unit is either indifferent to the copulation, or capable of reacting with great excitement—vocalizing, earflapping, head shaking, turning, backing and excreting.

Habits: Elephant calves spend several years dependent on and learning from the adults. There is nearly constant contact between mother and young. Calves suckle from the side, reaching for the teat just behind the mother's front leg. A young elephant that can walk under its mother's belly is probably less than a year old. A calf may suckle for up to five years. Young bulls stay with their family units until puberty, perhaps 10 to 15 years, after which they are encouraged to seek their fortunes. Young females in contrast may stay with the group for life. Young elephants in a relaxed, undisturbed and healthy population are a pleasure to watch playing with each other: chasing, play fighting, tug-o-wars and mounting. Their long childhood is only matched by higher primates, including man.

Group defence is common. If simply moving away does not work, the next line of defence is to draw up a formidable wall of adult females to face the intrusion with heads high, ears out, looking to and fro as if trying to make out the exact source of the annoyance. Young members of the group are pushed towards the rear. In full retreat, the group runs off in a tight bunch with the young in the middle.

A wounded or sick elephant causes great concern and excitement in the group. Faltering animals will be kept upright between two or more adults. Fallen animals will be fussed over with trunks and feet, and tusk- breaking attempts to lift a downed elephant have often been observed. The strange habit of burying a dead or immobilised fellow is also well-documented. Trunkfuls of dirt are tossed over the animal; branches are broken off and laid tenderly over the body, until it is completely covered. Human victims of elephant attacks have been treated in a similar manner. The reason is not known.

Elephants feed up to 16 hours a day, using their trunks from ground level to nearly five metres (16 feet). Feeding techniques are varied; nearly all vegetative material is eaten, from staples such as grass, trees and shrubs, to delicacies like fruits, seeds, herbs and creepers. Tree bark is stripped off *Acacia* trees in certain seasons. Elephants are attracted to over-ripe fruit trees and often gorge themselves to the point of intoxication on fermenting fruit.

The trunk is the principal feeding tool. Its main function is to reach downwards to harvest grass, the bulk of the elephant diet. The trunk evolved from a combining of nose and upper lip. It is an astonishingly mobile and dexterous collection of muscles. It has a "two-fingered" tip, used for smelling and for picking and plucking.

Tusks are also important tools, and are modified front biting teeth, not eyeteeth (canines), as might be expected. Tusks are used for chiselling, digging, prizing, levering and stabbing. Elephant ivory grows some 10 centimetres (four inches) per year, so the frequently broken tips are continually replaced by new growth. This has inspired suggestions of "live harvesting" of ivory to save elephant lives: complicated and costly, but feasible. No more than one or two percent of elephants are naturally tuskless.

Even while feeding on grassland, elephants detect the occasional herb or creeper by smell, and pluck them out. In this way, odd individuals break the feeding pattern as they come across an interesting "side dish" or a new bulk food. The entire group may then shift to the new delicacy.

The remaining elephant teeth, the molars, are also unique. An elephant has in its lifetime only six teeth on each side of each jaw, 24 in all. The teeth are large, 20 to 30 centimetres (eight to 12 inches) long, so only one, or two halves, end to end, are exposed on each jaw side at a time. Molars grow progressively forward, which provides scientists with a means of telling elephants' ages. Chewing is forward and backward: the lower jaw grinds against the upper in the forward stroke. Since elephants spend most of the day eating, they are almost continuously chewing.

An adult elephant can eat 150 to 200 kilograms (68 to 90 pounds) of vegetation a day. Water is essential—70 to 90 litres (123 to 158 pints) per day— both for helping digestion and for cooling. Water is sucked into

the trunk and then tipped and released into the mouth. Animals less than about six months old have generally not learned the trick; they drink from their knees, sucking water directly into the mouth. If water is not within a half day's march, elephants may use their trunks and tusks to dig for it, for example, in the bed of a sand river. Elephant wells, as much as a metre (three feet) deep, provide numerous other animals with access to water during dry periods. At such wells, or at a normal watering hole, elephants easily displace all other animals, including buffalos and rhinos, and have been known to kill them if it comes to a fight.

Left, family makes a river crossing. Above, play-fighting on the banks of the Ewaso Ngiro River.

RHINO

In our lifetimes, black and white rhinoceros (*Diceros bicornis* and *Ceratotherium simum*) are doomed to virtual extinction in the wild, outside small protected areas. Rhinos' horns are leading the beast to the brink of extinction. The alleged pharmaceutical qualities of rhino horn, as a nerve tonic and general restorative rather than an aphrodisiac, have supported relatively modest ancient markets in the Far East. Most of the recent poaching in East Africa, however, has been to fulfill the demand for dagger handles, a male status symbol in South Yemen. One well-wrought *Jambia* made from one life-long grown horn can fetch up to US $15,000 in Sana'a.

Horns are not bone, but tightly packed bundles of hair-like structures, similar to hooves and toe nails, mounted on roughened areas of the skull. Apart from this, rhinos are virtually hairless. The unabated and apparently uncontrollable problem of poaching is more than the rhino's naturally very low population density and slow reproductive rate can support.

Rhinos are odd-toed ungulates, like horses. Their footprints are unmistakable with three large toes. There are two distinct species of rhinos, the black rhino and the so-called white rhino. Both are in fact grey. The latter's name is a corruption of the Afrikaans word for "wide", referring to its broad upper lip, which is designed for grazing. Black rhinos have longer necks than whites which helps them to reach up into vegetation for browsing. The white rhino's relatively longer head enables it to reach the ground to graze.

African rhinos have two long horns, one set behind the other. They are distinct from their cousins, the Indian rhinos (*Rhinoceros unicornis*), which only have a single horn.

The rhinos' keen senses of smell and hearing compensate for their weak eyesight. They can turn their ears to locate the source

Face-to-face with *Diceros bicornis*.

of any disturbance. Like other large bodied animals, rhinos have potentially long lifespans of up to 50 years.

There are two distinct population ranges of white rhinos. The northern range (*C. s. cottoni*) extends from southern Sudan west through Zaire towards Lake Chad. The southern species of (*C. s. simum*) occurs south of the Zambezi. There are marked differences between the two subspecies in the concavity of the forehead. Populations have not been contiguous in recent historical times.

Some populations of black rhino developed characteristic traits. "Gertie" and "Pixie" of Amboseli were famous for their remarkably long, straight horns. The gene for such horns has now been poached out of existence.

Habitat: Black rhinos range from moist, montane forest to semi-arid bushland. The white rhino prefers the drier habitat of grassland and wooded grassland. Both species characteristically seek shelter and shade in dense undergrowth and thickets. The white rhino is partly nocturnal; the black rhino more strictly diurnal. Black rhino may become partially or totally nocturnal in regions of persecution.

Rhinos are never found far from a source of water. During drought, however, they have been known to go for up to five days without drinking. Rhinos are fond of rolling on their sides in mud or dust wallows. They cannot roll on their backs because of the elongated, blade-like protrusions on their spines. They frequently rub their belly, flanks or face on rocks or stumps and "polished" rubbing sites are dotted throughout rhino country.

Man is the rhino's main predator although lions and hyaenas may try to attack very young calves. Rhinos are able to rout or dispatch most disturbances after a short 50 kph (30 mph) charge. Disturbed rhinos are prone to attack, often before they have properly located the source of disturbance so the initial charge may not be directly towards the intended target. Other charges may stop short of target, as though the real purpose is to get close for enough to identify and intimidate the disturbance. In such a case the rhino is more likely to wheel about and run off than follow through with the attack.

Rhinos are not very sociable, especially black rhinos. White rhinos may form small, family groups including several females and immatures, but the most commmonly seen group is a female with her calf.

The territories of mature males vary from three square kilometres (just over one square mile) in forested areas to 90 square kilometres (35 square miles) in open grasslands. Old territorial males will only rarely stray from their familiar area. Territory is marked both around the edges and throughout the middle with large, conspicuous dung piles, or *middens*.

Middens may be a couple of square metres in extent. They are approached and sniffed by most passing rhinos, but only the dominant male defecates and scatters his latest addition on to the pile with his hind feet. Horn rubbing in the pile is common. A white rhino territory may have 20 to 30 middens around the boundaries. Scientists think that middens might be a sort of range "mailbox", allowing all rhinos in an area to keep track of who is in the neighbourhood and their reproductive state.

White rhino females often engage in friendly nose to nose greetings when they meet. Males however, especially black rhino males, tend to be less amiable. On encountering each other, serious fighting can result if one rhino does not give way. Sometimes gaping wounds are inflicted by the upward sweep of an opponent's horn.

White rhinos appear to be more phlegmatic than black: two males may spend an hour staring at each other from a short distance, sometimes nose to nose, as if greeting, occasionally wiping their horns on the ground. They then turn around suddenly and trot back to the centre of their respective territories.

Mating: Their territories ensure that white rhino males have access to receptive females. Once a female is found, she is continuously, but gently, herded within the boundaries of the territory for up to two weeks, often accompanied by her most recent calf. This may in part explain the length of time the male spends manoeuvring and tagging her— as much to get away from the "teenager" as to get close to her. Once close, the male prods the female gently with his horn, rests his chin on her back, rubs his face on her flank, and generally softens her up for the final approach.

A courting male and female may consort for several days. Mating may last as long as an hour, during which the male ejaculates several times. This does little to dispel the popular perception of the qualities of rhino horn.

Single calves are born after a gestation of some 16 months. Newborn animals are very small at birth, only one-twentieth of the female's weight. Females seclude themselves at the time of birth and female white rhinos, after calving, may isolate themselves from other animals for a month. It is said that white rhino calves run in front of their mothers; black rhino calves run behind.

Births are most common during the rains. Calves stay with their mothers from two to four years, depending on the birth of the next offspring.

Left, pondering his fate? Above, taking a sugar cane break.

The shape of the mouths of the two rhino species indicates the differences in their feeding habits. Black rhinos are browsers and have pointed, nearly prehensile upper lips, which allow them to choose and pluck small twigs, leaves, fruit and vines. Most plants, except grasses, are eaten by black rhino, nearly 800 species in some areas. White rhino are grazers with broad mouths and upper lips and teeth suitable for grinding grass. Both have lost the front biting teeth altogether and rely on their lips to gather vegetation.

BUFFALO

The African buffalo *(Syncerus caffer)* is traditionally known as the meanest beast in the bush, prone to launch a killing charge at the drop of a hat. It is an understandable reaction if one is being shot at, and solitary males are loath to be disturbed. But the majority of buffalos in cow-calf herds are nearly as docile as cattle.

There is the one main species, a very large blackpelted, grassland dweller, weighing up to 800 kilograms (1,764 pounds); one subspecies, the forest buffalo (*S. c. nanus*) which is smaller, reddish, with less robust horns; and at least two intermediate forms, *S. c. aequinoctialis* and *S. c. brachyceros,* found in grassland areas where large forests merge with grasslands. In East Africa, however, animals seen in forest clearings, or along grassy road verges, will certainly be African buffalos.

Buffalos in general favour open grassland, wooded grassland and bushed grassland. They are most active in evening, night and early morning, both for feeding and moving from place to place. The rest of their time is spent lying down and ruminating, in shade, if available, rather like cows in a field. Buffalos must drink daily, so are never found more than 15 kilometres (nine miles) from water. As with domestic cattle buffalo probably sleep not more than an hour a day.

The massive, bossed horns and exceptional size of the animal afford considerable protection. This has allowed blind, lame, even three-legged individuals to survive longer than could have been expected. Solitary bulls, without the redoubtable protection of numbers, commonly fall prey to prides of lions. It is not uncommon, though, for lions to be fatally injured during a prolonged battle with a wounded buffalo. A buffalo herd, when in danger of attack by lions, will form a defensive semi-circle, protected by bulls on the other flank, with the cows and calves grouped in the centre of the formation.

Delicate morning mist makes for a peaceful setting in the Maasai Mara.

An explosive snort heralds alarm. This is followed by a nose up posture orientated at the intruder, who should begin looking for the nearest tree if the buffalo is a lone male. An alarm in a cow-calf herd will bring the others to attention, and those close to the intruder may even move forward in a short-sighted manner, as if to have a better look. If the intruder stands his ground or even advances a few paces, the buffalos will invariably turn tail with high head tosses and the entire herd will run off.

Herd mentality: Buffalos live in herds which have a relatively stable composition, changed only by births and deaths. Herds may be as large as 2,000 for the African buffalo but rarely larger than 20 for the forest sub-species. A figure of 350 was estimated as an average in Serengeti National Park between the wet and dry seasons. Large herds tend to fragment during the dry season and regroup in the wet. This spreads the grazing load when grass is in short supply. In the dry season, there is a tendency to concentrate along the rivers. In the wet season, spacing over the habitat is quite regular. Each herd has a particular, fairly constant range, and there is little overlap between adjacent herds.

Despite their herding instincts, the basic social unit appears to be an adult female, her suckling calf and her two year old from last season's breeding. When any one of the three is incapacitated, the other two will stick together. Vocalizations play only a small role in social encounters: calves bleat; cows grunt to call calves. Otherwise buffalos are rather silent without the typical bovine lowing sounds.

Cow-calf groups do not appear to have obvious leaders. Decisions about which way a resting group should move next seem to be taken by a form of voting. During the period of resting and ruminating, individual females ocasionally stand up, face a particular direction for a few moments, and then lie down again to carry on chewing the cud. After a couple of hours, the whole herd moves off in the direction most buffalos faced when standing. Voting seems to pay off, since it appears to take the herd in the direction of the greatest amount of grass in the neighbourhood.

A lot of bull: Bachelor groups of 10 to 15 are common, and may consist either of old, retired bulls who no longer bother to keep competing for females, or younger bulls nearing their prime during the off season for matings, usually when it is dry. Solitary old males or small bull groups are the ones likely to charge intruders.

Fighting males circle each other for long

periods, head tossing and pawing the ground. The major threat is a lateral display with the head lifted and nose pointed to the ground. Presented side-on, the thickness and power of the neck and shoulder muscles are shown to full advantage. Seen from the front, the posture emphasizes the size and raised surface edge of the horns. Two males squaring off will accentuate their horns with tossing, hooking movements, and by thrashing nearby bushes. Such displays usually end with one animal giving up by simply walking away. Fights are therefore rare which saves everyone potentially serious damage. When fights do occur, they consist of terrific head-on clashes.

Males test for females in heat by sniffing their urine and genitals. Male competition for females does not entail much fierce fighting. Posturing and mock fights serve as substitutes for conflict, which helps keep these large, powerful animals from injuring one another. Courtship entails a temporary male-female bond which lasts only until soon after mating.

The principal calving period is between December and February. Calves are carried for eleven and a half months, and are dropped in the midst of the herd, usually in the rains. The afterbirth is eaten, the calf is licked clean, stimulated to defecate and suckled. On completion of this procedure the calf goes back to join the grazing herd.

Wallowing in mud holes appears to have a social function as well as keeping animals cool and discouraging skin parasites. A particular wallow, apart from being foul-smelling in its own right, may take on the scent of the bull which lays claim to it. The wallow thus serves as a passive territory marker. Cattle egrets can often be found in the company of wallowing buffalos. These birds give away the presence of buffalo concealed within a swamp.

Left, making a pit stop at the aptly-named Buffalo Springs. Above, oxpecker and friend.

Buffalos are strictly grazers. Different species of grass and even their parts— leaves, stems, inflorescences— are selected on a basis of smell, taste and protein content. The amount of protein is related to the tenderness of the grass so grazers weed out the nutritious from the coarse on the basis of the effort taken to pull the plant apart.

Seasonal changes in grass availability and nutritional value dictate the local movements of buffalo in semi arid areas. In the rains they will feed on open plains; in the dry season, they retreat to woodlands, hillslopes and river fringes.

KING OF BEASTS

Lions *(Panthera leo)* are the largest of the big cats, two to four times the weight of their cousins, the leopard and cheetah. Lions are far more social and also show greater differences between the sexes than other cats. Males have manes which are fully developed by four years old, and they are up to 50 percent heavier than females.

Young lions have a spotted coat which gradually, over the first two years, becomes a nearly uniform lion tawny. Like many of their prey, lions have a slighty counter-shaded colouration: a pelage darker on top grading to lighter underneath. This tends to neutralize the three-dimensional shadowing created by overhead lighting from the sun, as the shadow is lightened by the whitish belly fur. The overall effect is to enhance camouflage by flattening form.

Lions are widespread in wooded and bushed habitats. Although they are often seen in completely open grasslands, they prefer areas which have cover for hunting and hiding young. They appear to succeed in maintaining healthy populations in most game parks and reserves.

Lions are predominantly active in the evening, early morning, and intermittently through the night. They tend to spend nearly all daylight hours resting or asleep in the shade.

Roaring success: Both lions and lionesses roar, the males louder and deeper. Roaring typically consists of long moaning grunts followed by a series of shorter ones, the whole lasting 30 to 40 seconds. Roaring is most common at dawn and dusk or during the night. Its purpose appears to be definition and maintenance of territories, although it may well also be used to keep in contact on dark nights: a roar can be heard over two or three kilometres. Individual lions can recognise one another's roars. Cubs may make noises while older lions are roaring nearby. Lionesses often encourage their cubs by making soft moaning roars to them.

The high grasses help to hide lions from their intended prey.

Cool cats: Lions are the only really social cats. Prides are built around two to 15 related lionesses. These are accompanied by a coalition of males, many of whom are probably brothers unrelated to the females. The pride also contains dependent offspring. Young females mature and join their mothers and aunts as breeding pride members; young males emigrate and seek unrelated prides to attempt to take-over.

Lion prides are ever changing: males only hold sway over a group of females for an average of 18 months, before they are ousted, sometimes even killed, by stronger or more numerous newcomers. Only large male coalitions can aspire to control a pride over a period of relative stability, measured in months rather than years.

Females may produce as many as six litters during a lifetime. Lionesses come into season sporadically; the periods between heats vary from a couple of weeks to months. Oestrus lasts about a week during which time males compete over receptive females. Male coalition partners however, who test females as they wander over the pride's range, are rarely aggressive towards each other in competition over potential mates. They seem to operate an agreed first-come-first-served system.

Either partner may initiate mating. The male initiates with the so-called mating snarl, which has been described as a sneeze-like grimace. The mating snarl may or may not convince the female to stop and crouch. The female, if she initiates, keeps unusually near the male and may rub her head on his shoulders and sides, emitting a deep, sensuous rumble, walking sinuously around him and flicking her tail. She may even back into the male and crouch to stimulate his interest.

Copulation is unmistakable, if perfunctory. The female yowls and the male often bites her neck in a similar manner to the subduing neck bite of domestic cats. After ejaculating, the male leaps off, often to avoid a blow from the female who may spin around

when his weight is removed. Both animals then lie down. During one consorting week, the pair may copulate over 300 times, an average of once every 15 minutes during the waking hours. A successful pride male may mate 20,000 times in his lifetime: the King of Beasts indeed!

Gestation is relatively short: three and a half months. In a secluded, well hidden spot, often among rocks or dense riverside vegetation, the lioness gives birth to two or three cubs on average. She suckles and remains with them for long periods, only occasionally returning to the pride to hunt and feed. Litters are sometimes lost at this stage. After a month the cubs are led to the pride.

Lionesses appear to synchronise breeding activity; it is not unusual for several females in a pride to have litters at the same time. Unlike other mammals, lion mothers will commonly tolerate suckling by cubs of others in the pride, because they are generally all related. By providing milk to cubs belonging to her sisters and half sisters, a female feeds young lions that carry some of her own genes. Females do however more readily nurse their own offspring, who tend to approach boldly, mewing, purring, and pushing their mothers' faces before proceeding to suckle noisily. Cousins are more prone to sneak in from behind.

Nomadic males regularly kill cubs, presumably to eliminate from the population the genetic material of rivals. Cubs are also at risk from other predators such as spotted hyaenas and leopards. Lionesses keep their brood well hidden and move the hiding places if disturbed.

Getting a good grip: Lions' predatory habits have enhanced the evolution of a large head and powerful jaws equipped with long canines. Feet are large and retractable claws unsheath to present a catching tool as wide as a squash racket. Lions are the biggest African carnivores and routinely tackle prey, such as buffalos, that are beyond the ability of other predators. They stalk and ambush rather like cheetahs and leopards, but like hyaenas and hunting dogs, they also hunt collectively.

The majority of their prey are medium to large ungulates, such as wildebeest and zebra. There is, however, considerable seasonal and locational variation. In the Serengeti, for example, lions often hunt warthogs in the dry season when more favoured wildebeest and zebra have migrated north towards Kenya. At all times lions are opportunistic and take rodents, fledgling birds, ostrich eggs, etc. Males need to eat approximately seven kilograms (20 pounds) of meat a day; females five kilograms (11 pounds).

Females in the pride do most of the hunting but males, with their superior strength and weight, gain first access to a kill once the dirty work is done. There is a reason beyond sheer laziness: males have a distinct disadvantage in stalking with their manes, which appear like small hay stacks moving though the grass. Hunting tactics depend on prey and habitat. In open habitats lions tend to hunt at night, though they can be active during the day when there is enough vegetation to hide their approaches. River crossings and watering places are favourite sites for ambushes.

Stalking and rushing prey demands that lions get as close as possible. When several lions hunt together they tend to try to encircle prey to cut off lines of escape. Experts disagree on whether or not lions pay attention to wind direction, though hunts made upwind are usually more successful. Although lions can eventually reach nearly 60 kmh (37 mph), most of their prey can sprint faster. So lions must get within 30 metres (100 feet) before charging, overtaking, slapping down and grabbing the victim.

Once grabbed, the victim is subdued and suffocated with a relatively quick neck bite or a sustained bite over the muzzle. Larger prey may be overcome by several lions together and members of the pride may begin to open the victim while one lion is still suffocating it.

Left, the mating game. Above, the intense gaze of an alert lioness.

LEOPARD

Leopards *(Panthera pardus)* are the largest of the spotted cats. Their heavy build, pug-mark spots and thick, white-tipped tail distinguish them from the more slender cheetah. They are the most elusive of the large cats, principally because of their nocturnal and secretive habits. Leopards have been little studied so not much is known of their behaviour in the wild. They are active day and night, but veer strongly towards nocturnalism, due to harassment.

Leopards are found in all except the driest African habitats—woodland, bushland, wooded grassland and forest. They are the most widespread member of the cat family, even commonly occurring in suburban areas. Large rocks and *kopjes* and large trees along rivers are favourite resting sites. As long as there is an adequate food supply and a minimum of persecution, the leopard is at home. Despite their adaptability, persecution by those involved in the fur trade and competetion with man for living space, have reduced leopard numbers drastically.

Their characteristic call is a deep, rough cough, repeated 10 to 15 times, sounding like a saw cutting wood. Males have distinctly deeper voices than females. The sawing call serves to advertise presence and to discourage other leopards from trespassing into defended territory, thereby avoiding destructive territorial fights.

Greetings are often accompanied by a short growl. The beginning of an agressive charge may be heralded by two or three short coughs, and those foolish enough to corner a leopard never forget the beast rearing up on its hind legs and uttering a blood-curdling scream.

Blending in: Leopards have few natural enemies and their skill in climbing trees assures them protection from all but the most aggressive lions. Anyone who has scanned the branches and canopy of a tree for a leopard knows just how well they blend into the blotched light and shade. Even the cer-

This proud female displays a magnificent coat of fur.

tain knowledge that a leopard is in a particular tree is no guarantee that it will be discovered. The switch of a hanging tail is sometimes a giveaway.

Leopards are solitary animals. They occupy and defend home ranges which vary from one to 30 square kilometres (one to 12 square miles) depending on the availability of food. Males and females defend their own, often overlapping, territories from members of their own sex. Female territories tend to be smaller and several may be encompassed within one male territory. Males often fight over their space and mark trees and logs throughout their area by clawing bark and spraying urine.

Good breeding: Breeding can occur at any time of year. Leopards, like all other cats except lions, are solitary breeders. The only long term social bond is between a leopardess and her cubs. Females come on heat for about a week every 20 to 50 days. They advertise their receptiveness with the sawing call which soon attracts the nearest territory holding male. A pair will then consort during the week of heat when matings are frequent. Males court, consort and mate, but there the honeymoon ends: they leave and take no part in cub rearing.

Gestation lasts around 100 days. Between one and six (average three) young are dropped in solitary retreats such as rock crevices and caves. At birth the cubs are blind and do not emerge from their birthplace to follow the female until they are six to eight weeks old. Young are weaned after three months and independent after two years.

Most small to medium herbivores, large birds, rodents and primates, as well as smaller carnivores, such as servals and jackals are fair game to leopards.

Leopards use mainly stealth and surprise to capture their prey. Like cheetahs and lions, they are stalkers, but their tree climbing habit adds a third dimension to their hunts: it is a common tactic to leap out of trees onto prey. If the prey is not secured after a rush of a few metres, it invariably gets away. Long chases are avoided.

Out on a limb: leopard and its kill are far away from unwelcome intruders.

CHEETAH

Cheetahs (*Acinonyx jubatus*) are lean, muscular cats, approximately 40 to 60 kilograms (90 to 130 pounds) in weight. They have spotted coats and a "tear stripe" running from eye to cheek. Their silhouette is long and lanky. Unlike the rest of the cat family, cheetahs do not have retractable claws.

You have a reasonable chance of observing cheetahs in Tanzania's Serengeti and Kenya's Amboseli, Maasai Mara and Nairobi National Parks. They are most common in savannah parks and wherever there are sufficient stocks of their preferred prey — species such as Thomson's gazelles. They are the most endangered of the three large cats perhaps because, unlike leopards, they are unable to adapt easily to the changes wrought to their habitat by man, who is gradually forcing them into marginal areas.

Cheetahs are diurnal animals with activity peaks at dawn. In game parks where they are hassled by tourist cars, they have taken to hunting during the heat of the day, at noon when tourists return to the lodges for lunch.

Cheetahs are generally silent, solitary animals except for consorting pairs and females with dependent offspring. Young animals which have just left their mother tend to stay together for a time and males sometimes band together temporarily to defend a territory.

Males and females only socialize whilst the female is on heat. They select breeding areas that have a reasonable number of gazelles, with good hiding places for cubs, perennial water and relatively low densities of possible cub predators. Males congregate in these areas which allows them to mate with local females.

Born to run: After a gestation period of three months cheetah cubs are born blind, naked and helpless. Litter size varies from one to eight (usually three) cubs. Females hide them away for two to three weeks in dens, often in dense vegetation or among rocks on kopjes. Mothers hunt for food leaving the cubs in hiding, periodically returning to suckle them. Every few days she moves her litter between dens. Although adults are not very vulnerable to predators, cheetah cubs are preyed on by hyaenas and other big cats, including lions and leopards.

Cubs may be born at any time of year, but

there is a peak which coincides with the appearance of gazelle fawns, which are easy prey for females. In the Serengeti most litters are born between January and April, coinciding with the long rains.

At five to six weeks cubs venture out after their mother to join her at kills. This in an important period for cubs since, unlike all other cats, young cheetahs do not know instinctively how to stalk, chase, catch and kill their prey. A cheetah cub, presented with, say, a mouse for the first time, will stare stupidly at it, or perhaps even run away. A young leopard or serval cat, in contrast, will pounce on the mouse without hesistation. So cheetah mothers must bring dazed or half-dead young gazelles or hares to their offspring and take them patiently through the process of being a hunter. Not surprisingly, many young cheetahs that have recently left their mothers do not have much chance of survival.

Cheetahs prey on relatively small animals such as Thomson's gazelles, relying on sight to locate, stalk and initiate pursuits. Groups of males can take on larger animals such as wildebeest, especially calves which are often taken during the mass calving on the Serengeti plains.

They select the least vigilant animal on the edge of the group as victim. Although cheetahs are the fastest land mammal, they cannot maintain their top speed of 95 kph (56 mph) for more than 200 to 300 metres (200 to 300 yards) so unless they can get undetected to within about 30 metres of the prey, before starting their chase, they are rarely successful.

In a typical kill cheetahs use their forefeet to knock the running animal off balance then clamp tightly on to its neck to strangle it. The dead prey is invariably dragged into cover where the cheetahs will feed. Females will call their cubs to the kill with a soft, bird-like chirrup. Adults will only hunt about four days after a successful kill.

Left, the epitome of speed, power and grace. Above, youngsters at rest.

GIRAFFES

Giraffes (genus *Artiodactyla*) have three subspecies: Maasai (*Giraffa camelopardalis tippelskirchi*), reticulated (*Giraffa reticulata*), and Rothschild's giraffe (*Giraffa camelopardalis rothschildi*). These subspecies differ only in their blotch pattern, and their distribution. All other characteristics are the same. The pattern of their coats is fixed for life, making it possible for human observers to distinguish one animal from another. Animals tend to get darker with age. It is not easy to distinguish males from females although males tend to be a little bigger and seem to spend more time feeding from tree canopies than females, which prefer to feed on low-lying vegetation.

Giraffes have very long necks which curiously have only seven bones, no more than the necks of any other animals although giraffe vertebrae are elongated. Both sexes are born with horns which are covered by skin and topped with black hair.

Giraffes weigh up to 1,000 kilograms (2,200 pounds). They have a high centre of gravity and this together with their weight, may account for their strange gait. Walking is almost a "pace": both legs on one side appear to move at the same time making it look as if the giraffe is rolling. When galloping, the hindlegs swing forward together to plant in front of the forefeet. Giraffes can reach a maximum speed of 60 kph (37 mph).

They inhabit open woodland and wooded grassland. They may also be seen in bushed grassland and occasionally at a forest edge. Giraffes often frequent drainage line vegetation in the dry season. Riparian thickets are the only place you are likely to see them in dense vegetation.

Giraffes are diurnal, but also move about at night. They sometimes utter snorts and grunts, but are normally silent animals. Often the only noise to be heard when giraffes move by is the click of their hooves when the foot is lifted clear of the ground as weight is removed from them.

These long-legged animals have few enemies. They are most vulnerable when drinking when they splay their front legs and lower their heads often in the vicinity of thick, waterside vegetation. Animals will only drink after carefully looking around. Young animals may be taken by lions if the adults are not around. Females will defend their young against any attacker by kicking with their front legs. The mortality rate up to the age of three is about 8 percent—which is not so different from the 3 percent adult average.

Giraffes are not much hunted by man. They only rarely raid local maize farms and their feeding habits make them almost impervious to drought. This might be the reason why, as other wildlife species continue to disappear, the giraffe seems to have become the only remnant of a formerly impressive wildlife array.

The social group: The individual is the basic social unit in giraffe society. Animals are loosely gregarious, with a usual group size of between two and 12 (usually six or less). The composition of herds changes constantly as adults come and go. The home range of a female may be 120 square kilometres (48 square miles) but they spend most of their time in the central part of this range where they feed. Dominant males wander in and out of female home ranges.

Fights between dominant males are rarely seen. When they approach each other, they frequently adopt a threat posture—"standing tall". This is normally enough to persuade the less dominant male to leave. Serious fighting occurs only when the fixed dominance hierarchy breaks down, for example, when a new nomadic male comes into the neighbourhood.

A ritualized form of fighting, known as "necking", is carried out by young males, normally between three and four years old. The animals intertwine necks, often accompanied by light blows with the head. The "winner" of the bout often climbs on the back of the "loser". This frequently leads the observer to the erroneous conclusion that necking is a courtship ritual.

Giraffes breed all year round. Females reach maturity at five years, males only at eight. Dominant males patrol the home range looking for females that come into heat. Males test females by smelling their vulvas or sampling their urine to see if they are receptive. If a dominant male finds a female in oestrus he will displace lesser males and consort with her until mating.

During their lifespan, females may have up to 12 calves weighing approximately 100 kilograms (220 pounds) each at birth. Calves are usually two metres (six feet) high. Twins are very rare. Births normally occur in a calving ground. This might lead to the formation of "creches" in which the unweaned calves spend most of the day together on their own.

Feeding occurs generally between six and nine in the morning and three to six in the afternoon. Giraffes are ruminants and spend a good portion of the day resting and chewing the cud. They are exclusively browsers, with 95 percent of their feeding confined to the foliage of bushes and trees, mainly acacias. All parts are taken—leaves, buds, shoots, fruits. Very infrequently, grass, forbs and creepers are eaten. Acacia fodder is harvested in a couple of different ways depending on the strength of the thorns: young, flexible thorns are flattened and the leaves stripped off with a sideways sweep: older, more robust thorns may also be flattened, engulfed in mucous and swallowed along with the branch. Very woody branches may simply be nibbled at selectively. The lips are used as agilely as the tongue.

The black tongue is long, 45 centimetres (18 inches), and mobile enough to curl around branch tips. Giraffe saliva is especially thick and viscous which undoubtedly helps in dealing with thorny branches.

Left, Maasai giraffe necking in the woods. Above, reticulated giraffe amongst the acacias in Samburu.

Normally, giraffes drink once a week, taking up 30 to 50 litres (63 to 105 pints) at a go. If vegetation is particularly dry, they may need to drink every two days.

ZEBRA

Zebra in *Swahili* means "striped donkeys". There is quite a controversy about the functions of the stripes but the general opinion is that they serve as a form of visual antipredator device, either as a camouflage or to break up form when seen from a distance.

Two species are found in Kenya: Grevy's (*Equus grevyi*) and Burchell's zebra (*Equus*

Burchelli) Grevy's zebras are only found in northern Kenya. There are still about 10,000 left, but their numbers are rapidly disappearing due to hunting by man for their skins. They have narrower stripes and larger ears than Burchell's zebra. Burchell's or common zebras are found throughout East and Central Africa. There are more than 100,000 in the Serengeti alone.

Zebras are animals of open and wooded grasslands and are never far from perennial water. Grevy's inhabit more arid areas and are much more drought-resistant than Burchell's, which is rarely found far from a source.

They are active round the clock but will look for shade at midday, resting periodically at night. Zebras make a typical horse-like neighing sound.

Among their enemies are lions, hunting dogs, spotted hyaenas, and, of course, man. A common response to alarm is bunching. Zebra stallions are fierce fighters and kick back with great ferocity. Mares are as brave as stallions when their foals are involved.

The basic zebra social unit is a more or less permanent group of females with young looked after by a dominant stallion. The tightest social bond is naturally between mare and foal. Young females stay with the family group: as they approach maturity males wander off with bachelor groups.

Burchell's zebra live in groups with permanent membership: there is a lead stallion and a number of females with their offspring. The Serengeti groups get together and participate in the great migrations of East and Central Africa when the range begins to deteriorate in the dry season.

Grevy's zebra have temporary associations that rarely last more than a few months. Territorial males defend large territories during the breeding season and attempt to keep female groups within their boundaries. Outside the breeding season, they mix with other male groups.

Breeding in both species is linked to the rains. Sexual maturity in males occurs at about one to two years but social maturity—the ability to take and defend a territory or female herd—does not come until about six years. Courtship is followed by repeat matings at one or two hour intervals over two days. Gestation takes around a year.

Young independence: Young foals are surprisingly independent, even at one month old. They may be left alone whilst the mare grazes hundreds of metres away or walks several kilometres to water. Animals up to about the age of six months are reddish-brown rather than black.

Zebras have evolved stomachs which allow them to feed on coarse, stemmy grass largely passed over by other members of the grazing community. This enables them to survive when other grazers cannot.

HIPPOPOTAMUS

Hippos (*Hippopotamus amphibius*) are the second largest terrestrial mammals, weighing up to 2,000 kilograms (4,400 pounds), with barrel shaped bodies and short legs. Their heads are adapted for life in water with eyes, ears and nose all on the upper side. Hippo jaws can open up to 150 degrees wide, which makes a very impressive sight when this is all that can be seen of the semi-submerged animal.

Hippos live in still or slow-running water with frequent bends in the shoreline and deep pools with shallow sided banks, in the midst of open wooded or bushed grassland. The preferred water temperature is 18 C to 35 C (64 F to 95 F) and they have been seen along the sea coast. In the rainy season, males may take up temporary residence in seasonal water holes.

Hippos spend the whole day in water, totally or partially submerged, floating beneath the surface and bouncing from the bottom to come up to breath every few minutes. Dives generally last less than five minutes but can be as long as 15. Hippos come out of the water only at night to feed on the shore.

They have a variety of vocalizations, among them the dominant male's: "MUH-Muh-muh"; bellows and roars when fighting; a high-pitched "neighing" when attacked and a sort of snort when submerged.

Man is the hippo's only real threat although a pride of lions will attack a solitary hippo on land, and crocodiles undoubtedly take the occasional baby hippo in the water. Females defend their young by making use of their long tusks (canines).

Despite their benign look, hippos account annually for more wildlife-induced human deaths than any other animal, including lions and snakes. They specialize in capsizing boats which get too near, either drowning or biting the people inside. Hippos are also dangerous on land since they will run over anybody standing in their way to the water.

Gregarious groups: Hippos are gregarious and territorial. Group size averages 10 to 15 females with young, led by one dominant male. When water resources become scarce during droughts, groups of 150 have been

Left, Grevy's zebra have slimmer stripes and larger ears. Above, the more commonly seen Burchell's zebra.

observed. Dominant males space themselves out along the lake shore or river course advertising their dominance with the familiar "MUH-Muh-muh" call. Sexually mature males are kept on the edges of the group with threats and fights until, at eight to 10 years old, they feel strong enough to take on a territory holder. As the flow of river changes with seasons, territories may break down. In more stable lakes, territories may persist for years.

Hippos have a curious habit of spraying their dung around a two metre radius by rapidly whirling their stubby tails as they defecate. Defecation spots are communal and it is thought that there are used as a

display to other hippos.

Courtship and mating occurs in the water. In theory, territorial males have exclusive mating rights, but they are continuously challenged by other males. Fights are common, leaving animals visibly scarred along their back and flanks.

Females give birth in shallow water, suckling occurs on land or in ankle-deep water and lasts for eight months.

Hippos are grazers and feed at night, coming out of the water along well-worn trails. They do not feed on aquatic vegetation. During the rainy season when grass is abundant they feed near the shore. In the dry season they may walk up to 10 kilometres (six miles) inland in search of food. Their half-metre wide lips allow them to eat very short grass. Fallen fruits, like water plants, are rarely taken.

The total amount of food taken by an adult hippo is less then that ingested by other cloven hoofed animals (only 1 to 1.5 percent compared with 2.5 percent of body weight). This is only possible because of the hippo's undemanding life-style: floating around the pool all day does not take much effort.

PIGS

African pigs (*Suidae*) are medium-sized herbivores with compact bodies, large heads and short necks. They have coarse, bristly coats, small eyes, long ears, prominent snouts and tusks (elongated lower canines). The flattened face and broad snout, ending in the characteristic, naked pig nose, are related to both the search for food and fighting style. Despite their heavy bodies, pigs can swim and they are agile and quick. All male pigs are larger than females and have larger tusks and warts. Sight is, in general, the poorest of the pig senses.

A number of pigs can be found in East Africa, among them: giant forest hog (*Hylochoerus meinertzhageni*) covered with shaggy, dark, coarse hair; bushpig (*Potamo-*

choerus porcus) and red river hog members of the same species, varying in coat colour from reddish-brown in the forests to blackish-brown in the drier bushland, and warthogs (*Phacochoerus aethiopicus*) which are less hairy, but covered with sparse bristles with hairs along the neck and back that can be raised like a crest and serve as a signal during social interactions.

Pigs can be found in a series of different habitats ranging from grassland to bushed grassland (warthog), bushland to dry forest (bushpig), to moist evergreen forest (giant forest hog).

The warthog is an accomplished and compulsive digger and may dig its own burrow using its forefeet. Most other pigs seem to modify existing holes for their burrows. Bushpigs do not live in burrows.

All pigs are diurnal, except the forest hog which is mainly nocturnal.

Continuous, soft grunting serves as a contact sound to keep members of a foraging family group together. Tooth-grinding may be heard when the animal is aroused, presumably angry. Alarm and distress are expressed with porcine squeals.

The most important pig predators are lions and cheetahs. The largely nocturnal forest hog is susceptible to leopards. Pigs will defend themselves from predators by using their lower tusks (canines) which are pointed and very sharp. Nevertheless, their main line of defence is retreat at speed, usually towards the nearest hole.

Bushpigs are often very destructive to local agriculture. Warthogs live in a more pastoral setting and therefore do not cause much damage to agriculture.

Pig society: The main social group consists of temporary female-young bonds and more lasting female-female associations. Such associations account for the joining up of litters. There is no herd boar in a functional sense, although an adult male is usually attendant. Although not very territorial, males mark their 10- to 20-square-kilometre (four- to eight-square-mile) home ranges with secretions from lip glands, pre-orbital glands (warthog) or foot glands (bushpig) by rubbing them against tree trunks and stumps. Pigs are frequently seen rubbing their necks and spreading their saliva with their faces.

Greetings are mostly a matter of sniffing: nose to nose, to mouth, to pre-orbital region or rear end. They sniff the ground in places where another animal has stepped, presuma-

Left, hippos spend all day in the water. Above, amorous warthogs.

bly to taste the gland secretion. This may lead to a friendly greeting.

Pigs are rapid breeders. Females are able to conceive at the age of 18 months, and the size of their litter can be as many as a dozen young either born in a grass "nest" constructed by the sow or, more commonly, in a borrowed aardvark hole. Males, meanwhile, mature at nearly three years. This may explain why breeding males usually ignore fully grown yearlings.

As there is no marked male dominance among the pigs, conflicts between males are fierce when associated with a female in season. The strongest pig is the one that will mate. Fighting between male bushpigs may involve upraised snouts and tusks slapped from side to side, accompanied by whirling around. Side blows may be followed up by biting. It seems that hierarchy is based on tusk length; if a tusks breaks, as frequently happens, then the pig loses status.

The male warthog's threat display is characterized by stiff-legged strutting, an erect mane and tail and head and shoulders raised with the snout pointed downwards and towards the opponent to present the tusks to full advantage.

Female fights are seen less commonly than male encounters and consist of rushes and display if two family groups happen to come too close together. As farrowing approaches, pregnant sows may become more solitary and quick-tempered, engaging in male-like head-on encounters.

Females farrow in burrows. The piglets develop rapidly and can leave the burrow and begin to eat grass after only one week. However, they suckle for as long as four months. Suckling is generally done while the sow is standing, although she has to lie down in the burrow.

The previous young may try to rejoin their newly-farrowed mother but will be roundly chased off. This is a vulnerable time for yearling males since they have problems finding burrows to share with adults of either sex: mortality from predation is correspondingly high. If, however, the female fails to conceive, then the young will stay with her until she conceives the next time.

Shared burrows are uncommon, especially for farrowing and only occasionally occurs with first-time breeders. Males and females take up companionship again after farrowing, to culminate in mating the next season.

African wild pigs are basically vegetarians. They eat a wide range of plants, grasses, forbs, ferns, fungi and consume almost any part of the plant. Small vertebrates and invertebrates are also eaten when found. All pigs are dependent on water and will always be found near a source.

The manner in which different species search for food is largely dictated by habitat. Warthogs are the most specialized feeders and eat mainly grass. While feeding, they walk from tuft to tuft, or move slowly forward in a characteristic "kneeling" or "bowing" posture, by lowering their short-necked front end and settling on their fore carpals. Tusks are used only occasionally to root out food: most digging is done with the rough upper and leading edge of the nose.

Bushpigs spend more time rooting by digging with the nose. It can dig several centimetres deeper than warthogs by taking full advantage of its more flexible snout. Giant forest hogs feed mainly in grassy forest clearings where they eat mainly grass. In the forest, however, they like to eat a variety of herbs and plant parts. They rarely root with their snouts like other pigs. The combined length of their head and neck allows them to reach the grass sward usually without kneeling.

Left, sinister grin of the spotted hyaena. Above, one hyaena pack feeds while another waits its turn.

SMALLER PLAINS PREDATORS

There are four distinct families or subfamilies of smaller plains predators: hyaenas; the dog family (*Canidae*), broken down into African wild dogs and others such as jackals and foxes; and small cats.

HYAENAS

There are two distinct species of hyaena in East Africa: spotted hyaena (*Crocuta crocuta*) and striped hyaena (*Hyaena hyaena*).

Spotted hyaenas have powerful forelegs and shoulders, long necks and heavily built skulls with formidable teeth. Ears are round and the coat is spotted. Young hyaenas have darker spots then adults which appear almost spotless.

Striped hyaenas share the powerfully-built shoulders and head, but they have longer legs and the forefeet are much larger than the hindfeet. They have a crest of hair running down the back to a very bushy tail. Their ears are more pointed. The body is light coloured, with the outer surface of the legs striped boldly. Senses are very well developed in both species.

Spotted hyaenas are more widespread than striped hyaenas. They are found in dry acacia bushland, open plains and rocky country where there is abundant wildlife. They are not common in heavily wooded country or in forests. They can live at high altitudes, up to 4,000 metres (13,200 feet).

Striped hyaenas inhabit the drier parts of East Africa. They have been found in deserts in areas where water is not available. for many miles.

Both species are very vocal. The characteristic "whoop whoop" howl can be heard throughout the African night. When fighting, they issue a hoarse "ahh ahh" sound. When excited, the hyaenas' uneven-pitched howl is eerily reminiscent of the laughetr of a demented soul. Man and lions are the hyaenas' prime enemies.

Spotted hyaenas are usually seen alone or in pairs, but in areas of concentration of

number of males, of which one eventually mates with her. Young are born in dens shared by several females. After a gestation period of three to four months, one to three (usually two) cubs are born. Food is not carried to the den, so the young hyaenas depend entirely on their mother's milk for approximately eight months and are only weaned at 12 to 16 months when they can feed by themselves.

Striped hyaena females form a temporary bond with the male with which they have most recently mated. Gestation is three months and one to six (normally two to four) cubs are born in dens. Both parents will bring food to the cubs.

available prey they may form temporary or even permanent groups which share, patrol and defend a hunting territory against other clans.

Spotted hyaenas are not individually territorial. They make and break bonds with other hyaenas very easily. Once a group has been formed hyaenas become territorial.

The family: Striped hyaenas are basically solitary, forming a pair during the breeding season. Both parents help to rear the young and a family unit may be formed once young are mobile. Striped hyaenas occasionally congregate at a kill but do not form groups.

When in season spotted hyaenas attract a

Both species are opportunistic hunters and scavengers, taking advantage of wastes left by other animals or man. Hyaenas in general will take weaker or sick animals in preference to healthy ones. Spotted hyaenas show a primitive form of cooperative hunting which is generally more successful and is better able to resist the consequences of a counter attack by another wildebeest, for example. Both species will eat the dead of almost any mammal, bird, reptile, fish, irrespective of size or species. Striped hyaenas are more omnivorous than spotted hyaenas. Insects, birds, reptiles and fruit seem to be their main diet.

CANIDAE: AFRICAN WILD DOG

African wild dogs (*Lycaon pictus*), also called hunting dogs, are medium-sized carnivores. Adults are approximately the size of a labrador dog but are slimmer and light for their size, weighing an average of 25 kilograms (55 pounds).

Wild dogs have brindled coats of brown, black, yellow and white. They all have similar black face masks and white-tipped tails. The rest of the pattern is distinct for each dog. Large, rounded ears allow dogs to hear over long distances. They have very sharp shearing teeth.

Wild dogs are typically found in savannah grassland and woodland. They have also been seen at altitudes of 5,600 metres (18,480 feet) in the snow of Kilimanjaro or in the heat of deserts. They are most active in the early morning and evening and lie in shade during the day.

Wild dogs are surprisingly vocal. A bird-like yittering is often heard when the dogs greet one another. Alarmed dogs will bark: distressed puppies occasionally give a squeaky hoo-call.

Lions generally ignore wild dogs but will feed on the dog's kills after chasing them off.

Packs consist of a dominant breeding pair, five to six other adults and dependent young. With litters averaging 10, and bitches capable of producing 16 puppies, packs can build up to number as many as 50. Young females emigrate at between 18 months and three years of age: young males normally remain in their natal pack.

Like their hunting behaviour, the breeding system of wild dogs is an example of remarkable cooperation. In each pack only the dominant pair breeds: the other dogs help to rear their offspring. This is for the good of the pack since all dogs are related but only the strongest genes are passed on. Breeding coincides with the rains.

Gestation lasts approximately 10 weeks and the dominant female whelps in a den. For the first fortnight the female suckles the litter, spending long periods underground with them. After two weeks in the den, puppies make their first unsteady forays outside and begin to feed on meat regurgitated to them by all the pack members. Within a month or so, they are feeding on meat alone.

Wild dogs kill a wide range of animals but specialize on small to medium-sized antelopes such as Thomson's gazelles and wilde-

Left, wild dogs crowd round a kill. Above, grooming time for these silver-backed jackals.

beest. In bushed and wooded habitats their preferred prey are impalas. Such ungulates represent the vast majority of their kills, although they opportunistically snap up gazelle fawns, hares, hatchlings and other small prey they chance across. Antelopes as large as eland are only rarely taken.

Packs normally hunt once a day—more often if the group is large or there are puppies to be fed. Prey size also influences the number of hunts: a wildebeest will obviously satisfy more dogs than a gazelle.

CANIDAE: JACKALS AND FOXES

This group comprises smaller, dog-like carnivores: *jackals*—side-striped *(Canus adustus),* golden *(C. aureus)* and black-backed or silver-backed *(C. mesomelas)*—and the relatively common bat-eared fox *(Otocyon megalotis).*

All of them have a striking similarity with domestic dogs in the way they move, lift their legs, raise their hackles, scratch, bury food and roll in something rotten. The senses of *Canidae* are all very well developed.

Small *Canidae* are fairly common and may occur in densities of around 10 per square kilometre (a third of a square mile).

Jackals frequent open wooded and bushed grassland. Golden jackal and the bat-eared foxes prefer more open and arid habitats, from grasslands to semi deserts.

Small *Canidae* are active 24 hours a day; their tendency to be active at night is reinforced by human persecution.

All are vocal to a degree: black-backed jackals yapping is a familiar sound on the plains.

They are prey to larger carnivores and to rock python and large birds of prey, such as martial eagles. Alertness and quickness are their main defences.

They are very sociable animals. The basic social unit is a pair—either permanent or for a few seasons. Occasionally they pair up with the young of that year or with some of the previous year's offspring. Several breeding and non-breeding adults form more or less permanent social groups within a home range, but essentially a pair marks and may defend a small territory which includes one or more subterranean dens.

Pair formation begins with consorting and mutual grooming some months before actual mating. Behavioural observations indicate that pairs tend to persist beyond one season, at least six years in the black backed jackal, and probably longer. Bat-eared foxes, for example, pair for life.

The young are born helpless, just like dog puppies, and will stay in the safety of the burrow, suckled by the mother. When they emerge from the den, they are still suckled and begin to be fed on regurgitated food by adults.

"Nannies": Black-backed jackals, golden jackals and probably the other species, have non-breeding "helpers", usually the young of previous seasons, which assist in care of the current young. This ensures a higher rate of survival of the pups.

Although perhaps not as sophisticated as with lions, hyaenas and wild dogs, cooperative hunting is important in jackals. In general, jackals and foxes are opportunistic carnivores. They will feed on almost anything they can catch or unearth—small vertebrates, invertebrates of any size, young animals, eggs, carrion, even some fruits. Only bat-eared foxes show a tendency to some specialization: 80 percent of their diet is insects, mainly *Hodotermes* termites.

Scavenging is one of the main elements of the jackals' search for food. As many as 30 jackals can be seen, often along with vultures, at the fringes of a lion or hyaena kill, waiting for a chance to dash in and grab a piece of meat from the carcass.

SMALL CATS

This group of the cat family *(Felidae)* includes a number of small cats which are relatively common but infrequently seen. These include: wild cat *(F. s. lybica)*, sand cat *(F. margarita)*, caracal *(Caracal caracal)* and serval *(Leptailurus serval)*.

Their sizes vary from the two- to three-kilogram (four- to six-pound) sand cat to the 15-plus kilogram (over 33 pounds) serval and caracal. In general, cats do not have a very good sense of smell, but their hearing is excellent and keenly directional; eyesight is acute in dim light. All species have retractable claws.

Small cats are generally found in all habitats with perennial water. Only caracal range in semi arid country to the edge of deserts. All species are active during the night or early hours of the evening and morning.

As with domestic cats, small wild cats communicate with a number of variations on the "meow" theme. They snarl and spit when warning and purr when content. Servals have a characteristic little bark when calling to conspecifics.

Their main enemies are large raptors and other big predators. Their basic defence is alertness and avoiding open areas in which large birds of prey can fly.

All small cats are solitary except for short periods of consorting with a mate and during suckling and weaning of young. Individuals have a hunting range of 1.5 to three square kilometres. Males are territorial to other males within their range but permit females to wander in the territories unhindered.

Small cats are overwhelmingly carnivorous. Small living prey is stalked, pounced upon, hooked with extended claws, and then killed. Killing techniques depend on the size of the prey: very small animals—insects, lizards, mice—are simply bitten to death or squeezed in the jaws until suffocated. Larger animals—rats, hyraxes, large birds—are characteristically killed with a prolonged bite to the nape of the neck, which effectively breaks it and eventually severs the spinal cord.

Caracal, serval and other large cats are able to bring down large birds, such as bustards, and the kids of gazelles and other small antelopes.

In some cases wild cats take domestic

Left, the hard-to-spot serval cat. Above, Ethiopia's equally elusive Simien fox.

stock. It rarely becomes a habit, however, and such small depredations are more than offset by their importance in controlling rodent pests.

PLAINS GRAZERS

The open and wooded grasslands and bushlands of East and Central Africa support the most splendid variety of small, medium and large herbivores in the world. All are members of the family *Bovidae* to which our domestic livestock also belong. "Small" in this context means about the size of a large dog, "medium" means about the size of a pony; "large" means like a small horse; "very large herbivores" are the really big

ones, like elephants, buffalos and giraffe.

Alcelaphines are long-faced, rather foolish looking, medium sized antelopes with S-shaped, ridged horns and sloping backs. They include the spectacular Serengeti migrating wildebeest or gnu (*Connochaetes taurinus*), hartebeest or kongoni (*Alcelaphus buselaphus*), topi or tiang (*Damaliscus lunatus*), hirole or Hunter's hartebeest (*Beatragus hunterii*) and impala or swara (*Aepyceros melampus*).

There are some 20 species and subspecies of gazelles in the *Antilopini* family. All have the same general appearance with slender bodies, long legs, fawn-colouration on top, whitish underneath, and a white rump. Males are larger than females, have better developed S-shaped horns, a black-white blaze on their flanks, and a black-tipped, constantly wagging tail. The main species are Grant's gazelle (*Gazella granti*), Soemmering's gazelle (*G. soemmerringi*), Thomson's gazelle (*G. thomsoni*), Dorcas gazelle (*G. dorcas*), gerenuk (*Litocranius walleri*), dibatag or Clarke's gazelle, (*Ammodorcas clarkei*), and the springbok (*Antidorcas marsupialis*). Dibatag and gerenuk differ from the rest in having conspicuously long, slender necks.

Hippotraginae include the medium to large, thick necked, "horse-like" antelopes that sport impressively long, swept-back or straight horns. They include: roan (*Hippotragus equinus*), sable (*H. niger*), oryx (*Oryx gazella*) and the virtually extinct addax (*Addax nasomaculatus*). The broad chested *Hippotraginae* run with powerful, sure-footed strides, hence the name "Horse-like antelopes". Roan's Latin name belabours the point: "horse-like horned horse".

Reduncinae are medium-sized antelopes whose males have somewhat lyre-shaped horns. Their coats are usually coarse, brown to reddish-brown to beige, usually with a white chevron at the throat. This group includes the waterbuck (*Kobus ellip-*

siprymnus), kob (*K. kob*) and reedbucks (*Redunca spp.*) . They range in bulk from heavy set waterbucks to spritely reedbucks. Waterbuck horns have a smooth, open sweep; those of reedbucks are shorter, forward pointing at the tips; and those of kobs have the classical impala-like lyre shape.

Tragelaphinae are medium to large, slender, round-backed, longish-necked antelopes, russet to grey-brown in colour, often with vertical white stripes or a line of dots on their sides and flanks. Males in particular have well-developed spiralling horns, and are larger and often darker or greyer than the females. The only real plains grazer is the eland (*Taurotragus oryx*). Sitatungas (*T. spekei*) live in a very restricted habitat in the swamps and have evolved long, splayed hooves to assist in walking over mud and boggy vegetation. All *Tragelaphinae* have a delicate, high-stepping gait, even the two-metre- (six-foot-) high eland.

The relatively smaller, round-backed, dainty antelopes are members of *Bovidae* in the subfamily *Antelopinae*. Among them are duikers (*Cephalophinae*); dwarf antelopes (*Neotraginae*), including dikdiks (genus *Madoqua*), steinbok, oribi (*Ourebia ourebi*) and klipspringer (*Oreotragus oreotragus*). They look more or less similar and are all less than 80 centimetres (31 inches) high; they inhabit areas of dense cover, are territorial and live in very small groups centred around life-long pairs; all mark their territories with secretions from conspicuous glands in front of their eyes. Their short-necked, arched back body frame assists in rapid movement through thick bush. Steinbok and oribis which inhabit open grasslands have a more upright posture, as do klipspringer which must balance upright as they spring from rock to rock. Males have horns; females are either hornless or have weak and poorly formed horns.

Left, chaotic annual migration scene. Above, distinctive facial markings and straight horns of the oryx.

Habitats: Antelopes live in nearly all East African habitats. Where man has not extensively broken the soil wild herbivores can be seen. Wildebeest and their *Alcelaphine* relatives favour open and wooded grassland; bushland and thickets are generally avoided where grass growth is relatively weak and there is more cover for predators. Impalas, however, prefer woodlands, riverine strips and zones between vegetation types (transition zones). Gazelles have adapted to a wide range of habitats, from arid to semi arid country in the case of gerenuk and some

Grant's subspecies, to Thomson's gazelles which prefer better watered grassland. Roan and sable inhabit grasslands with good bush and tree cover and they both frequent well-watered grasslands and wooded valleys. In contrast, oryx prefer very arid habitat and can live in near-desert conditions.

All species of *Redundinae* antelopes prefer wetlands or tall, tussocky, marshy grasslands. Even hillside-living reedbuck are also associated with wet grasslands and hill marshes. Lechwes are almost semi-aquatic, and spend most of their time in marshes, swamps and inundated grasslands.

In general, *Tragelaphinae* live at low densities in arid to sub-humid areas, in regions of thick cover, such as forest (bushbuck, bongo), bushland (bushbuck, lesser kudu, nyala) and hill thickets (greater kudu, mountain nyala). The two notable exceptions are elands which frequent open grasslands often roaming as high as 4,500 metres (14,850 feet), and the long-hoofed sitatunga, which splashes around in swamps and marshy lakesides.

Very small antelopes exploit a wide range of habitats: forests (duikers, suni) to thickets (dikdik), on kopjes or rock outcrops (klipspringers) to open grasslands (steinbok, grysbok, oribi).

All species are diurnal, with preference for the cooler mornings and evening for feeding. Except for impala and some of the very small antelopes, most breed in a very narrow period, usually dropping the calves just before the rains.

They rely on alertness and flight to escape their enemies—big predators such as lions, leopards, cheetah, hunting dogs, hyaenas.

An eland group is usually made up of a few females and young with an adult male in attendance. This male is the one that will eventually mate with the group's females. Females reach sexual maturity at three years. Males are loosely territorial. During conception peaks, small groups aggregate into big groups of hundreds of elands which can move over great distances and can be considered to a certain extent migratory.

Steinbok and oribi are strongly territorial and pair for life. Their territories are small, about 50 to 500 metres (165 to 1,650 feet) in diameter. This small size allows territory holders to explore every bit of their turf so that they come to know of all the best bolt holes and special food plants. Sexual maturity is reached in less than a year and gestation is about six months, allowing two births per year.

Waterbuck, kob and reedbucks have small, loose associations of adult females and young, moving through a world of male dominated territories. Groups are rarely bigger then 10 to 15 animals.

Kobs exhibit a variation on the territorial theme. They establish a *lek* or territorial breeding ground, where a group of dominant males display and mate. Gestation is about nine months and a single calf is dropped.

Breeding season: Some male *Antelopinae* (gazelles) and *Alcelaphinae* (wildebeest, hartebeest, topis) are territorial during the breeding season. Males that have set up their territory try to prevent females from leaving. Females tend to move on when the grass supply is diminishing.

These herbivores are rarely seen alone, except for the occasional territorial male wildebeest forlornly standing his ground outside the breeding season. Most herds are usually made up of less than 100 animals, with the exception of large, migratory populations of topi and wildebeest which may be seen during the migrations in tens of thousands. Their herds spread over the landscape, often without any apparent structure. Breeding is polygamous with one male fertilizing many females. Births are synchronised, particularly in wildebeest, into a calving peak which lasts only a few weeks. This floods the predators' market and, even though they eat their fill, the percentage of calves that they can take is much smaller than if calving were spread throughout the year.

Hippostraginae (roan, sable and oryx) have a system of matriarchal hierarchy, resulting in small herds of five to 20 animals with varying degrees of male participation. In sables, female herds range over adjacent male territories: in roan, a single male accompanies the females; in oryx, male groups satellite female ones for most of the year. Only the dominant male in a group of oryx mates.

Left, long-faced Coke's hartebeest in Tsavo West. Above, the "horse-like" roan antelope.

Herbivores are of four basic types: those which mainly graze (wildebeest, hartebeest); those which graze very selectively, allowing the animal to take a higher plane of nutrition (steinbok, oribi, waterbuck, reedbuck, roan, sable and oryx); those which both graze and browse (topi, impala, eland, gazelles); and those which predominantly browse (gerenuk). The relative shapes of their mouths makes the distinction clear. All are ruminants, that is, they have several stomachs for fodder in varying stages of digestion and they reprocess already swallowed fodder by chewing the cud, like domestic cattle.

BUSHLAND AND FOREST ANTELOPES

These antelopes can best be divided into two separate groups, the medium to larger ones, known as *Tragelaphinae* (eland also belong to this group) and the dwarf and small antelopes.

Tragelaphinae are slender, round-backed, longish-necked antelopes, russet to grey-brown in colour, often with vertical white stripes or line of dots on their sides and flanks. They have short-cropped pelage and a delicate, high-stepping gait. An erectable crest occurs on the back of the bushbuck. Males have horns and their colour is often darker or more grey.

Members of this group are: eland (*Taurotragus oryx*), bushbuck (*Tragelaphus scriptus*), lesser kudu (*T. imberbis*), nyalas (*Tragelaphus buxtoni*), greater kudu (*T. T. strepsiceros)*, and bongo (*T. T. euryceros*).

They live at low densities in sub-humid areas, in regions of thick cover, such as forest (bongo), bushland (bushbuck, lesser kudu, nyala) and hill thickets (greater kudu, mountain nyala). They are all diurnal, although bushbucks will become nocturnal in areas of persecution.

All species are preyed upon by the obvious large carnivores and the smaller species are also vulnerable to smaller cats, such as servals, caracals, golden cats and other predators such as rock python and crocodile. When threatened, animals emit a sharp bark as an alarm. Most rely on flight for protection.

The basic groups are made up of females and immature animals, with an adult male in attendance. Group numbers vary from two to three in bushbucks reaching a maximum of about 10 to 15 in lesser kudus.

The onset of female sexual maturity is around 18 months. Males continue to grow beyond sexual maturity, so that the differences in body size between the sexes becomes gradually more striking.

A single calf is born after a gestation of about six months. Calves are weaned at about three to four months.

Tragelaphinae are very selective browsers and pluckers. This allows them to have a highly nutritious diet. Fruits, seeds, pods, flowers, bark and tubers are taken as well as leaves. All species take grasses, but not as a bulk food.

DWARF AND SMALL ANTELOPES

This group is made up of small, round-backed antelopes with short necks and conspicuously large ears—except dikdiks (genus *Madoqua*). It includes duikers (genus

Cephalophinae), dwarf antelopes (*genus Neotraginae*), beira (*Dorcotragus melanotis*) and klipspringers (*Oreotragus oreotragus*). The males have horns; females are either hornless or with very weak and variable horns. The smallest is a *Neotragine*, the royal antelope at a height of some 25 centimetres (10 inches); the largest is a *Cephalophine*, the yellow-backed duiker which may approach a height of 80 centimetres (31 inches). The rest, however, are 40 to 60 centimetres (15 to 23 inches) at the shoulder.

Females can be larger than their males by as much as 20 percent in height and weight. The relatively large size of the females may arise from the fact that these antelopes form a permanent pair bond, meaning the male does not have to fight continually with others for a mate. Females also share in the defence of territory and greater size in territorial conflicts is an advantage.

They are found in forests (duikers, suni), in thickets (dikdik) or on *kopjes* or rock outcrops (dikdik together with klipspringers and rare beira).

Small antelopes are generally active all day with peaks in early morning and evening. Their voices, used principally as an alarm, are high-pitched "zick-zick" sounds (hence the name dikdik) or a tiny snort. Most other species have an alarm whistle.

Their enemies are all large predators, plus an array of medium and small predators, including cats, snakes, birds of prey, ratels and baboons. Their only defence is to take to flight.

All these animals pair for life and are very territorial. Territory sizes may vary from 50 to 500 metres (165 to 1,650 feet) in diameter, depending on season and local conditions. Such relatively small territories allows the animals to know precisely both the location and season of food plants, the best escape routes and the most effective hiding places.

Sexual maturity may be reached in less than a year: gestation is about six months. Two births a year are therefore possible, given a relatively constant food supply.

At less than two years old, the young animal will leave its parents' territory.

Dwarf and small antelopes are almost exclusively browsers and nibblers on fine-structured vegetation. They take the most nutritious plants and parts which provide a high plane of nutrition. They are not normally seen drinking: moisture comes mainly from food plants.

Left, with its long neck, the gerenuk can stand on its hind legs to graze the higher branches. Above, a pair of impressive impala.

Smaller Mammals

There are many groups of small mammals in a number of families, ranging from small carnivores, such as genets, to herbivores such as hyraxes, to monkeys. They can be grouped into classes according to similar characteristics.

WEASELS, ZORILLAS, RATELS AND OTTERS

These small, carnivorous mammals belong to the family *Mustelidae*, which includes African weasels, such as the African striped or white-naped weasel (*Poecilogale albinucha*), zorilla or African striped polecat (*Ictonyx striatus*), ratel or honey badger (*Mellivora capensis*), spotted-necked otter (*Lutra maculicollis*) and the cape clawless otter (*Aonyx capensis*). They have elongated, flattened bodies, powerful jaws and relatively large brains. Except for otters, all these mammals have a striking horizontal black and white pattern.

They inhabit woodland, bushland and grasslands and tend to use one or more dens (holes in the ground) as bases. Otters are rarely far from river banks. Weasels and zorilla are mainly nocturnal; ratels are active most times except noon; otters are diurnal.

In general *Mustelids* are quite vocal, especially when annoyed. Otters are very vocal and have a range of twitters and chirps which serve as contact calls.

Few other animals will attack them due to their repulsive anal gland secretion and, in the case of ratel, its ferociousness. Weasels may be at risk from large owls. Otters are taken by crocodiles and rock pythons; zorillas are often run over by cars.

Otters live in small family groups of less than 10. Not much is known about their breeding behaviour except that a pair may consort temporarily for several months during the breeding season. Gestation is presumed to be two months.

Mustelids take almost any living prey smaller than or equal to their own body size. The zorilla's diet has been measured to consist of nearly 50 percent insects, 25 percent small mammals and the rest made up of birds, amphibians, spiders and plant material. Rodents are the main item in the ratel's diet although it may catch small antelopes. Otters feed mainly on aquatic food such as crabs, fish, frogs, molluscs and insects which they catch during dives of about one and a half minutes each.

GENETS, CIVETS AND MONGOOSES

These small mammals are low slung, long tailed, largely nocturnal carnivores, with a variety of striking spottings, stripings or conspicuous tails. All members of the group have large eyes, facing front, and outstanding night vision.

The family *Viverridae* contains three subfamilies: mongooses (*Herpestinae*), the true civets and genets (*Viverrinae*), and the palm civet (*Paradoxurinae*). There are some dozen species of more or less solitary mongooses, a dozen genets, one African civet (*Civettictis civetta*), and one two-spotted or African palm civet (*Nandinia binotata*).

Their habitat stretches from forest edge to semi-desert. Genets and palm civet are semi

Left, a large-spotted genet in the Aberdares. Above, eyeing the greater galago.

arboreal; true civets and mongooses, with the exception of the slender mongoose, are more likely to seek refuge in crevasses, holes in the ground tree trunks and tree roots.

Genets, civets, and most mongooses are nocturnal, occasionally crepuscular. Both Egyptian and slender mongooses are active by day, in the evening and on moonlit nights. The marsh mongoose is largely diurnal.

Vocalisations are used for contact (genets: "uff-uff-uff"; civets: "tsa-tsa-tsa"), as well as for alarm, excitement or threat, when they growl, spit and hiss.

Their enemies include medium sized predators such as African wildcats and large owls. When threatened they rely on stealth and early warning. Genets have very good eyesight, especially at dusk. If disturbed, they will take to the nearest tree.

Viverridae are generally solitary and only occasionally seen in pairs. (The exception is social mongooses). Their striking black and white markings on faces, body and tail are undoubtedly used in sexual and social signalling.

Breeding occurs throughout the year with some indication of seasonal peaks. Courtship and mating has been described as catlike. Genet matings last three to five minutes; gestation is some three and a half months and two or three young are born in tree holes. They are weaned after two to three months and become independent after nine.

Viverrids are carnivorous, almost omnivorous. They are generally opportunistic feeders but also function as quick and efficient predators. Their diet includes all vertebrates and invertebrates up to the size of antelope calves and domestic cats (in the case of the African civet).

Social mongooses: This is the second major division of the *Viverridae* and includes two main social mongoose species: dwarf mongoose (*Helogale parvula*) and banded mongoose (*Mungos mungo*). They are grouped together because they share ecological characteristics and are highly social, unlike the rest of the family *Viverridae*.

They are small, lively mammals, often seen moving in through the bush with weasel-like gait, pausing from time to time to stand upright and look around.

Both the dwarf mongoose and banded mongoose frequent wooded and bushed grassland; the latter prefer a somewhat more open habitat. The pack's home range includes a number of burrow sites, such as termite mounds or loosely aggregated rock piles which are used for night-time dens, as breeding sites and as lookouts.

Social mongooses are very vocal, and much communication significance is attached to their squeaks and twitters. Their main enemies are larger carnivores and birds of prey. Alertness, speed and the propensity to dive into nearby holes are their main defences.

Social mongooses occur in packs averaging around a dozen animals, but up to as many as 30 in the dwarf mongoose. Dwarf mongoose packs are led by a dominant female which is the oldest adult female of the group. She is normally the only one that conceives in a pack.

Both dwarf mongooses and banded mongooses synchronize their breeding within the pack which ensures that the season's brood can be fed and protected together. In sub-humid regions, such as western Uganda, social mongooses are capable of producing four litters a year. Mongoose young are born underground.

Mongoose packs tend to change dens every few days. This has a dual function: it ensures that a particular predator cannot focus its attention on restricted areas and it allows the neighbourhood food supply to recover from the depredations of the ravening hoard.

Social mongooses eat virtually any small terrestrial living creature, both vertebrate and invertebrate, as well as eggs and occasionally fruits. Poisonous snakes are frequently killed by the pack.

AARDVARK

The aardvark (*Orycteropus afer*) or ant bear is a peculiar looking, nocturnal animal with a humped back and a long snout and tail. An adult's body may be 1.3 metres (4.6 feet) from snout to rump, with the tail adding another 60 centimetres (two feet). They can weigh up to 65 kilograms (143 pounds). Males and females are the same size but can be distinguished by the female's slightly lighter colour. Local soil colour often masks their true yellowish grey. Since they are entirely nocturnal aardvarks are very difficult to see except when caught in the glare of car headlights. Then, they appear to be a light grey colour.

Left, the African civet out looking for a meal. Above, dwarf mongooses are playful characters.

Aardvarks are accomplished diggers, and can completely bury themselves in less than 10 minutes. Their eyesight is poor, but their scent and hearing are very good. They inhabit open grassland, woodland and bushland but are rare in forests. When moving, they make a snorting grunt sound.

Aardvarks are relatively vulnerable to

predation from all large mammalian predators and rock pythons. They invariably escape by running into a hole.

Breeding appears to be timed so that young are born at the onset of the short rains in East Africa. One or two naked young are born in burrows which consist of several metre-long runs with living chambers at the end. Although aardvarks are basically solitary, only coming together to mate, several animals may sleep in one burrow. Most likely these are related: a female and nearly grown young or a courting pair.

Their staple diet is termites, either looped up from the ground with their long, sticky tongues, or dug out of the ground.

gland which is active during states of arousal. Hyrax feet are shod with rubbery pads which sweat when running. The tacky surface allows them to climb up near vertical rock faces or tree trunks. Hyraxes have rudimentary tails and numerous long, tactile hairs over their bodies.

They inhabit *kopjes* in bushed and wooded grassland, although cliffs and rock faces form the centre of the rock hyrax's home range, from where they graze on the grassland surrounding the *kopje*. Tree hyraxes live in evergreen forests.

Hyraxes are diurnal with the exception of a forest species of tree hyrax. At night they sleep together for safety and mutual warmth.

HYRAXES

Rock hyraxes (*Heterohyrax brucei*) are found in most of East Africa.

Unbelievable as it may seem at first glance, these rather dull looking, medium-small, brownish-grey mammals are the closest living relatives to elephants and dugongs (sea-cows).

Many hyrax characteristics are elephantine: toenail-like claws which are really hoofs; two teats between the forelegs and another four in a more anterior position; internal testicles, a long gestation period of seven months which is remarkable for such a small beast; and a somewhat mysterious

Territorial males may call on moonlit nights.

Their most dangerous predator is Verreaux's eagle, which feeds almost exclusively on hyraxes. Other large birds of prey, rock pythons, civet cats, and baboons are also a threat. At an alarm call—an unmistakable, high-pitched shriek emitted by any member of the group—all hyraxes of both species, dive for cover in rock clefts.

Society: Hyraxes are extremely social and gregarious. Social organisation seems to be dictated by the size of kopjes. On small kopjes, both species live in family groups rather like harems, with one adult male overseeing three to seven (up to 17 in the bush hyrax) adult females and a number of juveniles of both sexes. The male is usually old and large, intolerant of immigrating adult males, and prone to frequent, raucous territorial display calls. Dominant males chase subordinant ones away.

On larger kopjes groups of females move within overlapping circles around core areas, which are dominated by a territorial male and younger, peripheral males. The numbers of hyraxes on a kopje is determined by the number of sleeping holes and retreats from predators as well as the quality of the local food supply.

Breeding is year-round, with a slight preference for the rainy season. Females are receptive once or twice a year and have a gestation period of seven or eight months. All females in a family group usually give birth within three weeks. Up to six (usually two to three) open-eyed young are born which are fully praecocial (capable of feeding themselves immediately) and able to run about as soon as they are dry. Sexual maturity in both sexes is around 18 months. Like elephants, females stay with the group: newly-matured males wander off.

Rock hyraxes eat mainly grass and other vegetation. Bush hyraxes are predominantly browsers. All species can eat, without apparent harm, many poisonous plants distasteful to other herbivores . They are rarely seen drinking and it is thought that most water is taken from leaf surfaces during feeding.

Left, rabbit-sized rock hyraxes are related to elephants. Above, playtime for this baboon and its young.

BABOONS

Baboons are the largest and most terrestrial of the *Cercopithecidae* family. They are heavy shouldered with rounded heads and protruding muzzles. Their arm bones are relatively longer than other *Cercopithecidae*. Their coat is shaggy and the colour varies from the yellow baboon's (*Papio cynocephalus*) light sandy yellow to the almost olive green of the olive baboon (*Papio anubis*). They have short tails.

There are five main species of baboon but only common baboons—*Cynocephalus*—are found in East Africa. Yellow baboons live in lowland areas of East and Central Africa: olive baboons prefer highland regions of East Africa.

Baboons can be seen virtually anywhere in grassland and wooded grassland where perennial water and adequate forage can be found. They are strictly diurnal and will sleep in trees at night.

They exhibit a wide vocal repertoire, from sharp barks in alarm or warning, to squeals and titterings when siblings play together, to screams as subordinant animals are chased off by superiors, to murmuring between mother and young.

They are sometimes preyed upon by large carnivores but their main enemy is man, with whom they compete for space and agricultural produce—baboons tend to depredate crops. They are also susceptible to human diseases, such as tuberculosis and yellow fever.

The family: They are very gregarious and live in family troops from 10 to 150 strong, with a mixture of all ages and sexes. At the centre of the groups are females and daughters who stay with the family group as long as they live. Males leave the family group on reaching sexual maturity.

There are generally several reproductive males in an average troop but one is dominant until killed by an outside agent or displaced in a fight by a younger, stronger male. Certainly the largest male gets most, but not all, opportunities to mate.

Baboon society is based on enduring bonds between individuals, especially core friendships which are probably formed between females as they grow up. However, young baboons and even males exhibit close ties to one another. Although males are generally promiscuous, adult females in a troop will have close ties to just two or three males, normally mating with the most dominant. These males will defend her offspring against predators.

Baboons breed the whole year round. New-born infants have short, nearly black soft fur which changes to adult colour at three months of age.

Baboons are omnivorous. The hand is their main foraging tool, and with its highly manipulative, opposable thumb and fingers, it is used to pluck, pull, strip and tear edible bits from the environment.

Predation on small animals is common in order to add protein to the diet. Baboons sometimes catch and eat larger prey such as hares or young gazelles and impalas.

VERVETS

These archetypical monkeys belong to the sub-family *Cercopithecinae.* Vervets (*Cercopithecus aethiops*)—also known as green monkeys, grivets, guenons, tantalus monkeys—are slight of build, agile and long tailed, and are always seen in noisy, bickering, family troops with lots of young animals. Both sexes are generally alike, although males are about 40 percent heavier, with conspicuous red, white and blue genital colouration.

They generally inhabit well-wooded and well-watered grasslands but can also be found from semi arid regions (generally near rivers or swamps) to evergreen forest edges at altitudes from sea-level to 4,000 metres (13,200). They are strictly diurnal animals although some feeding may occur on moonlit nights.

These monkeys are much less vocal then baboons. They are generally silent, except when defending their territory or when warning one another against any danger.

Vervets are prey to large raptors, snakes and cats ranging from servids to leopards. Their basic form of defence is alertness and flight.

Vervets are very gregarious and live in troops of between six and 60 animals, sometimes reaching as many as 100. Troops are comprised of one or more adult males, adult females and young of all ages and sizes. Troops are territorial and defend their range against neighbouring troops with noisy group displays at the territory boundary.

Breeding occurs all year round, although local peaks may be apparent depending on local nutrition conditions. Gestation is relatively long—180 to 200 days. A young vervet starts eating solid food at a very early age by reaching up and taking morsels from its mother's mouth whilst she is feeding. It will ride on her back and suckle until six months or until the arrival of the next sibling, often a year later.

Infant mortality from disease or predation is relatively high which keeps populations reasonably stable, although they get larger in periods of good rainfall and diminish during droughts.

Vervets are omnivorous with a predilection for vegetable matter. Their preferred tastes include fruits, flowers, grass seeds, shoots and bark, both wild and cultivated, as well as insects, reptiles, small mammals, young birds and eggs.

GREATER GALAGOS

Greater galagos (*Galago crassicaudatus*) are long, woolly-tailed, nocturnal primates with thick brownish or silver grey fur. They measure around 65 cm (25 inches) and their tails add about 30 cm (12 inches) to their length. Their large eyes are surrounded by a ring of darker fur and stare out from a pointed muzzle: their ears are cupped and rounded and their senses of hearing, sight and smell are excellent. They have small "hands" with long, thin fingers with flat nails and an extended claw on the second toe. Greater galagos are sometimes mistaken for bushbabies (*Galago senegalensis*) which are much smaller—only 40 cm (15 inches) long—and have comparatively longer, slimmer tails.

Greater galagos are arboreal and live in a range of habitats including rain-forested mountain slopes up to a height of 4,000 metres (13,200 feet), bamboo thickets, wooded savannah grasslands and eucalyptus, mango and coffee plantations. They can sometimes be seen in suburban gardens. They establish territories which vary in size according to season with between 70 and 130 animals per square kilometre. Territories are marked out with excretions from breast, anal and foot glands which are rubbed on the ground, trunks and stems of trees and bushes. Neighbouring territories may overlap and family groups often share the same sleeping hole with as many as 12 sleeping holes in each territory, located in dense foliage.

Left, primate portrait: a contemplative vervet monkey. Above, young vervet checks out the view.

The enormous variety and concentration of birdlife in East Africa has been sadly neglected. So far 1,293 bird species have been recorded and scientists believe that many unrecognised birds are yet to be discovered, especially in remote areas. Compare this figure with the 250 or so recorded bird species in Great Britain or the 850 in Canada, Mexico and United States and you will understand why Kenya, with over 1,100 different birds, is an ornithologist's delight.

Nature has provided East Africa with a tropical environment with every conceivable type of habitat. Snow on the equator from volcanic mountain ranges, cool lush forests on their slopes, vast open temperate plains, harsh dry deserts and lowland equatorial forests, sea shores and mangrove swamps providing perpetual food supplies, all contribute to this unique region. The vast majority of birds live and breed here year round, but several hundred species come from northern latitudes, when harsh winters destroy their food sources.

From as far east as the Bering Straits and as far west as northern Scandinavia it is estimated that up to 6,000 million birds make the annual journey, with some birds flying as far as the southern tip of the continent. Those that survive make the return journey each spring to breed in their chosen latitudes.

In East Africa there are only wet and dry seasons and even these vary dramatically from place to place and year to year. Driven by some age-old instinct birds know when to leave their breeding areas to go to find a better place to live.

Migration: This is instinctive, not learned behaviour. Many species that breed in the northern latitudes and migrate to Africa each year, actually leave their young to find their own way south, or perish in the attempt. A perfect example of this is the Eurasian cuckoo (*Cuculus canorus*) which lays its eggs in the nests of foster parents. The young cuckoo, even when totally blind, instinctively and forcibly ejects any other egg or even young chick from its nest. Foster parents spend the next 20 or so days feeding this voracious monster until it can fly and feed itself. By now its parents have long since left for Africa, often weeks before the European weather turns miserable. The young cuckoo starts the long journey south with no guidance, following its instinct to go or die.

Many ducks and geese exhibit similar behaviour. Once breeding is over adult birds go into *eclipse* when they moult their flight feathers and cannot fly until new feathers have grown. In the meantime their new brood has learned to fly very well and they disappear south, well ahead of their parents.

Those that survive the long journey over harsh deserts, flying mostly at night, find rest and feeding grounds in the amenable climate of East Africa. There is a sudden influx of birds almost overnight as huge numbers appear in the bush country, forests and even surburban gardens.

Bird migration has fascinated man for centuries. Their ability to navigate over thousands of miles with none of man's sophisticated technology, and to return repeatedly to the same nest site to breed is a remarkable achievement.

Bird enthusiasts throughout the world have cooperated to study this extrordinary phenomenon. They set up mist nets mainly at night to trap birds which are then weighed, identified, and have their wing length and other data recorded. Then a small numbered ring is attached to one leg and the bird is released. In this way, if ever a bird is recovered or seen again, its route, flight pattern and time taken to cover the distance can be roughly estimated.

Numbers of birds recovered are very small but these studies add to the sum of man's knowledge and some startling data has come to light. A ringed shore bird was picked up dead in Kenya's Rift Valley. Investigations showed that it had been ringed just west of the Bering Straits in the Soviet Union by Russian enthusiasts, only 18 days before.

The sophistication of the migratory phenomenon is just beginning to be understood. We know why birds migrate but much work needs to be done before we understand the complexities of birds' highly-tuned navigation systems.

Preceding pages: close-up of a bateleur eagle. **Left**, black-headed weaver bird and nests. **Right**, the distinctive saddle-bill stork.

Conservation: Birds play an important role in the lives of man, not least for their aesthetic value. We all know how much pleasure can be derived from watching them in our gardens, but birds also perform an essential function in keeping the number of insects under control. Perhaps most important though is the way birds serve as an indication of man's destructive activities on the environment.

The classic example is the effect of DDT (dichlorodiphenyltrichloroethane) on the

eggs of birds of prey. This phenomenon was widely published and alerted governments throughout the world to the perils of using long term insecticides. Birds of prey are at the end of a food-chain. Their prey, be it mice, rats, lizards or other birds, all feed on grains or insects which, in this case, had been treated with DDT. The compound built up in their bodies, without serious effect, to very high levels. But the effect on the birds of prey was to weaken their egg shells, effectively destroying their ability to reproduce. Once this was realised and conclusively proven DDT was banned.

The East African climate is controlled by

two major factors: a meteorological phenomenon known as the *intertropical convergence zone,* which produces the two main rainy seasons with specific wind directions, and the various ranges and altitudes of mountains in relation to these winds, at different times of the year.

The water birds: On the coast the climate is tropical year round, and the beaches with wide tide differentials (up to four metres or 13 feet) provide massive food supplies for migrating waders or shore birds. At low tide from September to March, thousands of these birds can be seen feeding along the beaches, coral pools and mud flats.

Sanderlings (*Calidris alba*), Whimbrels (*Numenius phaeopus*), Ringed Plovers (*Charadrius hiaticula*), Turnstones (*Arenaria interpres*), Oyster-catchers (*Haemantopus ostralegus*), Greenshanks (*Tringa nebularia*), and other migrants live and feed here storing up energy for the long flight back to their northern breeding grounds in the spring.

Resident birds are also much in evidence. Grey herons (*Ardea cinerea*) feed in the shallow pools, gulls of several species are ever present, and in the evening large flocks of terns come to roost on the coral cliffs. Breeding colonies of the roseate terns (*Sterna dougallii*) establish themselves on off-shore islands, and there is a confusing variety and large numbers of egrets, with all their oddities.

In shallow water without coral cliffs, mangrove swamps develop. Here, the mud attracts mangrove kingfishers (*Halcyon senegaloides*), night herons (*Nycticorax nycticorax*), and other species of heron, including the strange black heron (*Ardea melanocephala*), with its unique feeding behaviour. It paddles with bright yellow feet and then brings its wings up over its head in umbrella fashion to shade the water underneath.

There are also crab plovers (*Dronius ardeola*), and yellow billed storks, (*Ibis ibis*) which feed by sticking their partly opened, long bills into shallow water and, with sweeping action, snap them shut when they touch something edible.

Along the rivers: The great rivers flowing from the mountains hundreds of miles away, across semi-desert country down to the sea, create a third environment, known as riverine forests. Birds take advantage of this narrow strip of permament water where food supplies are always available. Weavers of all kinds nest in the overhanging trees.

Tawny eagles (*Aquila rapax*), martial eagles (*Polemaetus bellicosus*), Wahlberg's eagles (*Aquila wahlbergi*) and others nest in the tree tops and feed off small mammals, dry country game and birds such as guinea fowl and francolin which come to the water to drink.

Blacksmith plovers (*Vanellus armatus*) nest on the sand-bars and huge flocks of sand grouse come to quench their thirst and bathe.

On each side of these rivers stretch vast areas of semi-desert and scrub. This is harsh land at relatively low altitude, with scarce and erratic rainfall. For most of the year it is extremely dry. When rain does fall, however, the land blooms: every living thing from plants to elephants takes advantage of the vast increase in food supply. Insects flourish, plants flower and seed madly, and the bird life erupts to match.

Every tree is suddenly full of nesting birds: buffalo weavers (*Bubalornis niger* and *albirostris*), white-headed weavers, (*Dinemellia dinemelli*), thousands of queleas (*Quelea cardinalis*), hornbills of many types, yellow-necked francolin (*Francolinus leucoscepus*), and those that prey on this new abundance. Secretary birds (*Sagittarius serpentarius*) nest on the top of flat trees: black-crested snake eagles (*Circaetus gallicus pectoralis*) wait to snatch lizards or snakes.

Left, the bateleur is identified by its short tail. Above, enterprising Egyptian vulture uses pebble to crack an ostrich egg.

Arid regions: In Kenya, particularly the north, lie vast areas of almost true desert, most of it uninhabited by man. Here are the true dry country birds which have evolved to take advantage of their enivronment: large Heuglin's bustard (*Neotis henglinii*), sand grouse which fly 30 to 50 kilometres (18 to 30 miles) each day to scarce water holes, and the tiny, short-crested lark (*Galerida cristrata*).

Vast areas of East Africa are covered by savannah—great open grass covered plains with varying degrees of tree and scrub exist mostly at middle altitudes (1,000 to 2,000 metres or 3,300 to 6,600 feet). Rainfall is erratic, but usually good when it does fall. Water-courses, some permanent, others only seasonal, create bush and tree lined valleys which slice through the plains where larks (*Alaudidae*), and pipits (*Motacillidae*) of all types, plovers (*Charadriidae*), longclaws (*Macronyx*), and a vast variety of so-called grass warblers (*Cistocola*) breed and live. The ugly scavenging marabou stork (*Leptoptilos crumeniferus*) is often seen. Overhead soar almost every species of vulture. They nest in tree tops or on rocky cliffs many miles from the open plains that supply their food. Vultures can always be seen at the scene of a kill.

Along the valleys cutting through this region, heavier growth of trees and scrub provide shelter and nest sites for other birds, who feed on the plains: barbets (*Captonidae*) fruit and seed eating birds, bush shrikes (*Malaconotus*), francolins (*Francolinus*), guinea fowl and doves of all kinds.

Birds of prey, notably the chanting goshawks (*Melierax poliopterus, Melierax metabates*), find this environment much to their liking and bateleur eagles *(Terathopius ecaudatus)* effortlessly soar for hours at a time.

Rain forest birds: High altitude tropical rain forest covers all the mountain ranges. Winds are predominantly easterly, blowing from the Indian Ocean so eastern facing slopes tend to have a higher rainfall and more morning mist. Forests grow all year round so they are always green, lush and cool, and provide a permament home for bird life.

In the tree tops, insect-loving shrikes (*Laniidae and Prionopidae*) feed in noisy family parties, often accompanied by starlings (*Strunidae*) of several species.

At lower levels, nearer the moist, cool earth, plant and insect life are abundant. Robin chats (*Pycnonotidae*) of several species find this perfect. So do bulbuls (*Andropadus*) and greenbuls. From the forest floor to the treetops turacos with their brilliant scarlet wings and raucous calls feed on the abundant seeds and fruit.

On the forest floor, scaly francolins (*Francolinus squamatus*) scratch and worry the earth, while overhead the great crowned eagle (*Stephanoaetus coronatus*), possibly Africa's most powerful bird of prey, soars in display, sometimes only a speck in the sky, his piercing call drawing attention long before he is seen.

This is a place to sit quietly and watch. If the wild fig trees are fruiting, sit under one because the ripe fruit attracts green pigeons (*Treon australis*), more turacos (*Musophagidae*), olive thrushes (*Turdus olivaceus*), starlings (*Sturnidae*) and barbets (*Capitonidae*) of many kinds. Over-ripe fruit attracts insects, which are followed by a huge influx of insect-eating birds.

Mountain regions: The group of East Africa's mountain ranges on or near the equator, lies in an environment that is described by scientists as afro-alpine, or equatorial alpine. High altitudes of up to 6,000 metres (19,680 feet), coupled with latitude, create a peculiar habitat.

Permanent glaciers predominate above 4,700 metres (15,416 feet) but snowfalls, which are regular and heavy, usually melt fairly rapidly in the tropical sun making it seem like winter every night and summer every day!

From one of Africa's great birds of prey, Mackinder's eagle owl (*Bubo capensis*), down to the tiny scarlet-tufted malachite sunbird (*Nectarinia johnstoni*), or the hill chat (*Pinarochroa sordida*), birds confined to this alpine zone could probably not survive elsewhere. There are many other species of birds but the environment tends to keep numbers and variety down. Of great interest, however, is the way birds have evolved to survive in any environment: vultures have been recorded on the snowline of several East African mountains although there is no good explanation for their choice in habitat.

Low forest: True African jungle does not really exist in East Africa. Though there are remnants in Western Kenya's Kakamega Forest. But as this is at an altitude of over 1,500 metres (4,920 feet), it cannot properly be called lowland forest.

Along the Kenya coastline little remains of a once vast forest, created by local conditions of rainfall. Situated five kilometres (three miles) inland from the sea, this forest evolved to take advantage of the fertile coral-based soils, an erratic but heavy annual rainfall, and zero altitude. Much of this forest has now been destroyed by man, but in the Arabuku-Sokoke Forest, near Malindi, there are at least three species of birds that exist nowhere else in the world. The Sokoke Scops Owl (*Otus irenae*), the Sokoke Pipit (*Anthus sokokensis*) and Clarke's Weaver (*Ploceus golandi*) can all be seen with a bit of effort.

Swamps appear in deserts in years of unusual rainfall and immediately attract the attention of birds not generally found there. Since the swamp holds water long after the surrounding country has returned to normal, the birds will stay. Other swamps are more permanent. Water-loving birds always appear where there is food and disappear when the water dries up.

The rainy season in East Africa is the equivalent of spring elsewhere. Rain triggers food which in turn triggers breeding. So if you want to see bird-nesting sequences and behaviour you should plan your stay between April and May or in November and December. If you are more interested in seeing local birds rather than migrants then come between March and September.

Left, this early bird (superb starling) catches its worms. Above, the striking red and yellow barbet.

Big Bird

Ostriches (*Struthio camelus*) are the largest birds in the world, measuring up to 2.5 metres (eight feet) in height and weighing up to 135 kilograms (297 pounds). Males have attractive black and white plumage and are bigger than females which have the same dull, grey-brown feathers as their young.

There are two subspecies of ostrich in East Africa: Maasai ostrich (*S. c. massaicus*) can be distinguished by the male's flesh-coloured head, neck and legs which turn bright red during the mating season: Somali ostrich (*S. c. molybdophanes*) males have greyish-blue flesh.

Ostriches can be found throughout Africa although they prefer the lush, open grasslands of the savannah plains, dry thorn bush country and semi-desert. They are generally silent birds although during courtship rituals males sometimes make distinct reverberating calls.

Getting their kicks: Although ostriches still have flight feathers in their wings they cannot fly and have evolved long, powerful legs as their main form of defence. They have two toes on each foot and one kick is enough to kill a man. They can run up to 70 kph (45 mph) and can maintain speeds of 50 kph (30 mph) for up to 30 minutes. Their long necks enable them to sight enemies from a great distance and ostriches often serve as early warning systems for other plains animals. True to myth, they sometimes flatten their heads to the ground when approached.

The only real enemy of ostriches is man, who hunts males for their feathers and has decimated populations in certain areas. Ostrich eggs, however, are very vulnerable to predators, especially hyaenas and lions which may brave an attack by adults in search of this delicacy.

Ostriches live in family troups, small groups (up to 50 have been recorded in arid areas) or couples. They feed mostly off grass, bushes, leaves, succulent plants and berries. Their diet also includes small lizards and insects.

Birds of a feather: these long-limbed ostriches are easy to spot as they head across the wide open grasslands.

REPTILES

The richness of reptile fauna in East Africa compares favourably with any other part of the world. The region is a meeting ground for a number of zoogeographical zones, each with its own selection of species. Yet much of East Africa, including the whole of eastern Ethiopia, is largely *terra incognita* to the student of reptiles.

Fortunately, reptiles have no significant commercial value in East Africa. A few are killed for skins and smuggled out individually but, at least with snakes and lizards, the trade does not flourish. However, many reptiles are killed on sight by local people as a matter of principle. Sadly, the numbers of individual animals and even species are being rapidly reduced by habitat devastation and by the expanding population.

Lizards: Over 180 different species of lizard exist in East Africa, significantly more than the 150 to be found in the North American subcontinent. The variety of lizards is enormous, from the two-metre monitors (*Varanus niloticus* and *Varanus exanthamaticus*), to the tiny cat-eyed coral-rag skink (*Ablepharus boutonii*) which lives on outcrops of coral-rag and maintains an osmotic balance by having very saline blood.

There are many kinds of agama lizard (*Agama agama*), some with bright red or orange heads, others with steely purplish blue or shiny green heads.

About 40 forms of gecko are to be found here, including two transplants from Madagascar which probably arrived many generations ago with the dhow trade. These geckos are the brilliant emerald green common to the genus Phelsuma Gray. Native geckos are neither very large nor strikingly coloured. Giant plated lizards (*Gerrhosaurus major*) are handsome with skin that looks like chain mail, reddish brown in colour. Males are orange tinted on head and neck during the courting season. They are largely fructivorous but will eat insects and mice if opportunity allows. These lizards can be-

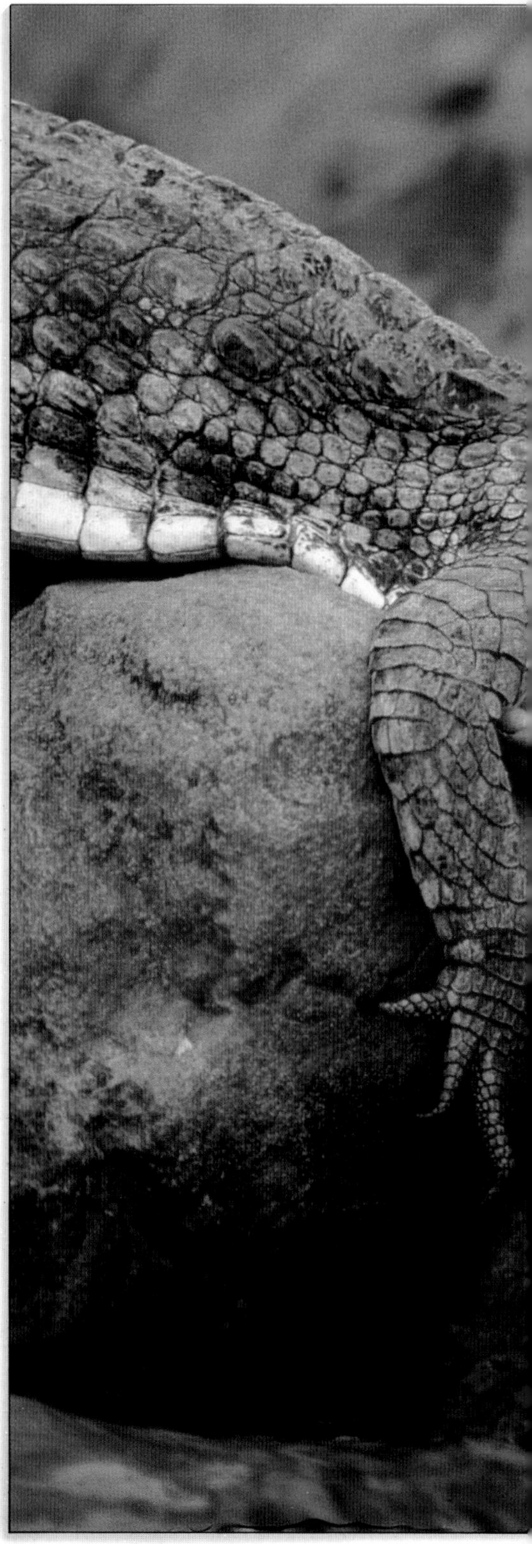

Green-eyed monster: the formidable Nile crocodile.

come very tame and sometimes will hang around campsites begging for scraps.

Chameleons are lizards, of course, but quite special ones. They come in many sizes and shapes, from the cat-sized Meller's chameleon (*Chamaeleo melleri*) which sometimes catches birds, to the tiny pygmy chameleon (*Rhampholeon kerstenii*), which is the size of a small mouse and lives on insects the size of a fruit fly. There are chameleons with three horns on the nose such as the dinosaur Triceratops, Jackson's and Johnston's chameleons (*Chamaeleo jacksonii* and *Chamaeleo johnstoni*). Others have two side by side protuberances looking like pineapples, such as Fischer's chameleon

(*Chamaeleo fischeri*). Still others have a single little spike on the nose and some even have plain, unadorned noses.

Of the standard lizard-shaped lizards there are too many to begin to describe individually but one deserves special mention: the serrated toed lizard (*Holaspis*). These are small lizards of the high primary forest, conspicuously marked with bright yellow longitudinal bars. They can glide from tree to tree like the Asiatic Dracos.

Turtles and tortoises: There are two freshwater varieties of turtles which have flattish rubbery shells and narrow pointed heads. Be careful when handling them as they have little sense of humour and bite like weasels. The first of the pair is the widely distributed Nile soft-shelled turtle (*Trionyx triunguis*). This occurs in the northerly end of East Africa and grows to a very large size. The other is the Zambezi soft-shelled turtle (*Cycloderma frenatum*) which comes from the southern regions.

Only three species of land tortoise can be found here. The leopard tortoise (*Geochelone pardalis*) is the largest and can grow to well over 45 centimetres (18 inches). This rotund animal of kindly disposition is a dull yellow colour with black flecks and lives in the savannah.

The forest, or hinge backed tortoise (*Kinixys belliana*) is quite carnivorous and can close up the back opening in its shell by a hinge more than halfway to the rear of its carapace. The back end closes down upon the plastron (under side of the shell) protecting the tucked in hind, limbs and tail. Other tortoises which can close their rear ends do so by having a hinge on the plastron which closes upwards. The forest tortoise's hinge is easy to see and looks a little as though the shell has been run over.

From the rocky areas comes the strangest of the three tortoises. Pancake tortoise (*Malachochersus torneiri*), so called because they are quite flat, have a flexible papery shell. Unlike most tortoises which when threatened retire into their shells, these fellows gallop off at a good speed and hide among the rocks like a lizard. Even when you find their hiding place they usually wedge themselves in so tightly that they are difficult to extricate.

Crocodilians: Where there was enough water the Nile crocodile (*Crocodylus niloticus*) used to be fairly ubiquitous at middle and lower altitudes. Although its range has been substantially reduced it is still far from uncommon in many areas. It grows to five metres (16.5 feet) and more. It has a very voracious appetite and, in spite of its reduced numbers, it accounts for many human lives every year.

Two other smaller species of crocodile occur in East Africa but both are from the western limits. One is the long nosed crocodile (*Crocodylus cataphractus*) which grows to a little more than two metres (6.6 feet) in length. It is reminiscent of the Asiatic gavial with its narrow nose and quite large teeth. It feeds almost entirely on fish and lives in Lake Tanganyika and associated waters. The other is the dwarf crocodile (*Osteolaemus tetraspis*), which comes from Uganda and points west. It seldom reaches two metres in length. Neither of the two smaller crocodiles are considered hazardous to man.

Snakes and adders: East African snakes are very varied with representatives from many families. The giant snakes of the area are rock pythons (*Python sebae*) which can reach a length of six metres (20 feet) and more. As they are heavy bodied, even one of medium length is quite massive. While they prefer to be near water, they can be found anywhere except at very high altitudes. A large python can be a dangerous adversary for man if disturbed but will not attack unless provoked.

The other members of the giant snakes are the boas. The family is represented in East Africa by a small relative, the sand boa (*Eryx colubrinus*), which rarely exceeds 45 centimetres (18 inches) and lives buried in the sand. Only the tip of its nose and eyes show and it ambushes any unwary, lunch-sized passing animal which it grabs and constricts before eating.

Some of the most venomous snakes in the world are to be found here. There are three species of mamba, the largest being the black mamba (*Dendroaspis polylepis*) which can

Left, the leopard tortoise is easy to spot. Above, monitor lizard lurks in the tall grass.

grow to more than five metres and is not black but a silvery olive colour. They have a nasty reputation for aggressive attack but experience shows that unless pursued or otherwise aggravated, they behave with the utmost discretion, which is just as well—black mambas can inflict a lightning bite, injecting immense quantities of exceedingly powerful venom.

The other two mambas rarely reach over two metres in length. One is the common green mamba (*Dendroaspis angusticeps*) seen along the coast; the other is Jameson's mamba (*Dendroaspis jamesoni*) from the west. Both are brilliant green but the Jameson's mamba has a velvet black tail.

They are quite deadly but their venom is only about one fifth as toxic as that of the black mamba.

Asia is often considered the home of cobras but Africa has many more and in more varied forms. In East Africa there are five types of true cobra and two closely related genera. Three of the true cobras can spit their venom quite a distance, aiming it accurately at the eyes of their antagonist. If the eyes are washed out quickly the effect is only temporary but acutely painful; if neglected, permanent damage to the eyesight can result. The three spitting cobras are the common spitting cobra (*Naja nigricollis*), Mozambique spitting cobra (*Naja mossambica*) and a subspecies, the red spitting cobra (*Naja mossambica pallida*).

Closely related to the true cobras is Storm's water cobra (*Boulangerina annulata*) from Lake Tanganyika and points west. Although it has adequate fangs and very toxic venom, it swims freely among fishermen waist deep in water, neither fishermen or snake giving each other much attention. The other large near-cobra comes from the forest canopy of the western primary forests of Uganda and western Kenya. It is Gold's cobra (*Pseudohaje goldii*) which is hoodless and only comes down to the forest floor to prey on toads.

Three of the four giant vipers of the world can be found in this area. They are the Gaboon viper (*Bitis gabonica*); rhino viper (*Bitis nasicornis*), and puff adder (*Bitis arietans*). These are all large bodied snakes with wide heads. In exceptional cases all three can reach a length of two and a half metres (eight feet) and the Gaboon viper can grow up to two metres. The colouring of Gaboon and rhino vipers is striking and beautiful. They can all inflict multiple doses of lethal poison in just one bite.

There are a number of small vipers but three merit special mention. They are all endemic to a small area of Africa around

Mount Kenya. The small Worthington's viper (*Bitis worthingtoni*) is related to the three giants. It is an attractive little snake with black, brown, lilac and white markings, horns over its eyes, a saturnine face and an irascible disposition. It comes only from the hills around Lake Naivasha. The mountain, or Hind's viper (*Vipera hindii*) is a diminutive snake resembling a tiny melanotic European viper. It can be found only well above the treeline in moorland on top of the Aberdare mountains where there is a deep frost almost every night.

The last of the trio is the Mount Kenya bush viper (*Atheris desaixi*) which was only discovered in 1967. This snake belongs to a

West African genus and nobody suspected that a species existed this side of the Great Rift Valley.

Other snakes: There are only two back fanged snakes in the world known to be deadly; the boomslang (*Dispholidus typus*) and the twig, bird, or vine snake (*Thelotornis kirtlandii*). The former is a medium sized snake with large eyes. Typically, males are green and females grey or brown. The latter is very slender with a large pointed head. The body looks exactly like a lichen covered twig and the head is white or off-white below and green, red or brown on top.

A third snake is suspect: Blanding's tree snake (*Boiga blandingii*) comes from the western forests. It is a long soggy-looking snake with a huge head which, when threatened, suddenly becomes very unsoggy indeed, expanding its neck and flattening its head into a very intimidating display.

There are too many harmless snakes to mention in detail, including specialist feeders such as centipede eaters, slug eaters, egg eaters and even a little shovel-nosed snake which lives off gecko eggs. Others are more general feeders, such as the sand snakes (*Psammophis*) which, amusingly enough, never live in sand. For some inexplicable reason Isis von Oken Boie named them this: *Psammos* is Greek for sand, *ophis* is Greek for snake. But that is not the only misnomer. The largest in the genus, the hissing sand snake (*Psammophis sibilans*)—*sibilans* is Latin for hissing—does not even hiss!

Where to find reptiles: Reptiles are difficult to find unless you know where to look. Fortunately each area in East Africa has its resident *Bwana Nyoka* (snake man) most of whom are excellent and well worth their hire. Reptiles can be found almost anywhere but places to hunt are by river or lake sides, forest verges, especially around weaver bird colonies and where one type of habitat merges into another.

Unfortunately there are few places to go to ask questions in Kenya. The National Museum in Nairobi (P.O. Box 40658, Nairobi) runs a long-established snake park opposite the main building with two additional satellites in Kisumu and Kitale. Kenya Crocodiles (Mamba Village, P.O. Box 85723, Mombasa) has an excellent display of reptiles and successfully combines an educational programme and sound conservation with commerce. Bio-Ken (P.O. Box 3, Watamu, Kenya) have a reptile farm and run a technical and advisory service for universities and museums.

The University of Dar es Salaam in Tanzania (P.O. Box 35091, Dar es Salaam) has a good herpetological section and Makerere University (P.O. Box 7062, Kampala) may be consulted by people wishing to visit Uganda.

Left, skeleton of a black mamba, one of the world's most poisonous snakes. Above, spitting cobra is ready to strike.

All continents are edged by waters not more than 200 metres (660 feet) deep, covering the so called continental shelves. In East Africa, this shelf is very narrow, not more than 75 kilometres (46.5 miles) at its widest at Zanzibar, and often not more than two to 3.5 kilometres (one to two miles). The East African coast has a series of continental islands, which are part of the continent, separated only by sinking of the intermediate land. Zanzibar is one of these islands.

Pemba and Latham islands are oceanic islands, rising straight out of the ocean. The flora and fauna of these oceanic islands is quite different from that of continental islands. On Pemba, for example, the fruit bat found is a genus found in Asia and Madagascar, but not on the African mainland.

Africa at one time is thought to have been part of a much larger continent, called Gondwanaland, which split into what is today known as India, Australia, Antarctica and South America. The existence of this larger continent helps to explain the presence of so many species common in East Africa which are also found on the Great Barrier Reef in Australia. Madagascar also has a great variety of species commonly found in the East.

The distribution of plants and fish is affected by the currents impinging upon the continent. The South Equatorial Current which flows westwards towards the East African coast turns right once it reaches a point on the Somalia coast north of the equator. The effect of the current changes with the season of the year and the winds. Between April and October, the south-east monsoon (*Kusi*, in *Swahili*) blows in the same direction as the current, strengthening the flow westwards and turning the current to push great water masses as far as Malindi. This effect is reversed between October and March when the north-east monsoon (*Kaskazi*) blows against the flow of the current. The temperature of the water is correlated to this water flow, being relatively lower in September—24 C to 25 C (75 F to 77 F)—compared with about 28 C (82 F) in March when the water is hottest.

Fish list: Open sea fish are the ones most affected by the currents. These fishes, among them marlin, sailfish, kingfish, tunny, runners, bonito, dorado, etc, are called pelagic fishes because they continually swim around the open water. They are most abundant during the fishing season from September to March off the north coast of Kenya as well as off Shimoni on the south coast. The narrow Pemba Channel is an excellent place for fishing during this time of the year.

Pelagic fishes are generally streamlined, with smooth bodies, slim tails, strong caudal fins and with dorsal and pectoral fins reduced in size. They swim mainly by rapid lateral movements of their tails and not with

the movement of their fins as is common in fishes that live in sheltered waters. They are generally dull coloured, blue and silver with darker bars or markings.

Sharks also occur off the East African coast. They present little danger to humans with the exception of those present in places like Kilindi or Dar es Salaam harbours where they live largely by scavenging.

Manta ray or devil fish and the basking or whale shark are two of the largest fishes. The ray may measure up to six metres (20 feet) across and weigh two tons. Sharks are several metres in length and can weigh a ton. Both are quite harmless if left alone. Despite their formidable size and appearances, they both feed on small floating or swimming organisms, known as plankton.

Reef life: All reefs off the East African coast are fringing reefs. These reefs grow in warm, clear, shallow water on platforms of the continental shelf. They follow the outlines of the land and enclose lagoons.

The reefs are corals which are colonies of animals (polyps) of the same species as anemones and jellyfish. Polyps resemble small sea anemones and belong to the same phylum *Coelenterata*. They extract calcium carbonate from the sea to form a skeleton of lime on which they perch. When the polyps open to feed, they cover the skeleton with a filmy mass of tentacles. Corals grow continually, adding to their skeleton as long as they are covered with water, even at low tide. Coral reefs are the richest ecological environment in the world.

A typical coral garden can support some

Left, coral fish off Malindi. Above, manta ray skims the ocean floor.

15 families of fish containing up to 60 species which, if added to those fish in the surrounding water and those in rock caves and those in the seagrass beds, would come to more than 200.

Odd fins: In an easy morning's snorkeling around an undisturbed coral head you can expect to find globe fishes and puffers (*Diodontids* and *Canthigasterids*), which inflate themselves with water when disturbed and are poisonous to eat unless prepared by an elite Japanese cook. Keep an eye out for the queer-looking file and trigger fishes (*Monacanthids* and *Balistids*) which anchor themselves into the coral with their dorsal spines to keep from floating away whilst asleep. Then there are parrot fishes, which graze on the living coral and help to convert the coral into sand. Further on, you might swim past shoals of snappers, and gaterins. Single, territorial damsel-fish (*Pomacentrids*) in a variety of colours and sizes, defend tiny parcels of the coral head. One genus, the *Abudefdufs*, are particularly belligerent and all 10 centimetres (four inches) of indignant fish will dash out to threaten a passing snorkeler.

Larger fish are kept parasite-free by cleaner wrasse (*Labroides*) which have conspicuous black and silver lateral stripes and approach potential customers with an undulating "invitation dance". A large angel fish (*pomacentrid*) may partially roll over to allow the wrasse to pick skin parasites off its belly. In a classic example of aggressive mimicry, sabre-toothed blenny have evolved an almost perfect imitation of the cleaner wrasse's colouration and invitation dance. This combination allows the mimic to get close to otherwise wary fish. Unfortunately for them, blenny eat fish flesh, not skin parasites, and before they know what has happened, the impostor has taken a bite out of the proffered flank and made good his escape.

Red, black and white soldier or squirrel fish (*Holocentrids*), jewel fish (*Anthiid*) and yellow and black butterfly fish (*Chaetodontid*) can also be seen. Surgeon fish (*Acanthuridae*) are oval and laterally flattened. They have a razor-sharp "scalpel" protruding from either side of the base of the tail and if handled or touched, they can flick it to make a nasty incision.

Serranids are generally duller but very tasty. The largest of the group, sea bass or groupers, may grow to 300 kilograms (660 pounds) with a mouth that one would feel nervous about swimming too near. Groupers wait quietly camouflaged in rock crevasses and when something edible swims past, they open their enormous mouths so rapidly that anything in the immediate neighbourhood gets sucked into the temporary vacuum.

The ocean floor: The bottom of the lagoon is white sand often covered with a dense growth of weeds, spotted with dead coral that is being eroded away. *Cymodocea ciliata* is the most common weed growing on sandy or coral rubble. The weed is actually not an algae but a marine Angiosperm. The bright green wrasse (*Cheilio inermis*) and a small olive green fish (*Leptoscarpus vaigiensis*), are some of the few that live here.

Where there is more sand and some coral rubble, a greater number of fish are found, among them the rabbit fish (*Siganus oramini*) and snappers (*Lutjanus fulviflamma*) with yellowish fins and a big black spot on their sides. The sandy bottoms are favoured by species like red mullet (*Pseudopeneus macronema*) and sting ray (*Taeniura lymna*). The ray is brown, with blue spots and a long tail with a sting which has serrated spines and is very painful. The ray will take to flight when you approach it, so it is not very common to be stung by one.

Far more dangerous is the stonefish (*Synanceja verrucosa*) which is found on the edges of lagoons and in pools. This brown, ugly and perfectly camouflaged fish lies at the bottom, often among rocks where it is very difficult to spot. If you should be so unfortunate as to step on one, its sharp dorsal spines that have poisonous sacs connected to them will penetrate any beach shoe and cause the most excruciating pain. Medical help has to be sought immediately as the sting can be fatal if not treated. Luckily, the fish will usually move out of the way of the unsuspecting walker.

Two marine mammals, dugongs and dolphins, live in the Indian Ocean off the East African coast. Dugongs (Dugong dugon) are heavy bodied, vaguely seal-like animals with almost atrophied hind limbs and a body that ends in a single flat flipper. Dugongs cannot move well out of water as their front flippers are weak and they must raise their bodies by the strength of their breathing muscles alone. They hide in mangrove swamps by day, coming up only to breathe. They feed on marine plants that grow on the bottom, normally some distance offshore. It takes some imagination and perhaps several months at sea to understand how dugongs could have given rise to the mermaid myth. Nowadays, dugongs and dolphins are threatened throughout their habitat.

Left, prolific coral growth in the warm waters of the Indian Ocean. Above, busy reef fish.

FLORA

The diversity of flora in East Africa is a result of the wide range of ecological and climatic conditions. Rainfall and altitude are the two major factors affecting the distribution and growth of different species of flora. The region rises from sea level to nearly 6,000 metres (19,600 feet),and varies in rainfall from 125 mm (five inches) to 2,500 mm (100 inches) per annum.

Geographical zones: Recognition of these geographical zones helps in identifying plant species. The coastal zone running along the Indian Ocean from north to south, extends approximately 16 to 24 kilometres (10 to 15 miles) inland, with a moisture index rarely below 10.

In semi-desert, often covered with arid bushland or dwarf shrub grasslands, rainfall is generally below 250 mm (10 inches) per annum. There are no true deserts in East Africa.

Bushland (*nyika* in *Swahili*) is generally found below 1,650 metres (5,445 feet) and is sometimes interspersed with grasslands. Rainfall varies between 250 mm and 400 mm (10 inches and 16 inches) per annum.

Grasslands are medium to dry rainfall areas with 400 to 600 mm (16 to 25 inches) per annum. They are found at altitudes of between 760 and 1,800 metres (2,492 and 5,904 feet).

Areas with medium to higher rainfall—625 to 1,000 mm (25 to 40 inches)—at altitudes of 1,100 to 2,000 metres (3,600 to 6,560 feet) are usually covered with wooded grasslands with *Acacia, Albizzia* and *Combretum* trees.

Highland areas between 1,800 and 3,650 metres (5,904 and 11,972 feet) contain moorland, upland grassy plains and higher rainfall forest which can be divided into two regions: on the main high altitude massifs, including mounts Kilimanjaro, Kenya, Elgon and the Ruwenzoris where there is considerable cloud cover and rainfalls exceed 1,000 mm (40 inches) per annum; and dry forests which, although evergreen, have an average rainfall of less than 750 mm (30 inches) per annum. Examples can be seen around Nyeri district, the Chyulu Hills and Langata near Nairobi in Kenya. There is no true rainforest in East Africa, except possibly a small area on the Usumbara Mountains in Tanzania.

Above 3,650 metres lies the alpine zone with its own species of flora adapted to the extreme conditions.

Altitude has a great influence on the distribution of flora throughout the region. Plants tend to extend their altitude range upwards as one moves westwards where the climate is influenced by both the mellowing effect of the great bodies of water in Lake Victoria and Lake Tanzania and their attendant satellite lakes, and by a reduction in the effect of glaciers on mounts Kenya and Kilimanjaro.

In the southern and extensive northern latitudes of the region, plants extend their range to lower altitudes due to the movement of the sun and its effect on mean temperatures. It must always be remembered that as altitude increases, the effect of rain is proportionately enhanced, and that on the high-

Left, tussock grass, high up in the Aberdares. Above, rhino grazes by a giant cactus.

est mountains the rainfall is often less near the summit than on the slopes.

Plant families: The 22 main families of flowering plants in the East African region—excluding the *Gramineae* (grasses) and the *Cyperaceae* (hedges)—are represented in the following families: *Acanthaceae, Amaranthaceae, Asclepiadaceae, Capparaceae, Combretaceae, Commelinaceae, Compositae, Convolvulaceae, Cucurbitaceae, Euphorbiaceae, Labiatae, Leguminosae, Caesalpinioideae, Papilionoideae, Mimosoideae, Liliaceae, Malvaceae, Orchidaceae, Rubiaceae, Scrophularaceae,* and *Tiliaceae.*

Many of these families such as the *Malvaceae, Compositae* and *Orchidaceae* have a wide tolerance of changing ecological conditions and can be found throughout the region. Others such as the genus *Caralluma* are generally found in semi-desert and bushland where rainfall is limited and temperatures are distinctly high. *Orchidaceae,* both terrestrial and epiphytic (growing upon another plant), can be found from sea level to altitudes around 3,600 metres (11,808 feet) in conditions ranging from high humidity and warm temperature ranges, to dry conditions at high altitudes with wide variations in the nocturnal and diurnal temperatures.

Where ecological zones have marked characteristics flora has adapted itself to meet them. In grassland areas often extending over a hundred miles or more, the genus *Acacia* has evolved to cope with fire and drought. In many species the seed germinates more easily after fire and leaves are thin, often turning their narrower margins towards the sun to limit transpiration.

In all the dryer areas the *Gramineae* (grasses) have an unusually plentiful supply of seed to enable them to survive long periods when no rain falls and seed germination is doubtful or impossible. Similarly, the flowers of many plants have a higher than average nectar content to attract bees to stimulate fertilization. Others such as *Loranthaceae* and *Aloes* have bright orange or red flowers to attract birds which can distinguish these colours since the dry conditions inhibit much insect life which normally performs the function of pollination. Many plants in semi-desert and bushland regions have grey and aromatic foliage: the colour limits transpiration and the scents attract moths and insects for pollination.

Higher ground: In high montane and alpine areas three curious evolutions can be noted: giant lobelia, tree groundsel (*Dendrosenecio*) and heaths (*Erica*). Both lobelia and tree groundsels grow to unusual heights, the former to four metres (13 feet) or more with an inflorescence of three metres (10 feet); and tree groundsels up to 10 metres (33 feet) with a flower panicle one metre long. Lobelia have evolved an exaggerated deep calyx in which the blue flower is almost hidden, a device which withstands the wide variation in temperatures which may range from 60 C (140 F) at midday to several degrees of frost at night.

Tree groundsels are selective in their altitude: none of them grow below 2,550 metres (8,364 feet) and most of them are found at 3,200 metres (10,496 feet) or more. They are also selective in their habitat and each mountain area, such as mounts Kenya, Kilimanjaro, Elgon and the Ruwenzoris, has evolved its own subspecies.

In these high altitudes giant heather (*Erica arborea*) can be found. This many-branched tree grows up to eight metres (26 feet) in height with white flowers clustered at the end of the branches. Giant heath (*Philippea keniensis keniensis*), is found up to 4,250 metres (13,940 feet) on Mount Kenya.

In the medium altitude zones of grassland and wooded grassland there are 14 genera and 137 species of *Malvaceae.* Notable among them are *Hibiscus*, *Abutilon* and *Pavonia* which can be found mainly in grassy plains and, strangely enough, in rocky terrain and lava flows.

Papilionoideae, the pea family, is also prominent in grasslands, wooded grasslands and highlands. Represented strongly by the genus *Crotalaria*, of which there are probably more than 200 species in the region, they are widely distributed and can often be seen in considerable drifts of colour, mainly with yellow or yellow and orange flowers though there are one or two species which are predominantly blue.

No-one knows how many species exist in this vast region but when botanists complete their international survey into flora of tropical East Africa, it is probable that more than 11,000 species will have been described.

Left, giant lobelia and tree groundsel in the Ruwenzoris. Above, golden showers of the *Cassia didimobotrya* plant.

Lilies in the field: Among the *Liliaceae* (lily family) are three outstanding species. *Gloriosa superba*, sometimes known as "the flame lily", is a particularly beautiful plant with a red, red and yellow or red and green striped flower whose outside petals bend abruptly backwards (reflexed perianth segments). It is widespread in the area below an altitude of 2,500 metres (8,200 feet). There is a singularly fine variant at lower altitudes with lemon coloured segments and a deeper violet meridian stripe. The plant grows from a V-shaped tuber and can reach five metres (16.5 feet).

Albuca wakefieldii or *abyssinica* is the most common lily in East Africa. This robust plant up to one metre tall has bell-shaped flowers rather widely spaced on the stalk (peduncle). Though they never open fully, these flowers are yellowish green in colour with a darkish stripe down the middle of each petal (perianth segment) with off-yellow on the margins.

Aloes are among the *Liliaceae* most widespread in the middle and lower altitudes. Red, orange or yellow with green spotted or striped leaves, they form dense groups of

beautiful colour especially in grassland areas where grazing has reduced competition. Elephants are particularly fond of the *aloe*.

In the family *Amaryllidaceae* are two beautiful species. *Scadoxus multiflorus*, the fireball lily, is an arresting sight. Growing from a deep rooted bulb in rocky places, riverine forest and open grassland, often in the shade of trees or on the side of antheaps, the flower spike arises before the leaves and is crowned with up to 150 small flowers making a single magnificent red to pink head, looking like a gigantic shaving brush. *Crinums* often flower at the same time,notably *C. macowanii*, named the pyjama lily after the pink stripes marking the long tubular flowers which curve at the end of a thick peduncle surrounded by heavy, dull green leaves. It grows throughout East Africa and is often seen at the sides of roads and in ditches.

The carpet flowers: A notable feature from sea-level to more than 3,350 metres (11,000 feet) is the *Convolvulaceae* family which have no less than 22 genera and 170 species in the region. Prominent everywhere is the genus *Ipomoea* whose myriad flowers scramble over coral at the coast, in semi-desert scrub, in wooded grasslands, forest glades and even on the moorlands.

Cycnium tubulosum tubulosum, with white "pocket handkerchief" flowers, a member of *Scrophuliaraceae,* is dotted all over the grassland plains, especially on black cotton soils. *C. tubulosum montanum* often grows alongside its near relative. It has large pink flowers and extends to slightly higher altitudes. Both are parasitic on the roots of grasses and speckle the countryside for mile after mile where grasses have been burnt or grazed heavily.

In dry open bushland is another parasitic plant from the *Orobanchaceae* or broom-rape family—*Cistanche tubulosa.* An erect, unbranched spike of yellow flowers like a large hyacinth springs out of a bare patch of soil drawing its nourishment from the roots of neighbouring shrubs or trees.

In the family *Iridaceae* the genus *Gladiolus* has three beautiful species: *G. natalensis* grows up to 3,050 metres (10,000 feet) throughout the region and south to South Africa. It has yellowish-brown to orange flowers and is a relative of the garden varieties derived from *G. primulinus*. Above that altitude, the finest species of them all, *G. watsonioides* with bright red flowers, grows in stony soils only in alpine and subalpine regions on mounts Kenya, Kilimanjaro and Meru. *G. ukambanensis* is a delightful species with white, delicately scented flowers, produced copiously but capriciously in wet years. It is restricted to stony soils in the Machakos district of Kenya and in the Maasai Mara and coast areas of Tanzania.

Throughout shady and damp places in higher rainfall regions, nestling in banks or decorating the sides of streams will be found members of the family *Balsaminaceae,* allied to "Busy Lizzies" of temperate gardens. There are more than 70 species in the region, ranging from near the coast, inland through the Ruwenzoris to Zaire and Cameroons in West Africa, and from the Red Sea Hills to the Drakensburg.

Left, bright flowers attract birds, which help in pollination. Above, this orchid plant is found in the higher altitudes.

The desert rose (*Adenium obesum*), is found in semi-arid areas, often among inhospitable rocks. From the family *Apocynaceae,* this plant has magnificent long red to pink tubular flowers and fat fleshy branches: it appears a glowing mass of colour in a semi-lunar landscape.

Flowering trees: East Africa also boasts beautiful flowering trees. The Nandi flame *Spathodea campanulata* (family *Bignoniaceae*) is a magnificent sight when in flower, with large open chalice-like orange flowers growing up to 18 metres (60 feet) in height. Originating in areas from western Kenya to Zaire, it is now widely planted as an ornamental tree.

Calodendrum capense, the Cape chestnut (family *Rutaceae*) is another beautiful tree, recorded throughout East Africa and as far south as the Cape. It grows in heavy stands which set the whole area alight with its cyclamen coloured flowers.

In dryer bushland areas *Cassia singueana* and *C. abbreviata* (family *Caesalpinioideae*) both flower in front of their leaves and look as if golden sheets have been thrown over their branches. In contrast, in the wetter forest areas in the middle altitude range *Cordia africana* (family *Boraginaceae*) is a resplendent tree growing sometimes to 24 metres (80 feet), with stalkless white flowers massed in panicles inside a strongly ribbed, soft, brown calyx.

When flowering, the *Acacia*, sometimes called the "umbrella thorn", is covered with a mass of highly scented globular to elongate flowers. There are more than 50 species in East Africa: in particular, *A. senegal* and *A. mellifera*, which grow in meduim to low rainfall areas and decorate the bush for miles. *A. Seyal*, a species found in colonies on stony ground or black cotton soils, has lovely spherical flowers which appear in great profusion before the leaves.

Those who delight in the grotesque must not miss the Euphorbia tree. Numbering more than 20 species and ranging up to 30 metres (100 feet) in height, these trees have twisted triangular, quadrangular or hexangular fleshy branches. They exude copious amounts of latex and their flowers look like blobs of squashed plasticine.

AFRICA AT THE MOVIES

Those who travel to East Africa immediately fall in love with the dramatic scenery, herds of wildlife and various ethnic groups. Each new visitor relives the exciting moments of explorers of old who first glimpsed snow-capped Mount Kilimanjaro or descended into the Great Rift Valley to see plains' zebra and *Maasai* herdsmen with their ochred hair and coloured beads.

To the traveller these images are forever imprinted on the mind, like stills from an old movie. So it is surprising that the romantic appeal of Africa went pretty much unnoticed until Sydney Pollack's award-winning *Out of Africa* in the 1980s. But filmmaking is a complex business involving audience tastes, timing and perhaps most of all, production costs. That very terrain which is so appealing to the eye is not always so conducive to the rigours of locational filming. Hence many old full-length feature films about Africa were studio creations.

Mention Africa to the average movie buff and the response might be *Casablanca*, that 1942 Humphery Bogart/Ingrid Bergman classic, or the *Tarzan* series (most of which was filmed in up-state New York) or even Elizabeth Taylor's *Cleopatra* (1962)—but that's a bit north on the Nile.

African mystique: The allure of East Africa lies in its mountains, rain forests, jungles and deserts. There is the call of the wild. It is where the Big Five game animals dwell in simple, natural and beautiful environs and people likewise live in what might be described as rustic, even primitive, styles. Sunsets under the whispering palms of an Indian Ocean island or nights spent under a canopy of stars blanketing a campsite on a Nubian desert reiterate the mysteries of the universe. East Africa was the birthplace of mankind: now it is an Eden revisited.

Little wonder then that the Hollywood moguls ventured out to magnify the munificence of this continent, so dark in its ancient imagination and romance.

Preceding pages: Grace Kelly fans the flame in *Mogambo*. Left, Sigourney Weaver in *Gorillas in the Mist*. Above, scene from the 1949 film, *Sheena, Queen of the Jungle*. Note the tiger!

Adventure films tell of the lone hunter on a new frontier. And in the case of Robert Ruark's *Something of Value* (starring Sidney Poitier and Rock Hudson) or the more recent *Kitchen Toto*, the tale of human rights and freedoms is depicted in the pre-independence era of Kenyan history.

But at first it was the strange, exotic and unusual which drew the cinematographers to East Africa. Later, partition offered historic ethnographical subject matter for the creative filmmaker. When the Wild West was

won in American movies, cowboy heros gave way to new adventurers such as the Great White Hunter in the image of Ernest Hemingway, who added the *Swahili* words *safari* (journey) and *hatari* (danger) to the English language. These hunters were rugged he-men, not the playboys or sweltering, smouldering lovers in the vein of Rudolf Valentino's sheik.

In the 1930s Africa was still little known to filmmakers. In that star-studded era Marlene Dietrich came closest to the African theme with two films, *Morocco* and Richard Boleslawski's *The Garden of Allah.* Here she features with Charles Boyer setting out on a honeymoon in a sandstorm, portraying the fantasy and fatalism of love in the African desert. In *Morocco,* director Josef von Sternberg clothed her in mystery, portraying her as a shadowy café singer in a Foreign Legion town.

Perhaps the most memorable film to come out of Africa in the 1950s was John Huston's *The African Queen*, based on C.S. Forester's classic novel, and starring Humphrey Bogart and Katherine Hepburn. As with many of Huston's films, the story lived in him for a long time before he directed it.

Filmed in what was then the Congo (now Zaire), and on both the Congo and Kagera rivers, *The African Queen* combines a love story with both comedy and adventure. The two unlikely lovers are Rose (a prudish missionary) and Allnut (a gin-swigging riverboat captain). This is a love-under-the-mosquito-net plot that sees them commandeering the boat, *African Queen*, over dangerous rapids to torpedo a German battleship. Says Rose: "I never dreamed that any experience could be so stimulating!"

The script by Huston, John Agee and John Collier developed into comedy not readily apparent in the original novel. On location, Peter Viertel assisted with dialogue. He later wrote of the experience in his novel, *White Hunter, Black Heart.*

The African Queen is one of those rare films about an intelligent woman in love. The story was originally bought by Warner Brothers in 1938 and was to star Bette Davis and David Niven. Niven once revealed to Bogie that he had spent four weeks polishing up his Cockney accent and growing a beard which made him feel like a diseased yak before the whole thing was cancelled when Davis refused to be filmed out of doors. Hence the story was sold to Twentieth Century Fox.

The love stories: Lesser known films from the decade when television was beginning to rob cinemas of their big-screen-addicted audiences include *Mogambo* and a series by Warwick Films which produced *North of Mombassa, West of Zanzibar* and *Where No Vultures Fly*, most of which were filmed in Kenya and included such stars as Robert Taylor and Donna Reed.

Mogambo starred Grace Kelly, Ava Gardner and Clarke Gable. This 1953 film tells the story of a white hunter, whose world is invaded by an American showgirl. Together with an archaeologist and his wife, they all wander off on a gorilla hunt. The film also featured Kenya actor—David Makio.

Much of it was filmed in Amboseli National Park bordering Tanzania and at Lake Chala and Lake Jipe in Taita Taveta district.

The famed Grogan Castle, built by the Cape-to-Cairo explorer Captain Ewart Grogan at the turn of the century, was also used as a site. Director John Ford admitted that he had made the film because he didn't want to deprive himself of a trip to Africa.

His holiday mood was apparently in evidence on film, as one reviewer bemoaned the lack of direction, concluding that the cast was out-acted by the gorillas.

Leni Riefenstahl, famed for her German propaganda films of the 1930s, first fell in love with Africa in 1956 while touring Kenya and Tanzania (then Tanganyika). Since cinema had become a forbidden domain to her, she turned to photography. But during that first visit she started a fictional documentary, *Black Cargo*, about contemporary slave traffic, which she planned to submit to the London Anti-Slavery Society. However, her plans were drastically altered after an accident with her landrover in northern Kenya. She spent several weeks in Nairobi Hospital suffering from a skull fracture and several broken ribs. In 1961 she returned to the Congo, Uganda, Sudan and Kenya. Further financial and world crises hindered her progress but not before she completed some documentary footage in 16mm and several ethnographic photograph collections, one of which resulted in *The Last of the Nuba.*

The Hemingway mystique lives on in many corners of the globe and Africa is no exception. In East Africa he played the big game hunter himself. One hotel, the *Blue Marlin*, in the Kenyan coastal town of Malindi, boasts a plaque—"Hemingway stayed here". The Indian Ocean satisfied his desire to go deep sea fishing. In both *The Snows of Kilimanjaro* and *The Green Hills of Africa* (Susan Hayward and Gregory Peck) Hemingway's personal introspection and his quest for adventure are revealed on screen.

Left, *Born Free* told a lion's tale. Above, *Out of Africa* brought mass recognition.

Animals as actors: African animals have been cast as hundreds of thousands of unpaid extras in a number of films, mainly shot in a *cinema verité* style. With no union or guild to plead their cause this runs nothing short of exploitation! But films have played a part in conservation efforts. Many a big producer has left substantial sums to the World Wildlife Fund or the East African Wildlife Society in gratitude for the participation of this silent majority.

Some producers have chosen to transport their own trained animals to location although many animals become surprisingly stubborn and wilful in the climate of their original roots. Others used stand-ins in films such as *Sheena, Queen of the Jungle* where ponies were painted as zebras. Still others have relied on the real thing in its natural habitat. *Sheena*, a female Tarzan-type adventure, also included the zebroid stallion Mariko, Chango the elephant and chimpanzees Tiki and M'Bongo of *Animal Actors*, Los Angeles.

Films such as *Jumbo, The African Lion* (an early Walt Disney True-Life Adventure), *The Lion* and *The Last Safari* continued the tribute to wildlife. Most of these were filmed in Kenya in the 1960s. *The Lion* featured

William Holden (a long way from *Sunset Boulevard*), Capucine (an equally long distance from *Walk on the Wild Side*) and Trevor Howard. Holden went on to become a household name in Kenya as one of the founder members of the original, exclusive Mount Kenya Safari Club. The club was visited regularly by Hollywood Types and later went on to gain fare as a wildlife sanctuary.

Born Free: Perhaps the one film that did as much, if not more than *Out of Africa*, to put Kenya on the map, especially for American audiences, was *Born Free*. This was based on Joy Adamson's book of the same title and told the poignant story of Elsa the lioness and the Adamsons' decision to release her from their camp at Kora in northern Kenya into the bush.

Virginia McKenna and William Travers brought the story of Joy and George Adamson to the screen. Local actor Peter Lukoye played the assistant to Joy. The theme song was used by *Kenya Airways* when planes approached the then Embakasi Airport. And the film itself developed into a television series, *Living Free*, which later led to the popular Ivan Tors *Daktari* series for the BBC.

In *Born Free* Joy is characterised as an angular British woman with a strange penchant for lionesses. Husband George is a bit leonine himself (remember Bert Lahr's *Cowardly Lion*?) but seems to understand his wife's motives. In real life Joy actually became known much earlier for her painting. Over 600 ethnographic portraits of the peoples of Kenya and an equally impressive collection of meticulously rendered watercolours of plant life are her legacy to the National Museum in Nairobi. The world was shocked by her untimely and mysterious death at Kora Camp in the early 1980s. George tells his own story in his autobiography, *My Pride and Joy*.

Documentaries: *Born Free* is probably

also responsible for a rash of documentary films made in the last 20 years. Alan Root's work is the most widely known, having appeared on numerous television networks and is still a favourite evening's entertainment in the up-country lodges and hotels of Kenya. Root's vivid photography has captured lions, elephants, hippos, wildebeest, hornbills, termites and almost every imaginable ecological subject. Anyone familiar with Albert Lamorisse's classic *The Red Balloon* will have heart palpitations and flights of fancy when they view Root's balloon safari which gracefully records the annual July migration of hundreds of thousands of wildebeest from Tanzania's Serengeti to Kenya's Maasai Mara Game Reserve.

Further afield there have been Jane Goodall's studies of chimpanzees at Gombe in Tanzania and Dian Fossey's gorilla work at Karisoke, Rwanda (*Gorillas in the Mist,* filmed in Kenya and Rwanda). Equally as impressive is Romain Baertsoen's *Ibirunga* about the volcanoes and gorillas of Akagera National Park in Rwanda. The fascination in both films is the close proximity to those magnificent apes.

Kenyan poet, painter and filmmaker, Sao Gamba, won international awards with his study of the *Maasai* tribe, *Men of Ochre*.

In 1984 Hughes Fontaine, a young visiting French teacher at Pangani Girls School in Nairobi, finished his documentary film, *The Singing Wells*, for French television. This told of how the nomadic *Gabra* managed to find water during periods of drought. Tracing their wanderings from Maikona to Kalacha to Balesa, Fontaine's final shot features himself emerging from a Paris underground station, above which is a poster of Souleyman Cisse's *Vinye*.

Apart from strictly informative documentaries about the beauties of African wildlife or the strange and exotic habits of birds and people, other films have been made to publicise a just cause or plea. *Serengeti Shall Not Die* made by Bernhard Grzimek (director of the Frankurt zoo) and his son Michael, set out to prove that vast herds of zebra and wildebeest, as well as other game, were in grave danger from hunters and poachers, especially if government plans were finalised to limit the park's boundaries. Michael Grzimek lost his life while making the film when a vulture struck the small plane he was piloting.

Left, Clarke Gable plays a white hunter in *Mogambo*. Above, Susan Hayward and Gregory Peck in *The Snows of Kilimanjaro*.

Serengeti Shall Not Die was not a highly organised or expensive film. The message is simple and clear: it deals with animal extinction as a loss to all mankind, with destruction blamed on both black and white man. Africans are shown with their wire snares and poisoned arrows: in another sequence a warehouse of what looks like tree stumps turns out to be elephants' feet made into souvenir wastebaskets and footstools.

More recently, the camera work of Kenya's own award-winning Mohamed Amin has opened the world's eyes to the devastating famine in Ethiopia. Prior to that Amin filmed countless coups and other news events in Kenya, Zanzibar, Central African Republic and Uganda.

Kenya's archaeologists have also got in on the act. Richard Leakey's seven-part series, *The Making of Mankind*, begins and ends in East Africa, the palaeontologists' favourite hunting grounds for early man. Most notable are the scenes in Olduvai Gorge and Koobi Fora.

Making history: In recent years Kenya's colonial history has provided the storyline for scriptwriters. Danish baroness Karen Blixen's unsuccessful attempts at coffee farming in the 1930s coupled with her disastrous love life were woven into *Out of Africa*, based on her own enigmatic diaries. Meryl Streep and Robert Redford (playing Blixen's lover Denys Finch-Hatton) were filmed in Karen (the Nairobi suburb named after her), the Ngong Hills and the Karen Blixen Museum (formerly Blixen's home).

Woody Allen's *Annie Hall* brought Diane Keaton's "Kenyan basket" out of the bush and into the front windows of Bloomingdales, just as *Out of Africa* pushed rumpled khaki into everyone's closet.

Similar pages from the same chapter of Kenyan history resulted in *The Flame Trees of Thika* with Hayley Mills, a series based on Elspeth Huxley's childhood memories in Njoro and Thika, then a fast-growing agro-industrial town (pineapples, sisal, motor vehicles) to north of Nairobi.

Kenya's own flying ace, Beryl Markham, is brought to the big screen in *Shadow on the Sun* with Stephanie Powers, Clair Bloom and a host of other big names. Granada's *After the War* series has an episode filmed in Mombasa: *The Winds of Change* with Clair Higgins and Art Malik.

The unsolved murder in 1941 of playboy Josslyn Hay, the 22nd Earl of Erroll, was the subject of *White Mischief* by journalist James Fox. Produced by Michael White, Simon Perry and Michael Radford, the film version emphasises the hazy reality of the Happy Valley (*Wanjohi*) residents. The BBC continued this saga in *Happy Valley.*

Jumping back a century, *Mountains of the Moon* tells the story of explorers Richard Burton and John Speke. The title refers to the Ruwenzori Mountains shared by Uganda and Zaire and the *Unyamwezi* (people of the moon) of Tanzania. The story is loosely based on Burton's own accounts, *First Footsteps in East Africa* and *Lake Regions of Equatorial Africa,* and Speke's *Journal of the Discovery of the Source of the Nile.*

For filmmakers with panoramic vision, East Africa will provide a cinematic setting for a long time to come. Multicultural peoples, history, big game, adventure and mystery abound. And for subject matter, modern novels and short stories offer dozens of scripts the big producers haven't even considered yet.

Left, Rock Hudson and Sidney Poitier on location in *Something of Value*. Right, Hepburn and Bogart aboard *The African Queen.*

CUSTOMISED SAFARIS

When Robert Redford was filming *Out of Africa* he flew over Kenya's Maasai Mara Game Reserve in a small plane. Redford peered out of the window and saw a solitary male lion encircled by vehicles filled with camera-toting tourists. "I know just how he feels," Redford sighed.

As tourism became increasingly important to the Kenyan economy, so the old-style safaris under canvas beat a retreat, leaving in their wake conveyor-belt tours of the "today Naivasha, tomorrow the coast" variety.

Hauntingly beautiful wildlife areas have become so congested that the authorities are deeply concerned about the rapid deterioration of the environment. The game-drive rush hours—early morning and late afternoon—are turning grasslands into cauldrons of dust.

Zoologists report that some predators have developed neuroses because they can no longer stalk their prey without being trailed by a horde of onlookers. The cheetah, which normally hunts in daylight, is adapting to the crowds by hunting under the cover of darkness instead.

This is not what safaris are supposed to be like. There is a different type of trip to be had that is in tune with the slow, quirky tempo of life in the bush. On this voyage the true face of Africa—both majestic and intriguing—will unfold before you at a leisurely pace.

Customised safaris that take you off the beaten track and into the wild come in many different wrappings. Some cosset their clients with champagne and comfortable Range Rovers. Others are arduous and travel to the edge of reality. Deluxe or demanding, they both offer skilled guides who do what you want to do and take you where you want to go, ensuring that every moment is savoured before being thoroughly digested.

The guides are today's pioneers of Africa. The spirit of exploration prevails as they share a landscape close to their hearts. The alchemy of limitless horizons, hard physical

<u>Preceding pages:</u> floating above the Maasai Mara; going on camel safari. <u>Right</u>, Mount Kilimanjaro in your back yard.

exertion and encounters with lion, elephant and buffalo outside the confines of a car conjures up a heady and irresistible brew.

Tony Church, a third-generation Kenyan, runs trips on horseback through the virgin forests and undulating grasslands that are home to the nomadic *Maasai*. The sceptics predicted Church's horses would succumb to tsetse flies and lions, but they were proved wrong. He injected his ponies against sleeping sickness and employed *Maasai* warriors to guard the picket lines at night. Now a well-established holiday in equestrian circles, his safaris are always fully booked.

Those who sign up have the option of riding hundreds of miles for two weeks, moving camp each day, or splitting their time equally between game lodges visited by car, followed by a week of riding through the countryside that was so stunningly portrayed in the funeral scene from *Out of Africa*.

Whichever you choose, you will get plenty of close quarter game viewing from the saddle. The horses are accustomed to cantering alongside topi, giraffe, zebra and even rhino and buffalo but have learned to keep elephant, who are notoriously inquisitive, at a respectful distance.

Richard Bonham, the Kenya-born son of a British game warden, and Conrad Hirsch, a former mathematics teacher from Texas, typify the trail-blazing spirit that is quintessential to out-of-the-ordinary trips. They lead safaris on foot and by boat through the dark heart of Tanzania's Selous Game Reserve. The Selous has been likened to a Disneyland without people. With the exception of Arab slavers and ivory hunters, it has been virtually ignored by humankind. Its miombo thickets and palm-lined rivers are rarely visited except by the occasional poacher.

Bonham's two-week safaris, partly by boat, and mostly on foot, emulate those of the 19th-century explorers. Bonham likes to live off the land, just as they did, and "shoots for the pot". His limit of eight visitors at a

time endows trips with a feeling of truly personalised service. The meandering route varies, depending on whim. Walking time is kept down to a manageable three or four hours a day. Trekkers are followed by a snaking line of 30 porters, each of whom balances 18 kilograms (40 pounds) of equipment on his head.

What can rival the thrill of hearing an elephant trumpeting a few yards ahead as you walk along a river bank? Visitors who stalk these awe-inspiring creatures are rewarded with a front row view of cows and calves indulging in an impromptu shower as they suck up water in their trunks and spray their backs.

Bonham and others who take out foot safaris have a well-honed knowledge of bush lore that keeps their charges safe from harm. The secret of successful game viewing on foot is to approach downwind so that the animals do not catch the scent of approaching humans.

Conrad Hirsch prefers the tranquility of paddling down the Rufiji, East Africa's greatest river, in inflatable rubber boats. To reach his camp at the foot of the Shuguri Falls' precipitous red cliffs, visitors take the train from Dar es Salaam, then embark on a five-hour odyssey in Land Rovers. There are no roads and the tracks made by the four or so vehicles that pass through each year are quickly outgrown. Hirsch admits to quite a bit of "searching around and getting lost".

By day you shoot white water rapids and paddle along sandy flats rimmed by doum palms. In the evening you put up tents on the river bank and eat Hirsch's personal gourmet specialities—Chinese stir fries, Ethiopian *wat* and Indonesian kebabs.

Hirsch's chief concern is if the rubber boats should flip, but visitors soon learn another danger lurks in the muddy depths. Thousands of hippos lie submerged, seeking protection from the sun's rays. Sometimes they surface as a boat passes overhead, lifting it out of the water. On very rare occasions, enraged by the intrusion, they bite a boat as it floats by.

Despite the adventure of rubbing shoulders with nature, open-air safaris are comfortable and well catered for. In fact, there is so much good food on offer that any resolutions to become as lean and lithe as your guide are hard to keep.

The magic of these safaris is that they give you the time to become properly acquainted with East Africa's stunning visage and its intimate secrets: a porcupine quill half hidden in the sand; a valley stretching for 115 kilometres (70 miles) to a horizon of crystal clarity.

You will also get the chance to talk to local people such as the ochre-painted warriors who wear beaded bracelets in the shape of a watch, an irreverent comment on our preoccupation with timekeeping. All nomads can tell the time of day to within half an hour by the position of the sun, which is as close as you need to get when you have no appointments to keep. Safari-goers would do well to take a leaf out of their book.

Left, geared up for a 1930s safari. Above, these photo fiends have two willing subjects.

HORSEBACK SAFARI

Horses are not indigenous to East Africa south of the Abyssinian Highlands and the Horn of Africa. In the past, any horse which ventured southwards from Kenya's arid northern region was soon an unfortunate victim of the fatal African horse or sleeping sickness (*Trypanosomiasis*).

Early riders: When European missionaries, pioneers and hunting parties penetrated the interior, they travelled mostly on foot or by oxwagon. One of these early adventurers was Lord Delamere, sometimes known as the Red Baron from stately Vale Royal in Cheshire. He later became the flamboyant leader of the British settler community. Delamere first visited what is today the highlands of Kenya in 1897 by way of Berbera in Somaliland and south across the baking hot Chalbi Desert. His main purpose was to hunt big game. His caravan of bearers, trackers and *askaris* (guards) were supported by tough little Somali ponies indigenous to the Horn of Africa and the highlands of Ethiopia. Ponies that survived the long journey and reached the crisp, clean, disease-free air of the Kenya highlands were among the first horses to be brought to East Africa.

Subsequent Europeans who ventured to East Africa to take up land brought their thoroughbred horses from the British Isles, only to have most of them succumb to a host of African diseases. So settlers soon crossbred imported horses with hardy sure-footed Somali ponies and started what was to become a substantial herd of resilient country bred horses, adapted to conditions prevailing in East Africa.

At the outset of World War I in 1914 the British found themselves facing German settlers on their southern border with German East Africa (now Tanzania). When the colonial authorities realised the declaration of war in Europe was a serious matter they hurriedly sent despatches to the South African government for horses. These were shipped to the port of Mombasa to help the war effort. Settlers formed the East African Mounted Rifles, a Cavalry regiment, to patrol the border and pursue General von Lettow Vorbeck, the elusive and cunning German pioneer of guerilla warfare. Thousands of horses perished in this campaign as described in Charles Miller's *Battle for the Bundu*. But horses were obviously the most reliable source of transport during the Great War. Some early pioneers thought they could capture and harness wild zebras but this was a failure since zebras have weak hearts.

After the Treaty of Versailles in 1918, another wave of settlers under the Soldier Settler Scheme emigrated to the young colony. By now horses were firmly established on colonial farms for pulling pony traps, checking long fence lines and scaring away lion and other predators which in those days were regarded as vermin. Theodore Roosevelt enjoyed big game hunting from horseback during his visit to East Africa in 1913 despite the fact that horse flesh was known to be a favoured delicacy for a pride of hungry lions.

As motor cars became more popular the horse as transport gave way to these machines. Horses were bred more and more for the race track, polo and as hunters. Cross-country journeys on horseback were seldom undertaken except during the long rains (April and May) when roads became a quagmire, or when following up a gang of cattle rustlers. Nomadic warrior tribes, particularly the *Maasai* and *Samburu* relish cattle raids on a moonlit night as all cattle are considered a god-given right of the tribe. European farmers who were the target of most of these raids used their farm horses to follow up the marauders.

Short circuits: Mass tourism got under way in the mid-1960s with the advent of jet passenger aircraft. Fashionable safaris for the rich now came within the reach of anyone with a love of the wilds. No longer were expensive mobile deluxe hunting camps the only accommodation available in the game lands of Kenya. Shooting safaris with cameras rather than rifles became one of the most talked about experiences. Safari lodges and tented hotels were established at intervals along various tourist circuits.

Left, hunting on horseback. Above, still chasing giraffes, but only to shoot pictures.

Horseback trips were organised in the mid-1960s using Somali ponies and zebroids (a hybrid cross between wild zebra and a horse) as pack horses for parties climbing Mount Kenya. But long-distance riding safaris set up on a commercial basis were only started in earnest in 1972.

To outfit and escort safaris into the heart of game country for visitors from America or Europe was not easy. Horseback riding was regarded as a dangerous sport and carried with it real responsibilities and grave consequences should anything go wrong. Limited rides began on a daily and overnight basis across the Kitengela Plains and up on to the

Ngong Hills on the eastern edge of the Great Rift Valley. These wooded hills contain bushbuck, mountain reedbuck, eland, kongoni, Cape buffalo, colobus monkey, waterbuck and the occasional lion and rhino.

With picnic lunches carried in saddle bags, these early rides proved a great success. Riders returned to their Nairobi hotels after an exciting day in the saddle away from other tourists in minibuses. Horseback safaris endow an amazing sense of being on even terms with wild animals—no other sounds or smells except those of the bush. Of course, to enjoy such a riding experiences clients must be fit, mentally tuned in, able to face unpredictable situations and most important be confident on horseback at all paces.

Longer trips: As these safaris gained popularity longer and more ambitious routes were pioneered. In 1972 a five-day trail was forged from the Athi Plains, over the southern shoulder of the Ngong Hills and into the Great Rift Valley, dropping into the Lookariak lugga (dry stream bed) and beyond to the Kedong River. Then across the great Akira plain between extinct volcanoes mounts Suswa and Longonot, finishing with a spectacular day through Hell's Gate Gorge to the shores of Lake Naivasha.

Tentage, camp gear, groceries, horse grain and safari staff are carried by truck along bush tracks while riders go cross country covering between 25 and 40 kilometres (15 and 25 miles) a day. Picnic lunch and waterbottles together with a few personal effects are carried in saddle bags strapped to cavalry saddles.

Another spectacular riding safari leads from the wooded Nguruman escarpment, over the rolling Loita Hills down to Narosura spring before branching north west across the Loita plains always teeming with game to the Mara River. This ride finishes on the beautiful Esoit Oloolol escarpment which forms the western boundary of the Maasai

Mara Game Reserve.

Safaris Unlimited (Africa) Ltd are the outfitters and organisers of these horseback adventures, with stables and safari depot 16 kilometres (10 miles) outside Nairobi. Those wishing to undertake a horseback safari should book in good time since these safaris are organised by special arrangement or by joining a group put together by an overseas agent.

Riding today: On arrival in Nairobi you will be taken to one of the capital city's leading hotels for the night. The next day you will be driven to a very comfortable and picturesque camp set up in a glade of Podo and African Olive trees.

After a restful evening followed by a substantial English breakfast you head out in the African wilds with your own horse on a cross country trek that will take you through mountains, forests, grassy plains, rivers and escarpments filled with a variety of wildlife. There are no fences, telegraph poles or tarmac roads and the sense of space and freedom is quite overwhelming. You walk among huge herds of plains game, sometimes canter with giraffe or wade across muddy rivers, closely observed by families of hippo. You are always led by a highly experienced English-speaking guide.

Each day you head into the wild blue yonder while staff pull down the tents, drive round on bush tracks and then re-erect the whole camp at the next waterhole. After six or seven hours in the saddle (broken by a lunch stop) the party ride into the camp at about 4 p.m. for tea or cold drinks and hot showers before dinner.

The 10-day route leads from the swamps of Morijo, over the Subugo ridge to Narosura, Maji Moto (hot springs), Olare Lamun, and Olare Orok near the Maasai Mara where you rest for a day before the long ride across to Musiara, the Mara River and up on to the Esoit Oloolol escarpment to the site where Denys Finch-Hatton was buried in the famous scene from *Out of Africa.*

Finally, it's back to the Mara River for your last night under canvas in the riverine forest near Hippo Pool. Four-wheel drive station wagons are always nearby, enabling anyone to take a break from the horses and head into the game reserve to photograph animals at close quarters from the safety of a vehicle.

Left, riding on the dusty plains. Above, a popular trail leads through Hell's Gate Gorge, near Lake Naivasha.

UP, UP AND AWAY!

The *Swahili* word *safari* conjures up images of dust-stained travellers winding their way across the savannah followed by lines of African porters. But imagine floating gracefully through the air above the dirt and insects of the plains and you'll have some idea of what balloon safaris are all about.

Jules Verne's novel *Five Weeks in a Balloon*, published in 1862, was the first to mention ballooning in Africa and was the inspiration for English gas balloonist Anthony Smith's visit 100 years later. Using a balloon lifted by hydrogen he successfully crossed from Zanzibar to Tanzania. He also completed flights over the Serengeti and the Great Rift Valley.

Accompanying Smith as camerman during these early flights was Alan Root, now a world-renowned wildlife film maker. He realised that if problems of expense and manoeuvrability could be overcome a balloon basket was the perfect place from which the majesty of the African landscape could be fully appreciated.

Alan Root had a hot air balloon delivered to Kenya and with the aid of a trained pilot set about learning to fly it. Early flights were hazardous until European flying techniques were adapted to African conditions. The result of Root's efforts was one of his most popular films, *Safari by Balloon*.

While on location Root was asked on several occasions by passing travellers for rides over the savannah. It was these visitors to Kenya, who wanted to see the game from a different perspective who prompted him to set up Kenya's first balloon company, appropriately named Balloon Safaris. Based at Keekorok Lodge in the Maasai Mara Game Reserve the company has flown 30,000 passengers since its inaugural flight in 1976. The original five passenger balloons with their cramped baskets have now been superseded by balloons three times bigger with baskets containing seats for 12 passengers.

Much hot air: Since those first flights balloon safaris have now become a major attraction for many visitors in Kenya. Other balloon companies have sprung up in the Mara based at Governor's Camp, Sarova Camp and Fig Tree Camp. Visitor's staying at any lodge or camp in the Mara are close to a balloon base if they wish to fly. Although open year round the optimum time to balloon in the Mara is from July to October during the annual wildebeest migration, when over one million animals cross the plains.

The once-daily flights lift off with the rising sun for an hour long journey over an average distance of eight miles. Being highly manoeuvrable the balloons can skim tree tops or rise to over 300 metres (984 feet) for panoramic views of the rolling Mara plains.

The basket is an ideal platform for photography. Binoculars are a bonus for spotting game from higher altitudes. Passengers often forget it is a hot air balloon and wear unnecessary extra clothing. However, a hat is recommended for any tall passengers who find that their place in the basket is located under the burner!

The vast plains are ideal places for landing balloons four times bigger than their counterparts in other countries. All balloon companies serve a champagne breakfast wherever they land in the park, something as memorable as the flight itself.

Because balloons are moved by the prevailing winds they cannot return to their take-off point. After breakfast passengers are driven slowly back to the lodge by retriever vehicles. All companies return passengers to their respective lodges by mid-morning after presenting them with a certificate to mark the occasion.

In 1988, balloon companies started flying in two other game parks. In Samburu Game Reserve ballooners have the chance to enjoy the magnificent scenery with Mount Kenya in the distance. Ballooning in the privately-owned Taita Hills Game Sanctuary near Tsavo West Game Reserve offers views of Kilimanjaro as well as a variety of game.

A balloon ride cost US$250 per person in early 1989. Visitors are advised to book before arrival.

Right, great care is taken to prepare a balloon for a flight, which invariably becomes the highlight of any East African safari.

5Y-OGR

Going By Camel

Camel safaris offer a leisurely face-to-face encounter with the real Africa. Those who have the vision and energy to travel this way, emulating the Samburu and Rendille nomads of northern Kenya, will be rewarded with a memorable adventure.

It can be an unsettling experience to enter a world where there is no sign of a building for hundreds of miles and where waterholes can be 50 kilometres (30 miles) or more apart. But this unparalleled solitude is the attraction for those who travel with camels.

You meander along dry watercourses known as *luggas* and have the option of either walking or, when you get tired, riding. You cover up to 24 kilometres (15 miles) each day and rest a few hours during the midday heat.

Camel safaris wander through a vast and varied landscape in northern Kenya: the sacred slopes of Mount Nyiru, where the Samburu sacrifice bulls on an altar of giant rock outcrops; the arid plains of El Barta, stippled with mauve grasses; the Leroghi plateau, rimmed with cedar forests; the Suguta Valley, an alien moonscape of laval scarps moulded by erupting volcanoes. Take a camel safari here and you will spend delightful days becoming acquainted with at least some of this spectacular scenery.

By 6 a.m. trekkers will be awake and enjoying their first mug of tea around the campfire. In another hour trekkers will be perched atop the camels' humps or striding alongside the train, enjoying the brief cool that comes before the sun soars towards its zenith. By 9 a.m. they will be bathed in sweat and, for the first day at least, preoccupied with brushing off swarms of flies.

Trekkers should be reasonably fit and, of course, accustomed to walking long distances. The average daily distance is 24 kilometres (15 miles), even though camels can cover 40 kilometres (25 miles) with ease. On the first day, every joint and muscle in your body complains. After that you should become attuned to the pace.

You are on the move for four hours in the morning, before the heat becomes intolerable, and two hours in the evening when the sun is sliding towards the horizon. The midday break is essential for camels and people alike to browse and rest.

Clothes are a matter of individual taste. They should be cool and comfortable and should not constrict your legs when striding. Bring a bathing suit for plunging into rivers, and a sweater or light jacket for chilly evenings. All this should be packed into a kit bag tough enough to survive the wear and tear of being strapped to a camel.

Trekkers spend the evenings sprawled in exhausted abandon around the campfire. Sleep comes easily on camp beds beneath an indigo sky. Camps are set up beside water holes that herdsmen have dug for their cattle. Each water hole is protected from wild animals and is reasonably clean.

Often the evenings are punctuated by the chilling laugh of hyaenas. And sometimes dawn reveals the spoor of a herd of elephants or a solitary lion.

The camel caravans are tended by local herders, who handle the mercurial moods of their charges with good humour. Camels can be cantankerous and wilful as well as beguiling. They have been known to crush the kneecaps of their herders with one swift snap of their teeth.

By comparison, safari camels are well trained and usually obedient. To watch these silken-lashed beauties gliding eagerly over the sand is to be reminded of sailing ships coming into harbour on a stiff breeze. The impression, however, is erroneous. Once mounted, it feels more like being adrift in an Atlantic gale until you acquire your sea legs.

At first sight, mounting a camel appears a daunting task. However, once tried, it can be executed with ease.

Some safari-goers make the daily journey almost entirely atop their mount. Others prefer to walk. To be perched on these beasts is less precarious than it seems though. Riders are never dislodged—perhaps because they realise it is too far to fall.

Mounted or not, you are hostage to the slow, quirky pace of African travel. And to the panoramas and pitfalls that unfold along the way.

Therein lies the magic of camel safaris. Africa's intimate secrets, hidden to those who travel by car, are revealed to passersby with more time to spare.

Left, loaded up and ready to roll. Above, resting by the warm glow of the campfire.

FRESHWATER FISHING

If you consider your angling to be a contemplative and sedentary pastime, fishing in East Africa is not for you. The variety of indigenous and exotic fish and the range of wild and unexploited locations turn a casual day out into an angling adventure.

East Africa, with its wealth of animals, has long been geared to the needs of the game safari traveller. Fishing safaris, on the other hand, are still relatively unstructured and consequently you need a sense of initiative, and adventure in your soul, if this is to be the holiday for you.

Head for the hills: Just an hour's drive from Nairobi, detouring off Kenya's main North road, you can climb into the foothills of the rolling Aberdare mountains, a series of steep ridges separated by narrow valleys, each with a dashing mountain stream. The precipitous slopes are heavily cultivated by Kikuyu whose tribal customs, ironically enough, revile cold-blooded creatures, including fish. So fishing here is reserved exclusively for visitors.

The nearest river, the Thiririka, is reached via Gatundu, the country home of Jomo Kenyatta. Trout can be found in the river just short of the forest to between 2,000 and 2,300 metres (6,560 to 7,545 feet), and deep within the forest itself to about 2,750 metres (9,020 feet). However, the rugged conditions will limit access to only a few kilometres from the river banks. This is quite sufficient, given the allowable limit for the river of six fish per day. The hardy angler, after bagging his limit, can traverse the ridge for a second bag on the adjacent river. But for most people the heavy going demands retreat until another day.

There are numerous rivers and trout holding streams on this side of the Aberdares including the Gatamayu, Ndurugu, Karimeno, Chania, Thika and Mathioya. There are also late opportunities for big fish on the plateau of the Aberdare National Park. Streams on the south-eastern slopes are narrow, little more than two metres (six feet) wide and consisting of a series of pools connected by narrow rapids overhung by bushes. A rising fish is a rare sight since, in the absence of a significant insect hatch, trout feed on underwater aquatic insects and crustaceans. Casting is not easy on the forested banks but the narrow streams do not need a long cast. Use a sunken fly, of the attractor rather than the imitative variety—a *coachman*, *invicta*, *Watson's fancy* and *butcher* are successful patterns as are "local" specials known as a *Mrs Simpson* or the *Kenyan bug*.

One delightful surprise in Kenya is the availability of high quality trout flies at very low cost. Produced for the overseas market, they are also made to order for local fishermen by the expert fly tyers of Kenya Trout and Salmon Flies at Kikuyu, just outside Nairobi, who pride themselves on reproducing any pattern you care to name.

On the Thiririka, as on most Kenyan rivers, trout are of the short, fat rainbow variety. Neither rainbow or brown trout are native to Kenya but were introduced in ova brought from Britain by Major Ewart Grogan in 1905. These, and later fish from South Africa were introduced to the Gura, Amboni and Nairobi rivers near Nyeri. (Do not confuse the Nairobi River with Nairobi City: the two are a hundred miles apart.)

By turning left into the Aberdares at Thika about 45 kilometres (30 miles) from Nairobi, and then driving the same distance again, you will arrive at the government fishing camp at Kamakia. Sleeping huts and firewood are available at little cost. The camp is ideally located on a ridge with the Chania and Kamakia rivers on either side of it. These rivers are cleaner, faster and shallower than the streams a few miles to the south and they provide a large population of rainbow trout.

The forest area is largely unfished and the more intrepid angler can descend close to 300 metres (1,000 feet) on the almost vertical elephant tracks that join the valley to the ridge. Elephants are rarely seen but evidence of their presence is all around you. The wise angler talks loudly to his companions as he descends the track, thereby avoiding any

Left, the one that didn't get away: giant Golden Nile perch.

possible confrontation with ascending elephant, rhino or buffalo. The simple rule is that East African anglers do not dispute possession of path or pool with the local big game!

Back at Thika, the Blue Post Hotel offers both accommodation and scenic beauty. It lies at the confluence of the Thika and Chania rivers which descend through the hotel gardens in two mighty waterfalls. Both rivers are stocked with trout in their upper reaches but at Thika the warm foaming waters contain barbus (*Barbus Thikensis*). These silver fish, streamlined and strong from a life in the fast flowing water, can be angled for, African-style, with some ledgered maize paste.

In the rivers fish barely reach 500 grams (one pound) but in the pools below the waterfalls at the Blue Post, fish weighing up to 4.5 kilograms (10 pounds) are caught. The idea is to clamber down to the tail of the pool and then carefully work your way to the falls. On a ledge beneath the waterfall itself, the bait can be cast into the white water and allowed to swing through the pool. The strike is hard and fish fight without showing themselves until they reach the landing net. Unfortunately they have too many bones so are not good to eat. Pickling in vinegar to dissolve the bones has been recommended, but it is also possible to exchange larger specimens for pineapples sold by the local roadside traders. This is a more refreshing reward for a successful day's fishing.

Gone fishing: As you continue to Nyeri and Nanyuki you will pass about 30 more trout streams flowing from the Aberdares and Mount Kenya. Two in particular merit special attention—the southern and northern Mathioya rivers belonging to the Kenya Flyfishers Club. These beautiful, crystal clear streams with manicured banks and paths provide the very best of fly fishing. Both brown and rainbow trout abound and the clubhouse walls are decorated with plaster-casts of specimen fish weighing up to four kilograms (nine pounds).

On the banks of both rivers are elegant timber buildings, each with sleeping accommodation for eight. Cooks and other staff meet weary anglers on their return from the river and clean and prepare their fish for dinner. Unfortunately such comfort doesn't come cheap and no day tickets for visitors are sold. However, the determined angler may perhaps be able to wangle an invitation from a bona fide member.

The moorland streams of the Aberdare National Park are, however, open to the public and the Nyeri, Chania and Gura contain long lean trout. The open banks allow full reign to the elegant caster but on these rivers you should keep an eye open for wildlife. Elephant and buffalo are very common and usually resent your company. On at least one occasion the hunter has become the hunted! There are camping facilities for visitors although you will have to cook your own catch.

Nyeri and Nanyuki both boast first class hotels and the Ark Mountain Lodge game lodges allow visitors to combine fly fishing with a game safari.

As you descend from the Aberdare National Park on its west side you will reach the incredibly beautiful Lake Naivasha. Lying in the crater of a collapsed volcano, it is surrounded by the extinct caldera of Longonot with its hot springs and exciting walks.

Several fine hotels and camping grounds flank the lake and from the Safariland Hotel on the south shore boats can be hired by the hour. This very beautiful lake, roughly circular and about 10 kilometres (six miles) in diameter, holds both tilapia, Kenya's favourite table fish and large mouth black bass first introduced in 1928 from America. *Tilapia nigra* average about 500 grams and can be caught with a float, fished worm or bread paste. Bass, however, are wary of predators and fall most easily to the variety of plugs, spinners and plastic worms designed by the United Nations specifically for catching this very fish.

An early start from Safariland will take you right across Lake Naivasha to Hippo Point in about 20 minutes. At this early hour the sky and lake blend together in a single pewter hue. Other fishers are already about their work: pied and iridescent malachite kingfishers patrol the shallows; African fish eagles prey on tilapia, and pink and white pelicans drive shoals of fish inshore before dipping together at some silent signal to scoop up beakfuls of the tiny fish.

Left, fishing for river trout. Above, Ethiopia's Lake Chamo is well stocked.

At Hippo Point, so aptly named, casting and spinning will reap a rich harvest of bass of one kilogram each with a bigger catch coming often enough to confirm Naivasha as a high quality fishery.

Naivasha lies on the road to Uganda. On the way north there are more trout at Gilgil and at Kericho and in the tea dams above 2,600 metres (8,600 feet), Kenya's largest trout are found.

The great lake: Africa's great Lake Victoria is vast and barely explored. Although for generations it was the home of vast quantities of the ubiquitous tilapia, stocks in recent years have been severely depleted by the introduction of Nile perch. The latter may be good to eat but they run a poor second to the sweet-tasting tilapia. The angler however is presented with a bonanza because a perch of under 14 kilograms (31 pounds) is considered a baby: specimen fish begin at a 45 kilograms (100 pounds) and double or triple that weight is possible.

Heavy tackle now comes into play and boat and shore fishing will provide you with anglers tales to beat them all. Great silver fish with bulging, orange eyes and cavernous mouths fall easy victim to a trolled plug or the long cast dead bait. The cheeks of Nile perch are a delicacy without parallel and the great slabs of fillet, somewhat tasteless on their own, are delightful when curried or combined with more flavoursome ingredients in a pie.

Big fish : Twenty thousand crocodiles live in Lake Turkana but the quality of the fishing is so great that you'll forget your fear and wade waist deep in the lake in order to lengthen your cast. This lake, whose northern shore shares a border with Ethiopia, is fed by the Omo river. A fine highway runs parallel to its western shore or you can fly there by light aircraft. Fishing lodges at Ferguson's Gulf on the western side and Loyangalani Oasis on the east provide accommodation, boats and tackle for visitors.

Giant Nile perch is the favourite quarry but there are a whole host of other fishing delights available in the lake, to be caught with fly rod or light spinning rod. In a day you can expect to catch innumerable tiger fish, tilapia and a strange but delightful small-headed humpbacked fish, *citharinus gibbosus*, more popularly known by its *Swahili* name, *sahani*, which means "plate". This bream-like creature is attractive because it is never caught by its mouth. The spinner invariably lodges in the dorsal fin, possibly because it attempts to stun its prey before eating by striking it with its flat (plate-like) sides. Whatever the reason, the unconventionally hooked fish, weighing up to 4.5 kilograms, will not be constrained by a lure in its mouth. Only the greatest skill will keep the light tackle intact and eventually beach the fish.

The tiger fish is well named: each armour-plated jaw is fringed with needle sharp fangs. Their fight is as spectacular as any trout but wear gloves or hire an assistant to help remove hooks or you risk the loss of a finger-tip or worse.

There are other fish less frequently caught, some of which are sufficiently unusual to require a visit to Kenya's National Museum to identify the specimen.

Uganda's lakes: From Kisumu and Lake Victoria a short drive will take you through the Uganda border at Busia and on the road to Kampala. At Jinja, the pools under Owen Falls have always offered the best sport for barbus. Similar to the Thika fish, possibly the same variety, they nevertheless grow considerably bigger and fish to nine kilograms (20 pounds) have been recorded, although an average fish is nearer three kilograms (6.6 pounds). A long cast into the boiling water under the falls yields generous rewards but hidden rocks and debris tend to snarl up tackle.

The Nile at Murchison Falls yields tiger fish, catfish and Nile perch. At his most spectacular of scenes the full weight of the Nile roars over the narrowest of precipices. As with barbus, an accurate cast into the foaming waters beneath the falls is the way to take perch from the river. Use a wooden plug or large siver spoon as the lure. As it sweeps down the current the waiting perch grab and can be fought to the bank. Although generally smaller than fish in Lake Victoria, nevertheless 20-kilogram (44-pound) fish are common and 50-kilogram (110-pound) fish do exist.

A dead bait tiger fish can attract big catfish. Tiger fish are easily caught for this purpose by using a small spoon. Alternatively you can follow the local example and knock down with your landing net some of the thousands of dragon flies on the river bank. Then catch your tiger fish on a number eight hook.

Nile perch, locally known as *mbuta*, can be found in the whole Nile system below Murchison Falls and can be fished for as far as the Sudan border. Security and access will prevent you from venturing too far off the beaten track.

Left, casting a fly in Lake Turkana. Above, flamingo-filled Lake Bogoria.

Lakes Albert and Kioga contain perch and a large variety of other fish including several species of barbus, tilapia, catfish and lung fish. The latter are reputed to grow to two metres (6.6 feet) long and feed on frogs, crabs, snails and small fish, though they prefer carrion and are particularly susceptible to dead bait. They are also reputed to bite fiercely and should be treated with caution if encountered.

The deeper waters of Lake Albert hold larger tiger fish up to 10 times the weight of its smaller cousin of Lake Turkana and the Nile at Murchison. Inshore, small tiger fish give wonderful sport but no evening feast since their armour plating conceals a host of small bones.

Trout were introduced to Uganda in 1932 and the Mabuku, Dwimi, Namwamba and Namusagani rivers were all stocked. However, there has been no stocking programme since 1970 although breeding conditions are good and it is possible that there are large wild fish waiting for the more adventurous angler.

Other parts: Rwanda's Lake Kivu is home to several species of tilapia and other *chichlidae.* There are about 300 members of this fish family in Africa and it is more than possible that you will not be able to identify some of your catch without the aid of the larger reference books.

Burundi shares the northern part of Lake Tanganyika with Zaire and Tanzania. Nile perch are absent from these lakes divorced from the Nile, but many new species of cat fish can be encountered. Barbus of the type found at Jinja, in Uganda, are common and the streamlined *labes victoricinus* is often mistaken for a heavy tiger fish when first hooked. Its non-acrobatic fight and absence of teeth soon confirm otherwise. Two other fish, locally referred to as tiger fish because of their exceptionally fierce fangs, are actually of a different species. *Bathybates ferox* and *Bathybates fasciatus* have slim bodies, powerful tails and a long dorsal fin. Bluish-green on their backs and with yellow-tipped fins, they are respected as fierce fighters.

Trout can be found in the rivers of the West Usambara Mountains, particularly the Mkusu River. As you head north again towards Kenya, snowcapped Kilimanjaro offers an exciting diversion and the lower slopes host splendid fishing for rainbow trout. Arusha is an appropiate base and the Usa River offers the best prospects. A good tarmac road returns you to Kenya via Namanga.

East of Amboseli National Park you can explore the Athi and Bushwackers camp at Kibwesi for tilapia and catfish. The natural lakes at Hunters Lodge, 150 kilometres (93 miles) east of Nairobi contain barbus, small but obliging tilapia and giant eels, weighing up to three kilograms.

Ideally a fishing safari in one of the world's last wild places should take in lakes Naivasha, Victoria and Turkana, a trout stream or two on Mount Kenya or the Aberdares and a few hours on the rivers at Thika in search of barbus. Whatever the catch, it should be enough to guarantee a lasting reminder of an angling adventure in Africa.

Fishing Ethiopia: A fishing safari to East Africa would not be complete without a visit to Ethiopia. The 13 mountain streams of the Bale Highlands have become a fabled trout fishery. Stocked only in 1960 by brown and rainbow ova from Kenya, the progeny have turned into monster trout where a five kilogram (10 pound) fish is not exceptional. The most productive areas are the pools below the waterfalls of the Upper Webi and in the Dinka River. Accommodation can be found at Dinshu, the headquarters of the Bale Trout Fishing Club where there is a self-help lodge. There are also hotels in the nearby regional capital, Goba.

But trout are only one of more than 200 species of fish that can be found and angled for in Ethiopia. Nile perch, tiger fish, tilapia and catfish—including the one-metre-long electric *Maloterurus* with its 100-volt electric shock—all live in Lake Tana, the source of the Blue Nile.

Although wild and remote, the islands on this 3,600-square-kilometre (1,440-square-mile) lake are inhabited. Churches and monastries, some dating from the 14th century, cling to their rocky shores. Hotels at Bahar Dar on Tana's south shore provide a comfortable base from which fishing or exploration trips can be mounted.

Left, Nile crocodiles can be a menace. Above, fishing from a palm log raft.

Ethiopia is a country of mountains and lakes. Like pearls strung on a necklace, the great Rift Valley lakes of southern Ethiopia continue the chain running through Tanzania and Kenya. The largest are Chamo and Abaya and provide part of a breathtaking view from Arba Minch, a town perched on the forested ridge which divides the lakes. Appropriately enough, it is known as the "bridge of Heaven". Nile perch, tiger fish, catfish and tilapia all offer great sport and while you fish the presence of hippo and crocodile constantly remind of the wildness and adventure of Africa.

As you continue north into the Ethiopian lake district, the lakes of Zway, Langano, Abijatta, Shalla and Awasa can be found. The favourite for fishing is Awasa, about 270 kilometres (167 miles) from Addis Ababa. The 200-metre (600-foot) deep, turbulent Lake Shalla and the shallow, bird crowded Abijatta both have their individual fascination for the visitor.

Fishing gear: For a successful fishing safari you need to plan carefully. You must come equipped although, despite the wide spectrum of coarse and game fish you will encounter, you need neither great quantities of clothes or tackle. Not much, however, is available locally and what can be bought is expensive so provide yourself with a 2.5- to three-metre (eight- to 10-foot) fly rod; sinking line and casts to 2,725 grams (six pounds) test. A medium spinning or casting rod with a variety of spoons; diving, running or surface lures of 25 grams (one ounce) or less; a heavy casting rod complete with 140 grams (30 pounds) line; rig for live and dead bait and a set of larger lures designed for ocean game fish, even though your quarry are all freshwater fish.

Clothing should be lightweight and quick-drying as the climate is too warm for thigh and chest waders. Stout non-slip wading shoes can be used although many anglers find canvas tennis shoes more comfortable. Trousers can be long or short but beware of the fierce sun concentrated by low latitude and high altitude which can savage even the least susceptible of skins. Bring a hat, dark glasses and high factor sunscreen.

If you wish to arrange your own safari, the best place to start is Nairobi.

RUSINGA ISLAND

Rusinga Island in the middle of Lake Victoria has featured on the anthropological map since Louis Leakey discovered Proconsul ape there. It also has a place in the history books of modern Kenya as Tom Mboya's birthplace.

Lake Victoria itself, all 70,000 square kilometres (28,000 square miles) of it, plays a minor and under-rated role in Kenya's tourist circuit. This is despite its scenic beauty which has attracted praise since the days of the early explorers. Today the villages are sprawling townships with metallic roofs shining in the sun. The concealing trees have greatly diminished but the peaceful beauty remains.

African Explorations Limited have established a camp on Rusinga Island from which they run their half-day fishing excursions. Guests are usually collected from the Maasai Mara in the early morning, flown over the Kisii highlands—which on a clear day is a colourful patchwork of green tea, yellow maize and brown, freshly turned earth—and delivered to Rusinga airstrip which lies directly behind the camp.

Next, over a full English breakfast served in the dining area with a view of the lake and surrounded by well-kept lawns, trees and shrubs, the different activities available are discussed with experienced guides. These include Rusinga's most famous activity—fishing. However, bird watching comes a close second and for those with no particular yen for fin or feather there is a general sightseeing expedition by boat.

Nile perch is the big catch for fishing enthusiasts. These fish are abundant and all casters stand a good chance of bagging a big one—45 kilograms (100 pounds) or more—even if it is the first time you have held a rod! They have been recorded at twice this weight and the guides are always discovering new fishing grounds. Besides the perch, which were only introduced in the 1950s, Lake Victoria's wide variety of fish is regarded as an evolutionary phenomenon. These fish, particularly the brilliantly coloured cichlids which are so popular in aquariums, are now being seriously threatened by the predatory Nile perch.

Feeding fish eagles has become a main feature of the trip and guests have excellent opportunities to take action photographs of these majestic birds as they swoop down beside a boat to grab a fish. Fish eagles and hammerkops are among the many birds nesting in and around the camp, where sightings of over 100 species have been recorded.

The sightseeing trip takes you past picturesque fishing dhows to unspoilt, traditional fishing villages where there is the opportunity to witness the colourful scene of the day's catch arriving and being distributed for cooking, drying or smoking. There are secluded areas on the shoreline where families of the rare spotted-necked otter are often seen. Giant monitor lizards bask on rocks, slipping into the water as you approach.

At present the excursion is a one-day trip, returning to the Maasai Mara in the evening. Overnight accommodation will be available in the near future. It is also possible to visit the island from other parts of Kenya on arrangement with African Explorations Limited, P. O. Box 40075, Nairobi, Kenya. Tel: 540780/9. Telex: 24015 BLOCKMAN. Fax: 331422.

TANA DELTA SAFARI

Kenya's Tana Delta is flat and low, criss-crossed with tidal channels, savannah grasslands, stands of doum palms, swamp and thick forest. Some of the numerous and divergent tidal waterways are navigable for as far as 15 kilometres (nine miles) inland, either by small canoes through narrow channels or more slowly by larger dhow on the main waterway.

Guests are collected from Malindi and driven a couple of hours north along a rutted dirt track to a village on the river. From here the dhow *African Queen* travels at a leisurely seven knots down the Tana River on a four hour journey to the camp.

Sitting on a deckchair on the 11-metre- (36-foot-) long, high-decked, diesel-powered boat, one travels through channels the colour of café au lait with dense jungle on either side. Where the banks slope gently to the water's edge, herdsboys of the *Orma* tribe, akin to the more famous *Maasai*, bring cattle down to drink. Elsewhere, Pokomo women wash cooking pots or clothes as their children wade in the shallows—in spite of the crocodiles which claim at least one victim per week.

Hippos wallow in deep pools on the river bends, snorting and blowing bubbles before submerging again. In grass clearings reedbuck, topi, buffalo and sometimes the rare bushbuck or a lone bull elephant roam.

The Tana River is most famous for its prolific birdlife, featuring the exotic, the spectacular and the merely odd. There are huge flocks of egrets, pelicans, ibis and storks on every sandbank. Beautifully coloured bee-eaters, hornbills, kingfishers and many other birds are a constant delight.

By sunset, the *African Queen* arrives at the small tented camp, which lies at the foot of a giant, white sand dune. There are no roads or other camps in this remote area. The tents are roomy, insect proof and secure with a large attached verandah in front of each and a private shower and toilet near by.

The ideal safari lasts four days. This gives you time to explore and birdwatch in the vast floodplain, either by boat or on foot.

Left, sailing to Rusinga. Above, taking a leisurely trip along the Tana.

Deep Sea Fishing

The abundance and variety of game fish, together with well-equipped boats and professional crews make the coast of East Africa a paradise for sports fisherman.

From Pemba Channel on the Kenya-Tanzania sea border all the way north to Diani, Mombasa, Mtwapa, Kilifi, Watamu, Malindi, Lamu and Kiwayu, there are efficient charter operators available during the recognised eight-month fishing season (August to March). The high winds of the rain-bearing *Kusi* (south-east monsoon) make waters unfishable between late March and late July.

In Tanzania there are exhilarating fishing locations off Tanga in the north. From Dar es Salaam to Mafia Island and other spots along the extreme southern coast of the country there are numerous fishing opportunities.

The most challenging sport fish of the ocean—black, blue and striped marlin (*Makiara inoica, Makiara nigricans* and *Tetrapturus audax*) and Pacific sailfish (*Istiophorus platypterus*) all abound in these waters, as do the powerful yellowfin tuna (*Thunnus albacares*).

For both the enthusiastic beginner and those experts seeking the exciting billfish (marlin and sailfish), Kenya offers many possibilities. Pemba Channel Fishing Club owned by the Hemphill family is located at Shimoni on the southern border with Tanzania; at Watamu big black and blue marlin are caught every year; and Malindi is one of the world's most prolific venues for Pacific sailfish.

Kenya's sport fishing grounds rank in the world's top five areas for high annual average catches of billfish. But billfish are not the only attraction. Kenyan and Tanzanian waters are teeming with a wide variety of game fish including barracuda (*Sphyraena barracuda*), dolphinfish (*Coryphaena hippurus*), kingfish (*Scomberomorus commerson*), pacific bonito (*Sarda spp*), mako shark (*Isurus spp*), tiger shark (*Galeocerdo cavieri*), and hammerhead (*Sphvrna spp*). There are strong seasonal runs of yellowfin tuna and some dogtooth (*Gymnosarda unicolor*), especially from August to December.

The broadbill swordfish (*Xiphias gladius*), one of the most elusive of the ocean game fish, is found in the indigo waters off the East African coast.

Fishing tackle and boats run by Kenya's

dedicated professional operators—about 20 in all—are based at points along the 402-kilometre (250-mile) coastline.

Boats up to 15 metres (50 feet) in length include such popular "fishing platforms" as *White Otter* (13.5 metres/44 feet), the catamaran *Pingusi* (nine metres/30 feet), and *Broadbill* (14 metres/46 feet) at Hemphill's Pemba Channel Club. James Adcock owns a fleet at Mtwapa; *Ol Jogi*, a 10-metre (33-foot) Bertram and *White Bear*, an 11.5-metre (38-foot) Sport Fisherman, are both based at Kenya's newly-opened deep sea fishing, diving and sports venue—Hemingway's in Watamu. This 100-bed luxury resort is fully geared to cater for sport fishing. Boats are available for charter all year round—subject to weather conditions.

At Malindi—the sailfish "Mecca"—there are charter operators who have become especially skilled at light tackle and fly casting for sailfish and dolphin.

Fish are plentiful off Lamu and Kiwayu on Kenya's extreme north coastal waters bordering Somalia. Though not as accessible as areas further south, there are charter boats available both from the Peponi Hotel at Lamu and from the exclusive and remote Kiwayu Lodge. Several very large marlin have been caught in these waters.

Prices are reasonable by world standards with rates for the large boats averaging around Ksh 5,000 to Ksh 6,000 (US$270 to US$335) per day, with up to six rods and a maximum of three fishermen for marlin or four fishermen for smaller fish.

Left, heading for fishing grounds off Kenya's coral coast. Above, ready to reel 'em in.

The Kenya Association of Sea Angling Clubs (KASAC) is the recognised body that oversees the organisation of deep sea fishing on the Kenya Coast. It arranges a full programme of deep sea fishing competitions and events are attracting an ever growing number of competitors from overseas. For further information write to: Kenya Association of Sea Angling Clubs, P. O. Box 84133, Mombasa, Kenya.

Deep sea fishing enthusiasts going to Tanzania should contact the Dar es Salaam Yacht Club, which can arrange a trip to your liking.

The great snowcapped peaks of Africa rise on the eastern side of the continent. Kilimanjaro at 5,895 metres (19,340 feet) straddles the border of Kenya and Tanzania; Mount Kenya at 5,199 metres (17,058 feet), stands in central Kenya. The Ruwenzori Mountains at 5,106 metres (16,763 feet) dominate the border of Uganda and Zaire.

Few mountains in the world are as steeped in legend and mystery as these giants of Africa, and for over 100 years they have drawn the curious traveller to their slopes. Their very discovery proved controversial: the sightings of Kilimanjaro and Mount Kenya in 1848 and 1849 respectively by Church Missionary Society preachers Johann Rebmann and Johann Krapf, was disbelieved by the Royal Geographical Society in London. Snow, they said, could not exist so close to the equator, and what the missionaries had seen could only be "calcareous earth". Rebmann and Krapf would not be vindicated until the Scottish explorer Joseph Thomson witnessed the snows of both peaks in 1883. Henry Morton Stanley met no resistance from Europe's Geographical Societies when he sent word back of his discovery of the Ruwenzoris in 1889.

Kilimanjaro: Kilimanjaro is East Africa's loftiest landmark. It is the highest mountain in Africa and one of the highest extinct volcanoes in the world. Its name is synonymous with Africa itself and few mountains anywhere on earth have been so immersed in romance and folklore. Even the names of towns that grace the base of the peak have a dream-like quality to them—Oloitokitok, Rongai, Moshi and Marangu.

Great writers with no interest in mountains have written about Kilimanjaro. Songs have been sung about it. Empires fought for it. Many stories and myths have grown, some true, some false. The partially preserved skeleton of a leopard does exist on the icy crater rim at 5,666 metres (18,600 feet); but Queen Victoria did not give the mountain to the Kaiser as a birthday present.

However one looks at it, Kilimanjaro possesses an atmosphere, a personality, the type from which legends are easily born.

The base of the mountain is approximately 80 kilometres (50 miles) long by 40 kilometres (25 miles) wide, and boasts three definite peaks. Shira to the west is the least obvious and reaches a height of 4,002 metres (13,140 feet). Its long, gentle, whale-back appearance can barely be distinguished from the general downward roll of the mountain slopes.

Mawenzi to the east attains an altitude of 5,145 metres (16,890 feet) and is a beautifully sculptured masterpiece of a peak.

Kibo rises to the highest point of the mountain and is young, resplendent, a giant upturned bowl of a summit with an almost perfect circular crater 2.5 kilometres (1.5 miles) wide. It is covered by three large icefields from which hang 15 extremely steep glaciers. When viewed from either the north or the south the great sweep of The Saddle captures the eye as one of the dominant features of the massif. This vast plain at 4,569 metres (15,000 feet) is about seven kilometres (4.3 miles) across and separates Mawenzi from Kibo.

Shira, Mawenzi and Kibo were all individual volcanoes. Research during the last 50 years would suggest that Shira was the site of the first eruption, followed later by Mawenzi. Kibo, geologically speaking, is recent and most of its activity occurred during the Pleistocene period. It is possible, however, that the last major episode of activity took place within the last few centuries.

Green belts: As with other mountains of eastern and central Africa, there are definite vegetational zones on Kilimanjaro. A thick belt of montane forest encircles the entire mountain between 1,888 metres (6,200 feet) and 2,985 metres (9,800 feet) and is dominated by podos trees which are of giant proportions with buttressed roots. Cedar also grows in profusion on the lower northern slopes and the majority of these trees are festooned with thick, woody lianes. An array of large ferns and nettles makes passage in the forest difficult but, unlike other African

Preceding pages: breathtaking view of Kilimajaro's Mawenzi peak at dawn. Left, rock climbing in Kenya.

mountains, virtually no bamboo grows on Kilimanjaro.

Kilimanjaro forest is a silent place and although it is inhabited by mammals they are not in the numbers which can be seen on Mount Kenya. Elephant and cape buffalo are the most common larger animals of the forest; bushbuck and duiker can also be seen. But the chances of the visitor seeing or hearing them, or even noticing so much as a sign of them, is remote. Spotting a Colobus monkey is probably your best chance of seeing any wildlife on Kilimanjaro.

In the higher regions of the mountain there are a few deep sheltered valleys and the slopes consist of very porous lavas, so the little water that falls as rain or descends as melt-water from the glaciers does so underground for long distances and is virtually unavailable to plants. Compared with other East African mountains, Kilimanjaro is floristically poor.

The spoor of leopard can often be seen on the trails of the upper moorlands and alpine zone though sightings of the animal itself is rare. For many years The Saddle has boasted a large herd of eland; they are interesting because their fur appears to be longer than that of the lowland species and the animals are larger.

Reaching the top: Kilimanjaro was first climbed on October 5, 1889 by the German mountaineers Hans Meyer and Ludwig Purtscheller. This ascent to the summit of Kibo involved no great technical skills. Today thousands of visitors annually reach the lowest point of the crater rim, Gillman's Point, at 5,681 metres (18,650 feet). Some people continue clockwise round the crater for an extra 90 minutes to Uhuru Peak at 5,895 metres (19,340 feet), the highest point of Africa.

There are seven tracks up Kilimanjaro from both Kenya and Tanzania but at the time of writing all Kenyan routes had been closed. Ninety-five percent of visitors will ascend the peak by way of the tourist route from the little Tanzanian town of Marangu near Moshi. Moshi can be reached by rail or bus, but perhaps the most efficient way is by air to Kilimanjaro Airport, some 34 kilometres (21 miles) west of Moshi on the main road to Arusha. This modern airport is served by daily flights from Dar es Salaam, as well as international flights from Kenya and Europe. In Moshi there are two or three hotels and provision shops, but these shops should not be relied upon to stock sophisticated mountain food.

Marangu is 27 kilometres (17 miles) east of Moshi, and located here are two hotels which arrange for fully packaged mountain safaris up Kilimanjaro: Kibo Hotel, P. O. Box 102, Marangu, Kilimanjaro, and Marangu Hotel, P. O. Box 40, Moshi, Tanzania. These hotels will arrange porters, vehicles to the park gate and all food. Some trekking equipment can be hired. Hotel managers will also steer visitors through the tedious officialdom which most people encounter at the national park entrance.

Technically speaking, the ascent is no more than a stiff walk for those who arrive at the base in a state of reasonable fitness. The trick to climbing any large mountain is to go slowly. Mountain sickness tends to affect everyone to some degree and this can come in the form of headaches, loss of appetite, nausea and lack of energy. By walking slowly one can go a long way towards alleviating some of these ailments.

To make things easier the Kilimanjaro National Park has constructed an excellent series of huts at convenient intervals up the side of the mountain, and visitors are strongly recommended to spend four nights and five days for the ascent and descent of the peak. After leaving the park gate, which is at 1,800 metres (5,909 feet), the path works its way up through the lower forests to Mandara Hut at 2,700 metres (8,864 feet). Although this walk can be done easily in about five hours, spend the entire day over it and keep your body fluid level up.

On the following day the trail traverses the southern slopes of the peak, leading out of the forest and on to the stark moorlands. Horombo Hut at 3,807 metres (12,500 feet) is reached late in the afternoon. If it is possible for you to arrange a two-night stay at Horombo Hut it is a very worthwhile plan. This hut is located at a fairly critical altitude and acclimatising well at this point will ensure stronger chances of success in the two days to come.

After Horombo Hut the path climbs gradually to The Saddle at 4,264 metres (14,000 feet). This vast alpine desert is easily crossed towards the peak of Kibo, until the track begins to rise to Kibo Hut at 4,721 metres (15,500 feet). Not a great deal can be said for the location of this final hut on the way to the summit of Kilimanjaro. It is desolate and cold and the majority of walkers are in no mood to enjoy any small pleasures which the district may offer. Headaches and nausea are being nursed by many of the hut's occupants and most people are thoroughly intimidated by the prospect of the final 944 metres

Left, the southern glaciers of Kibo. Above, the summit of Africa—Uhuru Peak.

(3,100 feet) of scree, the ascent of which will begin at 1 a.m.

There are good reasons for starting at this early hour. The scree which is composed of small pebble-like ash cinders is frozen then and is easier to climb; when the sun rises it does so from behind Mawenzi, beautifully illuminating the serrated silhouette of this peak. The higher you are up the Kibo scree by the time the sun rises, the grander the view. But for many, the third reason for beginning so early is the one which makes greatest sense: it is dark and therefore one cannot see the apparently endless distances ahead, which in daylight never seem to come any closer!

Gillman's Point will be quite enough! Descent is then made to Horombo Hut for another night, before returning to the park gate on the fifth day.

For succeeding on Kilimanjaro, the best advice is to hedge your bets by following these rules. Insist on a good guide from the hotels; wear good, warm equipment (do not be fooled by the mountain's proximity to the equator for Kilimanjaro is as cold as any other 5,800-metre (19,000-foot) peak in the world, and the temperature at the top will be well below freezing), take the ascent easily—do not walk fast, and remember to drink plenty of liquids.

"The Ostrich hill": Mount Kenya is 5,199

There is little genuine pleasure in the ascent of the final few hundred metres of Kibo and it is only when you reach the edge of the crater that you will appreciate why you chose to suffer all the pain. The path zigzags up the side of the mountain. The average walker begins by climbing perhaps 100 steps before resting, and this number of paces becomes for a few hours the focal point of your existence. By the time 5,483 metres (18,000 feet) is reached you'll have reduced the distance covered between rests to a few stumbling steps. The aim of the exercise is to reach Gillman's Point although some people will continue to the main summit. But for most,

metres (17,068 feet) high and a little over 100 kilometres (62 miles) in diameter. It is the second highest mountain in Africa and is in many ways more interesting than its larger neighbour Kilimanjaro. It is an ancient extinct volcano whose period of activity was between 3.1 milion and 2.6 million years ago, when it probably rose to over 7,615 metres (25,000 feet) with a shape resembling that of Kilimanjaro.

Today the shape of Mount Kenya belies the fact that it was ever a volcano at all. There is little trace of a crater left, so beaten has the mountain become through time and countless eruptions, but its majestically battered peaks, draped by 11 glaciers, radiate an air of supreme elegance. From the plains, the peaks of Mount Kenya seem to float like a distant fortress in the sky. Or as the Wakamba people who dwell some 200 kilometres (125 miles) away used to say: like a "cock ostrich". The contrast between its white glaciers and the dark rocks do indeed look like the black and white plumage of the male ostrich. This could well be the origin of the word "Kenya" for in the Wakamba language *Kiinya* means "The hill of the cock ostrich". For the Kikuyu people who dwell and farm around the foothills of Mount Kenya the mountain has special meaning. Since the earliest of times their lives have been inextricably linked to this often uncompromising peak. It brings them their rain and therefore their livelihood, and during frequent periods of drought they would pray to it, and make sacrifices of lambs and goats. Even today, elder members of the Kikuyu rise early in the morning to offer prayer to *Ngai*—their god who lives among the summit peaks. It is said that by sunrise *Ngai* will have dispensed all of his blessings upon the early risers.

Left, resting on the crater rim on the way to the top. Above, view of Mount Kenya's Point Lenana.

Forests, moors and glaciers: The heart of Mount Kenya lies in the thick, semitropical rain forests which grace its lower slopes up to an altitude of 3,350 metres (11,000 feet). These forests begin between 2,132 and 2,437 metres (7,000 and 8,000 feet) above the uppermost point of the Kikuyu's fertile farms or *shambas*. Giant trees of camphor wood, pencil cedar, podocarpus and East African olive, draped with vine-like lianes, rise above a tangled profusion of dripping ferns, nettles and bamboo. It is an area difficult for humans to penetrate and because of this, it is home to herds of cape buffalo and elephant, as well as bushbuck and the occasional black rhino.

With dramatic suddenness the forests end at 3,350 metres and the moorlands begin. This region which extends upwards to 4,264 metres (14,000 feet) is more compromising, with its stark beauty akin to the Scottish highlands. Tussock grasses, with intermittent heather growing occasionally over three metres (10 feet) in height, gradually transforms into the drier world of the exotic tree-like giant groundsels and lobelia.

The area is studded with rock islands of porous volcanic ashes and agglomerates standing out like dark, incongruous sentinels. Big game is rarely seen here but its presence is always felt. The spoor of eland, cape buffalo, and leopard is always evident, and every now and again giant groundsel patches lie like withered carcasses, smashed to pieces by elephant.

There is a deceptive flatness about this area which at first appearance would seem to extend as high as the base of the summit peaks. But centuries of glacial activity have slowly eaten away the surface to the extent that the entire region surrounding the peaks has been carved into numerous valleys and gorges—some as deep as 609 metres (2,000 feet).

A vast carpet of lobelia and giant groundsels extend over these valleys, clustering

thickly along the edges of rivers which flow down them. But for the sounds of the rivers there is a profound silence in these valleys.

The valleys and gorges of Mount Kenya rise to a point at 4,417 metres (14,500 feet) where plants and vegetation can no longer survive. The green landscape becomes grey: rocks and boulders are suddenly of gigantic proportions, steep scree chutes of pebbles and cinders lead up to the snouts of glaciers and the final 609 metres of vertical rock and ice.

Climbing Mount Kenya: The summit of Batian, the highest peak on Mount Kenya, was first stood upon by Halford Mackinder, Cesar Ollier and Joseph Brocherel on September 13, 1899. It was the culmination of a four-month expedition and the successful outcome was an outstanding achievement. The easiest route to the top of the mountain had involved scaling a 457-metre (1,500-foot) vertical rock and ice face, and it would be 30 years before it saw a second ascent.

Today few people attempt to reach the highest point of Mount Kenya, but many trek to the summit of the third highest peak—Point Lenana at 4,982 metres (16,355 feet). It is worth the effort as it gives a good idea of the atmosphere of this beautiful mountain.

The base for climbing Mount Kenya is at Naro Moru, which is located 171 kilometres (106 miles) from Nairobi on the main road to Nanyuki. The Naro Moru track is the shortest way to the peaks, and can be undertaken in three days. Because of this the bulk of tourists ascend the mountain this way, neglecting the more involved routes, of which there are seven. Naro Moru can be reached from Nairobi by bus or by a number of local taxi services, and most of these are centred around: East African Road Services, Racecourse Road, P.O. Box 30475, Nairobi. Tel: 23476.

The journey to Naro Moru will take between two and three hours and visitors are recommended to make their way to the hotel which is the undisputed starting point for the majority of Mount Kenya attempts: The Naro Moru River Lodge, P.O. Box 18, Naro Moru. Tel: Naro Moru 23. This lodge will arrange fully packaged ascents of Point Lenana.

The recommended way of ascending the Naro Moru track is as follows:

Day One: Begin at the National Park gate which is 17 kilometres (10 miles) from Naro Moru township. Hike to the Meteorological Clearing at 3,046 metres (10,000 feet). Camp here or stay in the bunkhouses which can be booked through the Naro Moru River Lodge.

Day Two: Walk up through the forest which soon becomes boggy moorlands. These are crossed to the edge of the Teleki Valley. Descend round into the valley and climb it to its head. At this point is the Mackinder's Lodge at 4,143 metres (13,600 feet), which is a good place to stay. Good campsites also exist in this area. The distance to the lodge can be covered from the Meteorological

Station in about five hours, but the entire day should be spent slowly working your way up to the head of the valley.

Day Three: A day spent resting and acclimatising in the Teleki Valley.

Day Four: Begin the trek to the summit of Point Lenana at 3 a.m. The path contours up the west side of the Teleki Valley, then strikes up the screes on the south side of the obvious Lewis Glacier to the Austrian Hut at 4,788 metres (15,720 feet). Point Lenana is ascended by way of the ridge above the hut and takes approximately one hour more. The descent to the Teleki Valley can be made in two hours and many parties go all the way to the Meteorological Station in the afternoon.

Point Lenana is often ascended by way of the Chogoria Track on the eastern side of Mount Kenya. The town of Chogoria is located on the main road linking Nairobi to Meru and is 228 kilometres (141 miles) from Nairobi.

There are no hotels in Chogoria which offer fully equipped safaris up the mountain, but porters can be arranged through: Mr. Livingstone Barine, P.O. Box 5007, Chogoria. Tel: Chogoria 26.

The Chogoria track is a very beautiful way up Mount Kenya and three days should be allowed to reach Point Lenana. A good trek is to ascend the Chogoria track to Lenana then descend the Naro Moru track so completing a full traverse of the mountain.

The main summit peaks of Batian at 5,199 metres (17,068 feet) and Nelion at 5,188 metres (17,032 feet) should only be undertaken by experienced mountaineers who understand the intricacies of climbing steep ice and rock at altitude.

If approached in the correct way Mount Kenya can give the trekker a more rewarding experience than Kilimanjaro. It offers far richer flora and fauna than the latter and the scenery is more spectacular. It should be noted that although the equator runs through Mount Kenya, visitors should not underestimate the weather conditions.

Mountains of the Moon: No other mountain range in Africa, perhaps in the world, is as steeped in legend as the Ruwenzoris—Ptolemy's fabled Mountains of the Moon. A trek to the icefields of this range is no ordinary climb—it is a journey to the pulse of the African continent.

The Ruwenzoris lie along the western border of Uganda and rise approximately 3,960 metres (13,000 feet) above the western

Left, looking towards the main peaks of Mount Kenya. Above, camping on the "Ostrich hill".

Rift Valley. The range is some 112 kilometres (70 miles) long by 50 kilometres (30 miles) wide and was formed from a block which was tilted and thrust up during the development of the rift. Unlike Kilimanjaro and Mount Kenya it is not of volcanic origin.

In the centre of the range there are six major peaks all draped with permanent snow and glaciers. Mount Stanley at 5,106 metres (16,763 feet) is the summit of the Ruwenzoris and it is surrounded by a complex system of other mountains such as Speke (4,886 metres/16,042 feet), Baker (4,840 metres/15,889 feet), Gessi (4,712 metres/15,470 feet), Emin (4,788 metres/15,720 feet) and Luigi di Savoia (4,624 metres/15,179 feet). On each of these mountains there are several peaks and glaciers.

Early travellers acknowledged the Ruwenzoris as the source of the Nile and the first information to be brought back from the interior was by Henry Morton Stanley after his 1888-89 expedition. The first expediton to explore the mountains thoroughly was led by the Duke of the Abruzzi in 1906.

Unlike the other great mountains of East Africa, the Ruwenzoris rarely dazzle the surrounding countryside with their shimmering snowfields. In fact they are hardly ever seen from the plains, being almost continually wrapped in thick cloud. The range is arguably one of the wettest in the world but it has its worthy compensations in that it is a botanist's paradise. Giant groundsels and lobelia grow in thick forests above 3,046 metres (10,000 feet) and attain heights of up to 15 metres (50 feet). Heather can be found in clumps nine metres (30 feet) high.

For the traveller the Ruwenzori Mountains offer problems that other East African peaks do not. Merely reaching their base can be a mini expedition in its own right. Ascents to the peaks can be made from both Zaire and Uganda, but the former does not compare in either beauty or ambience. The approach from Uganda is the traditional way up to the icefields.

Climbing from Uganda: All expeditions begin from the small town of Kasese, which is located on the south-eastern side of the range. To reach Kasese, the easiest way is by train from Kampala and there is, at the time of writing, a daily service connecting them. One should always allow a few "buffer" days for any trip to these mountains for in Uganda there can always be delays. Driving between Kampala and Kasese is currently inadvisable. The road is long and badly in need of repair. Charter flights between Entebbe and Kasese can be arranged.

In Kasese there are two hotels, the Saad and the Margherita. The latter is government-run and expensive. At the Saad, on the other hand, the management was friendly and able to provide transport to the roadhead at Ibanda. They will also assist with porters although it is advisable to book these through: John Matte, Shop and Club Agent, Ibanda, P.O. Box 88, Kasese.

The porters are very friendly and eager for work, but they do expect blankets to be purchased for them in advance. Allow two days around Kasese and Ibanda for finalising porter arrangements.

It should be stressed that the Ruwenzoris are alpine in character and the warmest mountain clothing should be carried to the peaks. It is a very wet range and good rain-

proof equipment is essential. Knee-high gumboots are very useful on the approach to the peaks. There are currently no national park services of any kind and all visiting parties should be totally self-sufficient.

The majority of climbers will start from Ibanda, which is approximately 32 kilometres (20 miles) from Kasese. The trek to the peaks can be done as a "horseshoe" so that the same ground is not covered twice.

Left, men on the 'moon': scaling the Ruwenzoris. Above, attacking the south face of Batian.

The recommended route is as follows:

Day One: Camp at Ibanda at 2,437 metres (8,000 feet).

Day Two: Hike through the lower rainforests where you might be lucky enough to spot the Ruwenzori turaco flying amongst the branches. The walk today is about six hours long and in the late afternoon Nyabitaba Hut is reached at 3,046 metres (10,000 feet).

Day Three: The path continues up a narrowing ridge and after some three hours reaches the infamous Bigo Bog (3,350 metres (11,000 feet).

Day Four: The path steepens as the forests are left behind and the beautiful Bujuku Valley is reached. At the end of this valley is Bujuku Hut at 3,960 metres (13,000 feet).

Day Five: The path climbs into the relatively inhospitable alpine zone leaving all vegetation behind. At the edge of the Elena Glacier the Elena Huts are reached at 4,843 metres (15,900 feet).

Day Six: A climber's day, when the Elena Glacier is ascended to reach the Stanley Plateau. From here the summit peak of Margherita at 5,106 metres (16,763 feet) can be climbed in several hours.

Day Seven: In the morning a descent is made to the entrancing Kitandara lakes where Kitandara Hut is located at 3,960 metres (13,000 feet).

Day Eight: A trek is made across Freshfield Pass to descend into the Mobuku Valley. The Mobuku Caves are reached in the late afternoon. The altitude is 3,655 metres (12,000 feet).

Day Nine: The path traverses out of the Mobuku Valley and rejoins the ridge above Nyabitaba Hut, thus completing the "horseshoe" circuit.

Day Ten: The final descent into Ibanda.

To reach the highest point of the Ruwenzoris, Margherita Peak, requires some experience of snow and ice climbing. The majority of strong hikers wearing crampons and carrying ice-axes will be able to reach the Stanley Plateau, which in many ways is the most spectacular area of the entire range. This plateau, which is the largest single ice mass on the African continent, is the watershed of Uganda and Zaire and most hikers would be very satisfied to reach it. Margherita, which rises above the Stanley Plateau is ascended by its obvious East Ridge.

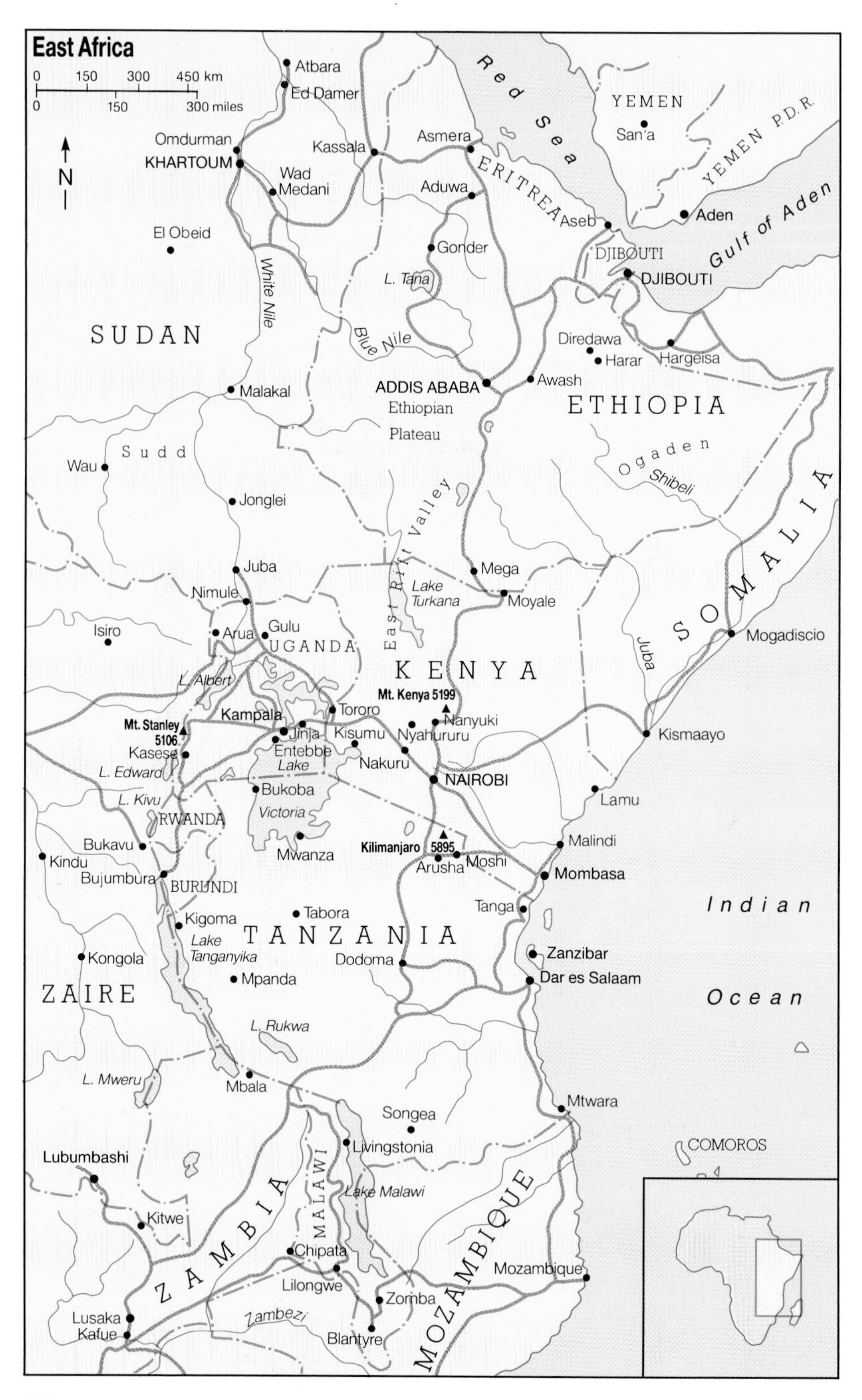
East Africa
0 150 300 450 km
0 150 300 miles
N
Atbara
Ed Damer
Omdurman
KHARTOUM
Kassala
Asmera
Wad Medani
Aduwa
Red Sea
YEMEN
San'a
YEMEN P.D.R.
ERITREA
Aseb
Aden
Gulf of Aden
El Obeid
Gonder
L. Tana
DJIBOUTI
DJIBOUTI
White Nile
SUDAN
Blue Nile
Diredawa
Harar
Hargeisa
Awash
ADDIS ABABA
Malakal
Ethiopian Plateau
ETHIOPIA
Sudd
Wau
Ogaden
Shibeli
Jonglei
East Rift Valley
SOMALIA
Juba
Mega
Nimule
Lake Turkana
Moyale
Isiro
Arua
Gulu
UGANDA
Juba
Mogadiscio
KENYA
L. Albert
Mt. Kenya 5199
Tororo
Mt. Stanley 5106
Kampala
Jinja
Kisumu
Nyahururu
Nanyuki
Kismaayo
Kasese
Entebbe
Nakuru
L. Edward
Lake
NAIROBI
Bukoba
Lamu
L. Kivu
RWANDA
Victoria
Bukavu
Mwanza
Kilimanjaro
5895
Malindi
Kindu
Arusha
Moshi
Bujumbura
BURUNDI
Mombasa
Kigoma
Tabora
Tanga
Indian
Lake Tanganyika
TANZANIA
Zanzibar
Kongola
Dodoma
Dar es Salaam
Mpanda
ZAIRE
Ocean
L. Rukwa
L. Mweru
Mbala
Mtwara
Songea
Livingstonia
COMOROS
Lubumbashi
Lake Malawi
MALAWI
Kitwe
ZAMBIA
MOZAMBIQUE
Chipata
Mozambique
Lilongwe
Zomba
Lusaka
Kafue
Zambezi
Blantyre

GAME PARKS AND RESERVES

There is nothing quite like the drama of wildlife, and the game parks and reserves of East Africa provide the best and biggest stage from which to watch. A herd of elephants trudging across a grassy plain, hippos happily wallowing in a mud pool, vultures riding the air currents over a group of grazing gazelles—all are a part of the animal kingdom's daily routine. East Africa's game parks give a fascinating glimpse of animal life in a series of pristine habitats. This can't-miss combination of birds, animals and scenery brings thousands of visitors each year to some of the most famous reserves in the world.

Kenya and Tanzania are at the heart of the safari experience, and the vast majority of safari goers will be heading for destinations within these two countries. They possess the tourist infrastructure, the facilities and the variety of wildlife to pamper those who like to temper their adventure with the trappings of luxury, although arrangements can also be made for anyone who prefers to rough it out in the less-explored regions.

Political and economic uncertainties in recent years, coupled with the problems of poaching, have complicated the wildlife picture in Uganda and Ethiopia. However, their respective governments recognise the value of tourism and have embarked on recovery programmes to protect the game parks there.

The tiny nation of Rwanda is best known for its population of mountain gorillas, whose habitat extends across the border to eastern Zaire. Neighbouring Burundi is only mentioned in passing in this guide as its wildlife has, sadly, been almost completely poached out of existence. The struggle for conservation continues in each of the other countries.

Despite these inherent problems, East African game parks well deserve their reputation as destinations that exceed expectations. This section discusses all the major game parks in East Africa, and some of the lesser ones as well. Whether it's a day trip to a nearby reserve or an extended journey through the bush, a visit to the land that spawned the word *safari* is one that few are likely to forget.

Preceding pages: tree groundsel near the top of Kilimanjaro; volcanic hills near Lake Turkana; the Mara River winds its way past a tented camp; room with a view: Mount Kenya as seen from the lounge of the Safari Club.

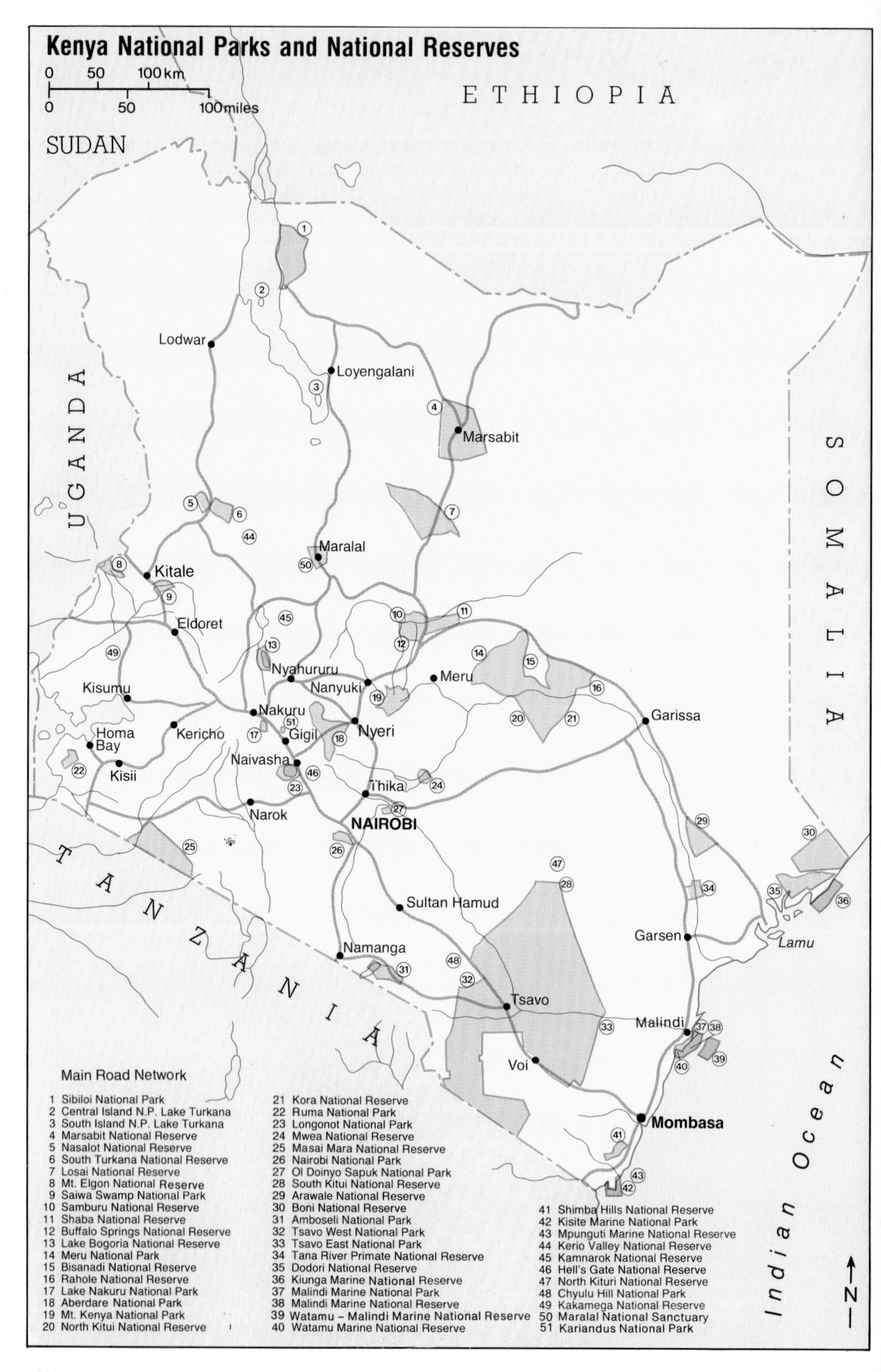
Kenya National Parks and National Reserves
0 50 100 km
0 50 100 miles
ETHIOPIA
SUDAN
UGANDA
SOMALIA
TANZANIA
Indian Ocean
N
Lodwar
Loyengalani
Marsabit
Maralal
Kitale
Eldoret
Kisumu
Nyahururu
Nanyuki
Meru
Nakuru
Nyeri
Gigil
Kericho
Homa Bay
Kisii
Naivasha
Thika
Narok
NAIROBI
Garissa
Sultan Hamud
Namanga
Tsavo
Voi
Garsen
Lamu
Malindi
Mombasa
Main Road Network
1 Sibiloi National Park
2 Central Island N.P. Lake Turkana
3 South Island N.P. Lake Turkana
4 Marsabit National Reserve
5 Nasalot National Reserve
6 South Turkana National Reserve
7 Losai National Reserve
8 Mt. Elgon National Reserve
9 Saiwa Swamp National Park
10 Samburu National Reserve
11 Shaba National Reserve
12 Buffalo Springs National Reserve
13 Lake Bogoria National Reserve
14 Meru National Park
15 Bisanadi National Reserve
16 Rahole National Reserve
17 Lake Nakuru National Park
18 Aberdare National Park
19 Mt. Kenya National Park
20 North Kitui National Reserve
21 Kora National Reserve
22 Ruma National Park
23 Longonot National Park
24 Mwea National Reserve
25 Masai Mara National Reserve
26 Nairobi National Park
27 Ol Doinyo Sapuk National Park
28 South Kitui National Reserve
29 Arawale National Reserve
30 Boni National Reserve
31 Amboseli National Park
32 Tsavo West National Park
33 Tsavo East National Park
34 Tana River Primate National Reserve
35 Dodori National Reserve
36 Kiunga Marine National Reserve
37 Malindi Marine National Park
38 Malindi Marine National Reserve
39 Watamu – Malindi Marine National Reserve
40 Watamu Marine National Reserve
41 Shimba Hills National Reserve
42 Kisite Marine National Park
43 Mpunguti Marine National Reserve
44 Kerio Valley National Reserve
45 Kamnarok National Reserve
46 Hell's Gate National Reserve
47 North Kituri National Reserve
48 Chyulu Hill National Park
49 Kakamega National Reserve
50 Maralal National Sanctuary
51 Kariandus National Park

KENYA

In 1985, the Academy-Award-winning film *Out of Africa* introduced a new generation of travellers to the beauty of Kenya. Cloudless, azure skies and golden savannah plains packed with animals fired the imagination of countless romantics and adventurers who wanted to experience this exotic Eden for themselves.

Long before that movie, though, a long line of writers, princes, poets and politicians had already enhanced Kenya's reputation as East Africa's leading safari destination, although cost and distance had hitherto prevented mass audiences of any sort.

The recent jump in visitor arrivals has been a welcome boost, and tourism is now Kenya's major earner of foreign exchange. Some of the increased revenue has been used to improve communications and tourist infrastructure. Nowadays, numerous highly organised companies offer scores of specialised safaris—horseback, walking, ballooning, mountaineering and bird-watching—to name just a few. Visitors may lodge at five-star hotels and dine on international-class cuisine, accompanied by views of unsurpassed beauty.

Ironically, as more people flock to Kenya, the once-vast animal stocks that draw them there have dwindled dramatically. In spite of international laws restricting imports of animal products, many animal species continue to be endangered. And animals are also at risk from Kenya's rapidly increasing human population—the fastest growing in the world.

Concerted government efforts to protect both wildlife and its habitats are showing slow, but mainly encouraging results and a wealth of conservation and wildlife organisations are contributing to the fights against ignorance and poaching. For the perceptive tourist, a visit to any of Kenya's magnificent parks and sanctuaries will be enough to convince him of the need to win the battle.

National Parks (in alphabetical order)

No.	Name	No.	Name	No.	Name
18	Aberdare National Park	13	Lake Bogoria National Reserve	47	North Kituri National Reserve
31	Amboseli National Park	17	Lake Nakuru National Park	27	Ol Doinyo Sapuk National Park
29	Arawale National Reserve	23	Longonot National Park	16	Rahole National Reserve
15	Bisanadi National Reserve	7	Losai National Reserve	22	Ruma National Park
30	Boni National Reserve	37	Malindi Marine National Park	9	Saiwa Swamp National Park
12	Buffalo Springs National Reserve	38	Malindi Marine National Reserve	10	Samburu National Reserve
2	Central Island N.P. Lake Turkana	50	Maralal National Sanctuary	11	Shaba National Reserve
48	Chyulu Hill National Park	4	Marsabit National Reserve	41	Shimba Hills National Reserve
35	Dodori National Reserve	25	Maasai Mara National Reserve	1	Sibiloi National Park
46	Hell's Gate National Reserve	14	Meru National Park	3	South Island N.P. Lake Turkana
49	Kakamega National Reserve	43	Mpunguti Marine National Reserve	28	South Kitui National Reserve
45	Kamnarok National Reserve	8	Mt. Elgon National Reserve	6	South Turkana National Reserve
51	Kariandus National Park	19	Mt. Kenya National Park	34	Tana River Primate National Reserve
44	Kerio Valley National Reserve	24	Mwea National Reserve	33	Tsavo East National Park
42	Kisite Marine National Park	26	Nairobi National Park	32	Tsavo West National Park
36	Kiunga Marine National Reserve	5	Nasalot National Reserve	40	Watamu Marine National Reserve
21	Kora National Reserve	20	North Kitui National Reserve	39	Watamu—Malindi Marine National Reserve

AMBOSELI

Amboseli is one of the oldest national parks in East Africa, having enjoyed more or less protected status for over 40 years. It was originally part of the Southern Maasai Reserve which also encompassed the **Kajiado** and **Narok** area where several clans of the nomadic, Nilo-hemitic Maasai people lived. The park became the Amboseli Reserve in 1948 when the right of the Maasai people to live there was recognised and a special area for wildlife was set aside. In 1961 the Amboseli Reserve was handed over to Maasai Tribal Control and became a Maasai Game Reserve together with the much larger Maasai Mara Reserve.

However, competition for grazing became such a problem that in 1970 a sanctuary around the swamp was preserved for game only and the *Maasai* were not allowed to enter. This aggrieved them so much that they killed many of the rhino population without even taking their horns. Consequently, a ring of bore holes around the park and a portion of the swamp was given back to the *Maasai* in exchange for an area to the north. Eventually, in 1977, Amboseli achieved full National Park status.

Left, this vulture has a great view of Mount Kilimanjaro.

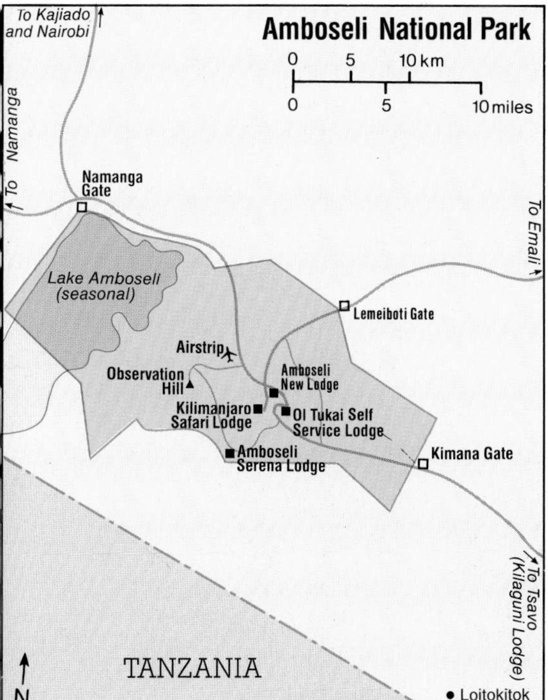

Amboseli National Park

0 5 10 km
0 5 10 miles
To Kajiado and Nairobi
To Namanga
Namanga Gate
To Emali
Lake Amboseli (seasonal)
Lemeiboti Gate
Airstrip
Observation Hill
Amboseli New Lodge
Kilimanjaro Safari Lodge
Ol Tukai Self Service Lodge
Amboseli Serena Lodge
Kimana Gate
To Tsavo (Kilaguni Lodge)
TANZANIA
N
Loitokitok
Mt. Kilimanjaro 5895m

Elephant tales: Lying at the foot of Africa's highest mountain, Kilimanjaro, Amboseli National Park is famous for its tranquil beauty and easily approachable wildlife. The Amboseli elephant population, only some 600 strong, is one of the few in all of Africa which has not been ravaged by poachers. It is also one of the longest studied and best researched by Cynthia Moss and her colleagues who know every elephant by face and name and have written about them in the book, *Elephant Memories.*

Clouds of soda dust which blow up from the perennially dry bed of the pleistocene **Lake Amboseli** provide a stark contrast to the lush vegetation of the swamps which form the heart of the ecosystem. The swamps are fed by the melting snows of Kilimanjaro which percolate through porous volcanic soils, forming underground streams which rise close to the surface in the ancient lake basin.

Lake Amboseli, from which the park takes its name, is a dry lake, some 10 by 16 kilometres (six by 10 miles), and is only flooded during the rare occasions when there are heavy rains. The maximum depth in the wettest years is about half a metre (two feet) but the surface is more usually a dry, caked expanse of volcanic soil. The fine, alkaline dust has a habit of creeping into every crevice, so photographic equipment should be protected in plastic bags.

Forests of towering yellow-barked fever trees used to surround the swamps but their numbers were gradually reduced by the elephant population which stripped off and ate the bark and were initially blamed for all the damage. However, it was then discovered that the naturally rising water table, induced by a period of good rains, was bringing toxic salts to the surface which were "pickling" the tree roots. This caused physiological drought because the trees could not absorb enough water to compensate for that lost from the leaves through transpiration. Even today you can see moribund fever trees which appear to be dying from the top down. Overall however, the park has a varied habitat with open plains, umbrella acacia woodland and the swamps and surrounding marsh areas.

Park life: Due to the open nature of most of Amboseli, lions are easily found and can occasionally be watched stalking their prey. Buffalo numbers have increased and plains game such as zebra, giraffe and gazelle abound. Small groups of gerenuk can occasionally be found in the arid bush standing on their hind legs to browse upon more succulent leaves on the higher branches.

Hippos live in the open waters and swamp channels formed by seeping waters from Kilimanjaro. Buffalos feed in the shore line swamps and elephants penetrate deeper, often emerging with a high tide mark on their flanks.

For years ecological and behavioural studies of these beasts have been carried out in the park, so animals are accustomed to cars and visitors will be able to observe these large mammals in close proximity from inside their vehicles. However, historical as well as recent encounters with *Maasai* warriors have left the animals particularly wary of people on foot. An elephant feeding peacefully three metres (10 feet) from your car will run off in alarm—or attack in a rage—if someone suddenly gets out.

The density of visitors has had negative impacts on wildlife. Cheetahs, for example, have been so harassed by crowding vehicles, that they have abandoned their usual habit of hunting in the early morning and late afternoon, and have taken to hunting at midday, when most tourists are back at the lodge having lunch and a siesta. Since this is not the best time of day to hunt, the result has been a reduction in the cheetahs' reproductive success.

Elephants ambling across Amboseli.

The swamps and marshy areas support a wide variety of water fowl with no less than 12 species of heron. Over 400 different birds can be found since the park encompasses both dry and wet habitats. Taveta golden weavers are very common. Birds of prey are also represented with over 10 varieties of eagle, as well as kites, buzzards, goshawks and harriers.

Just the facts: Amboseli can be reached from Nairobi by two main routes, the most common one being along the main Kajiado-Namanga road, turning left at **Namanga**, entering the park through the main gate near Namanga and following the road to **Ol Tukai Lodge**. The distance from Nairobi to the lodge is 240 kilometres (150 miles). The second access point is along the main Mombasa road, turning right just beyond the railway bridge past **Emali** and then following the Oloitokitok road for approximately 65 kilometres (40 miles), taking another right turn near the flat-top **Lemeiboti hill** and following this road for 32 kilometres (20 miles) before reaching the lodge. This route is shorter but the Namanga road is in better condition. Flights from Nairobi are also available.

The original camp at Ol Tukai was built as a film-set amenity in 1948 for *The Snows of Kilimanjaro*. These buildings remain as self-help *bandas* (grass-thatched, traditionally-styled houses), but nearby is **Amboseli Lodge** with international standards of accommodation and cuisine. Other lodges include: **Kilimanjaro Safari Lodge** with a seasonal swamp where game congregate, and **Amboseli Serena Lodge** which is built in the style of a *Maasai manyatta*, near the well-head of one of the springs feeding **Enkongo Narok** swamp.

Outside the park is **Kilimanjaro Buffalo Lodge**, about 15 kilometres (9 miles) from the **Kimana Gate** on the Emali road. Since this lodge is not restricted by game park regulations visitors are allowed to take game walks. **Kimana Lodge** is also outside the park, about 80 kilometres (50 miles) from Emali.

Zebra and wildebeest are common companions.

TSAVO

Tsavo is a vast arid region of roughly 21,000 square kilometres (8,400 square miles) comprising a series of habitats, ranging from open plains to savannah bushlands, semi-desert scrub, acacia woodlands, riverine forests, palm thickets, marshlands and even mountain forests on the Chyulu and Ngulia Hills. It is the largest park in Kenya. The park is divided by the Nairobi-Mombasa road and railway (the "Lunatic Express") into two sections: the north-east of the park is called **Tsavo East**, with headquarters near **Voi**; the part south-west of the road is called **Tsavo West**, with headquarters at **Kamboyo** near **Mtito Andei**.

Depleted herds: Tsavo was once world-renowned for its large elephant population. Only a few years ago, it was unthinkable to drive from Nairobi to Mombasa without seeing several herds. Tsavo elephants have always been characteristically red, taking on the colour of the soil from dusting and mud bathing. However, the elephant herds which once roamed freely through Tsavo have now been devastated by rampant poaching for ivory. In 1987 conservationists counted less than 6,000 elephants, compared with 35,000 in a similar aerial survey in 1973. The animals that remain are wary of people and ready to run at the sight of a car. The same sad fate has befallen the once-large, now non-existent rhino population.

Both Tsavo parks have been the study sites for a number of significant wildlife research projects. They were also the arena for fierce controvery over the "elephant problem". One school held that there were too many elephants that were destroying the wooded and bushed habitats in times of drought, and that they should be scientifically culled. Such culling would provide valuable data on the population dynamics of the beasts, whose numbers would, of course, be altered by the culling itself.

Cloud-covered Kilimanjaro from Tsavo West.

Another school argued that, since most known animal populations have the ability to self-regulate their numbers through habitat-induced alterations in birth and death rates, there was no reason to believe that elephants do not have the same ability. As it transpired, when the rains returned, so did the grass and trees, but the elephants all but disappeared. A few died of starvation but most were killed by poachers. The results of the research were inconclusive, but favoured the school of *laissez-faire*.

Tsavo West: Much of Tsavo West is of recent volcanic origin and is therefore very hilly. Entering from the **Tsavo Gate**, one comes across the palm-fringed **Tsavo River** from where the country rises through dense shrub to the steep, rocky **Ngulia Hill** which dominates the area. Volcanic cones, rock outcrops and lava flows can be seen, the most famous being **Sheitani**, a black scar of lava looking as if it has only just cooled, near Kilaguni Lodge.

The famous **Mzima Springs** are found in this volcanic zone. The springs gush out 50 million gallons of water a day of which seven million gallons are piped down to provide Mombasa with water. The rest of the water flows into the Tsavo and Galana rivers. The water originates in the **Chyulu Hills** as rain which percolates rapidly through the porous volcanic soils.

Hippos followed by shoals of barbels and crocodiles can be watched from an underwater observation chamber. The best time for viewing is early in the morning; during the day hippos move to the shade of the papyrus stand and remain out of sight.

East of the springs (downstream) is a stand of wild date and raphia palms, the latter with fronds of up to nine metres (30 feet). North of the Mzima Springs are numerous extinct volcanoes, rising cone-shaped from the plains. Mount Kilimanjaro dominates the western horizon.

South of Mzima Springs is a beautiful picnic site at **Poacher's Lookout** on the top of a hill. The view across the plains to Kilimanjaro is worth the trip.

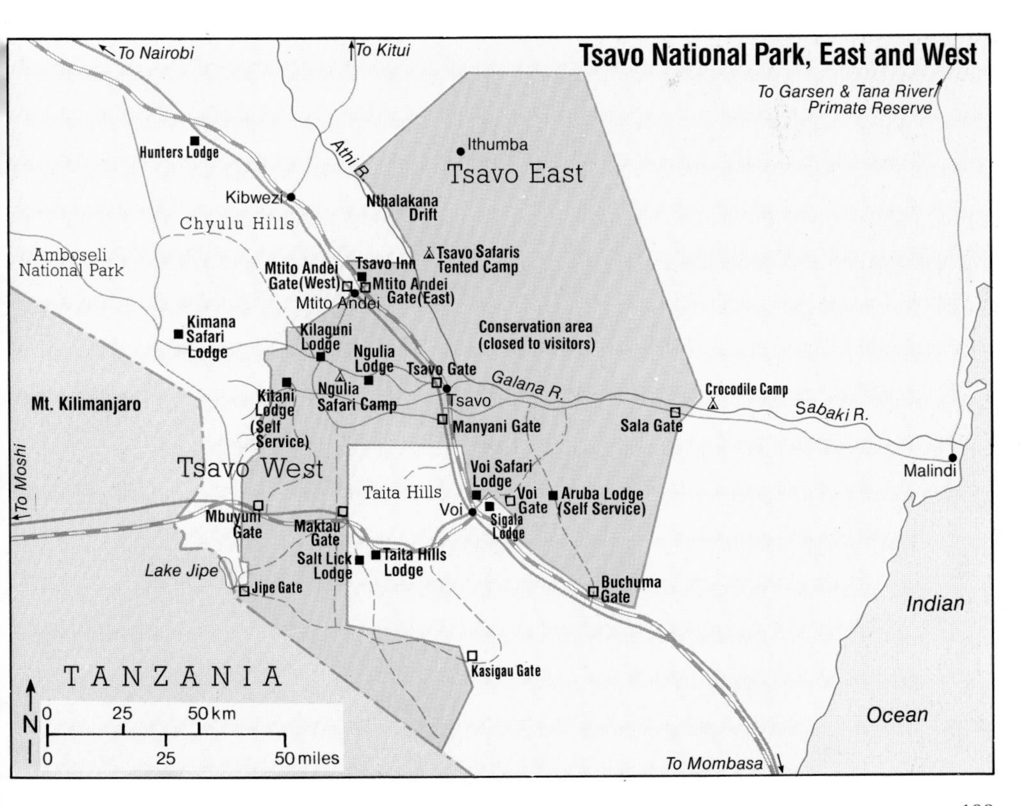

Tsavo West stretches further south to the **Serengeti Plains** which, despite their name, have nothing to do with the Serengeti National Park, although the landscape is similar. This part of the park is crossed by the road and railway from Voi to **Taveta** and lies at the foothill of Kilimanjaro.

Birds and baobabs: Tsavo West has spectacular baobab trees, which used to be far more numerous. In the mid-1970s, there was an enormous and as yet unexplained attack by elephants on baobabs. Some claim it was because of the drought, others claim that there were "too many" elephants. Whatever the reason, the remaining baobabs are quite safe today.

The variety and sheer numbers of birds in Tsavo are incredible. **Lake Jipe**, at the southernmost tip of the park, is surrounded by tall reeds and is one of the most important wetlands in Kenya, providing a sanctuary for a number of water and marsh birds, including migrants from Europe. Some of the birds commonly seen at the lake are knob-billed geese, pied kingfishers, white-backed night herons, black herons, palm-nut vultures and the African skimmer. Lake Jipe can be reached by following the road to Taveta and then turning south after leaving the park through the old **Mbuyuni Gate** or following a murram road which begins near park headquarters across the Serengeti.

If you take an early morning game drive accompanied by a park ranger you might catch a glimpse of another endangered species. The few rhino left in Tsavo are protected in a fenced sanctuary at the foot of Ngulia Hill. Other wildlife in the park includes lion, cheetah, leopard, buffalo, spotted hyaena, warthog, Maasai giraffe, kongoni, duiker, waterbuck, klipspringer, impala, Grant's gazelle, oryx, eland and zebra. The lions of Tsavo are legendary but after the rains, when the grass grows very long, they are difficult to spot.

Staying: Accommodation is available at a series of lodges within the park. The floodlit waterhole at luxurious **Kilaguni Lodge** attracts an incredible

The victor reaps his spoils.

variety of animals, especially in the dry season. All the rooms have splendid views and the food is first class.

Ngulia Lodge, sited on the edge of a great escarpment, is frequently visited by leopards, some of which have been carefully studied by scientists who put radio collars on them to track their movements. Only five metres (16 feet) from the verandah there is a waterhole and salt lick where elephants converge to within touching distance and dig at the salt-bearing earth with their tusks.

Taita Hill and **Salt Lick lodges**, situated in a private reserve on the outskirts of Tsavo West on the Voi-Taiveta Road, offer excellent wildlife viewing with a luxury resort atmosphere. **Ngulia Safari Camp**, not far from Ngulia Lodge, has six *bandas* on the side of a hill overlooking a small dam visited by elephants and an incredible view over the valley where elephants gather. There is another self-catering camp at **Kitani**, not far from Mzima Springs. At minimal cost these self-help camps provide lamps, gas, bedrolls and mosquito nets, although you may prefer to bring your own.

Tsavo East: The main physical feature in Tsavo East is the **Yatta Plateau** which runs almost parallel to, and is easily seen from, the Mombasa road. The plateau, which is between five and 10 kilometres (three to six miles) wide and about 305 metres (1,000 feet) high has its origins as a lava flow deriving from Ol Doinyo Sabuk east of Nairobi. Natural erosion over the millenia has exposed the flow to form the striking plateau seen today.

Around Voi, close to the road boundary, extends flat, dry, semi-desert thornbush country stretching as far as the eye can see. From Voi, running east, is the **Voi River**, which is partly swamp and does not flow all year round. It meanders slowly to **Aruba** where a large man-made dam, the remains of a defunct fish-farming scheme, makes an oasis for both animals and birds. Along the banks of the river is the dependent riverine woodland and numerous wildlife paths leading down to water holes.

Mom gets a grip on one of her cubs.

The road between Aruba dam and **Buchuma Gate** on the Mombasa road is heavily populated with weavers, starlings and lilac-breasted rollers with iridescent wings.

One of the most spectacular sights in the park is the **Lugard Falls** on the **Galana River**, 40 kilometres (25 miles) north of Voi. Here the river, which in its early reaches borders Nairobi National Park, rushes through water-worn coloured rock and at the narrowest point it is said one can step across the river. Perhaps the crocodiles downstream survive on those who try. A good spot to see them is **Crocodile Point** further along the river. Lesser kudu hide in the dry bushland along the river banks.

Mudanda Rock, a 1.5-kilometre-long rock between Voi and **Manyani Gate** is a water catchment area which supplies a natural dam at its base. It is a vital watering point during the dry season and therefore one of the best wildlife viewing areas in the park. Large numbers of elephants used to congregate there. Visitors can leave their cars at the rock and climb up to overlook the dam.

Feasting beasts: Tsavo lions were made famous by Colonel Patterson in his book, *The Maneaters Of Tsavo*, which records the havoc caused by marauding maneating lions to the imported Indian labour brought in to build the Mombasa-Nairobi railway during the early part of the century.

Grant's gazelles, zebra, impala, kongoni, giraffe and lion have replaced elephants as the most common animals in Tsavo East. Large herds of buffalo can also be found. Buffalo have enjoyed a well deserved reputation in the past as being extremely dangerous when wounded or hunted. But since the banning of hunting, buffalo no longer associate danger with man or vehicles so they are generally quite docile.

Some of the more rare and unusual animals include oryx, lesser kudu and klipspringers; the latter can be seen standing motionless on rocky outcrops. Rock hyraxes, the improbable first cousins of elephants, can be seen sun-

Salt Lick Lodge.

ning on rocks and chasing one another in and out of rocky crevices.

Conspicuous white-headed buffalo weavers (the most striking characteristic of which is arguably the red rump and not the white head) and red and yellow bishop birds are found everywhere. Some of the more unusual local birds include pale chanting goshawks, carmine bee-eaters, red and yellow barbets, palm-nut vultures, African skimmers, yellow throated longclaws and rosy-patched shrikes.

The roads north of the Galana River and east of the Yatta Plateau are closed to the public, except when special permits are granted by the park warden. The country is wild and woolly, and spotted with outcrops such as **Jimetunda** and seasonal rivers such as **Lag Tiva**.

Various accommodation is available. **Voi Safari Lodge** clings to the side of a hill overlooking the vast expanse of Tsavo and is literally built into the rock—many of the floors are natural rock. There are two water holes and even during the hottest times of day various wildlife, such as impala and warthog, come to drink.

Other accommodation can be found at the self-service **Aruba Lodge** near the Aruba dam and **Crocodile Tented Camp** on the road to Malindi, just outside the park beyond the **Sala Gate.** Every night, there is a ritual in which huge crocodiles are "called" out of the Galana River with chants and drums, up to the verandah to be fed on offal.

A dirt road from Mtito Andei runs to the west bank of the Athi River opposite **Tsavo Safari Camp** which is reached by boat. Don't miss the incredible view at sunset from Yatta Plateau. Plans are underway to re-open **Sheldrick Blind**, an overnight hideaway on the eastern wall of the plateau from where leopard and other nocturnal animals can be watched. It is named after David Sheldrick, the most famous of Tsavo park wardens.

Public camp sites with minimal facilities are available thoughout both Tsavo East and West.

Looking out from Voi Safari Lodge.

THE MAASAI MARA

Although not the largest protected area, the **Maasai Mara** must be one of the best for game viewing. The Mara area is an extension of the famous Serengeti Plains (*serengeti* in *Maasai* means extended place) just over the border in Tanzania. Animals don't recognise international boundaries so every year, in July and August, over a million wildebeest and thousands of zebra migrate from the depleted grasslands of Tanzania to take advantage of the fresh grazing after the long rains in Kenya.

Migration madness: They follow an established, circular route which begins in February with calving on the Serengeti plains 150 kilometres (93 miles) to the south. The route is inherited by instinct and crosses the **Sand**, **Talek** and **Mara rivers** at exactly the same place each year. The instinct that drives the herds is so strong that in southern Serengeti they swim or wade across small **Lake Ndutu** when they could easily go around it. They are naturally reluctant to enter the thick riparian bush but the pressure of wildebeest pushing from behind builds up until the front animals are forced to take the plunge, in mad, lemming-like, suicidal leaps which Alan Root has so spectacularly captured on film. Many animals get swept away and drown. After the territorital encounters, courtship and rutting season they migrate back to Tanzania following the short rains in November and December.

The Mara Game Reserve was established in 1961 and covers an area of some 1,800 square kilometres (720 square miles). The southern boundary lies on the border with Tanzania's Serengeti National Park. The **Loita Hills** mark the eastern boundary; to the west lies the splendid **Siria Escarpment** and the north is bordered by the **Itong Hills**. The wide horizons are unforgettable and wildlife are clearly visible.

Herd of Burchell's zebra.

Plains Game: The lush grasslands interspersed with silver and russet leaved croton thickets, hillocks and forested river banks provides a good variety of habitats for wildlife. There is a small resident population of roan antelope, many buffalo as well as herds of Thomson's and Grant's gazelles, topi and impala. Predators include large prides of lions, a fair number of cheetahs and leopards, spotted hyaenas and the silver- or black-backed jackal. African wild dogs are also found and two bedraggled groups of females which had originally been marked with radio collars by researchers in the Serengeti, showed up in the Mara in the late 1980s.

There are over 450 recorded species of birds in this reserve, including the large orange-buff Pel's fishing owl which is a common sight along the Mara River but rare elsewhere. Other common birds include kori bustards and various birds of prey.

The best time of year to visit the reserve is from July to October when the migration is at its peak and as many as two million wildebeest and 500,000 zebra are grazing, fighting, courting and mating.

The Mara is the archetypical arena of conflicts between man and nature in modern Africa. Wheat-schemes and livestock improvement programmes to the north meet the greatest remaining wildlife migration to the south. At the interface, conservationists and ecologists strive to reconcile the needs and aspirations of the *Maasai* landowners. Although fraught with problems, many of the results have been encouraging—apart from elephant and rhino poaching which is a blight thoughout Africa. Many tourism-based enterprises, such as tented camps, are run by local landowners who recognize that wildlife can be a resource worth husbanding.

The route to the Maasai Mara is via the Nairobi-Naivasha road, turning left after 56 kilometres (35 miles) towards **Narok**, 103 kilometres (64 miles) further away. The road continues through Narok then forks: four-wheel drive vehicles are usually necessary on the

.ifting off rom the savannah.

northern track leading to the west of the reserve. The road south leads to **Keekorok Lodge**, 106 kilometres (65 miles) away. Travelling from Nairobi can take anywhere between five and 10 hours depending on the season, but the travel time will get shorter as the roads improve nearer the reserve boundary.

Keekorok Lodge in the east was a traditional resting place on the long safari from the Serengeti to Nairobi. These days it is well laid out with cottages and good facilities including car mechanics—sometimes essential after the rough and bumpy roads!

Mara Serena Lodge in the west is set high on a saddle overlooking rolling grasslands and the far off Esoit Oloololo escarpment. Bedrooms are stylised mud *manyattas* grouped in outward-looking rings.

The Mara area has numerous tented camps including **Governor's Camp** on the Mara River where old colonial governors used to pitch their tents. It is now a very up-market retreat. **Kichwa Tembo** (in *Swahili*, "elephant's head") **Camp** and **Fig Tree Camp** offer romantic settings where you can lose yourself in the true safari atmosphere. **Cottar's Camp** specializes in night drives and game walks. **Mara Buffalo Camp**, with thatched *bandas* is situated near the (supposedly) best stocked hippo pool in Kenya. A number of camp sites are also available, many of which are good examples of local entrepreneurship where *Maasai* landowners have recognised the potential gains to be made from wildlife. For all accommodation it is recommended to book in advance.

Taking flight: Balloon safaris can be taken every morning from Keekorok, Mara Serena and Governor's Camp. For an hour or so, up to 12 passengers float silently across the plains watching the game from barely 100 metres (330 feet) above. Vehicles follow the balloon to serve a champagne breakfast in the bush and bring the passengers back. Early morning fishing trips to Lake Victoria out of Governor's Camp can also be arranged.

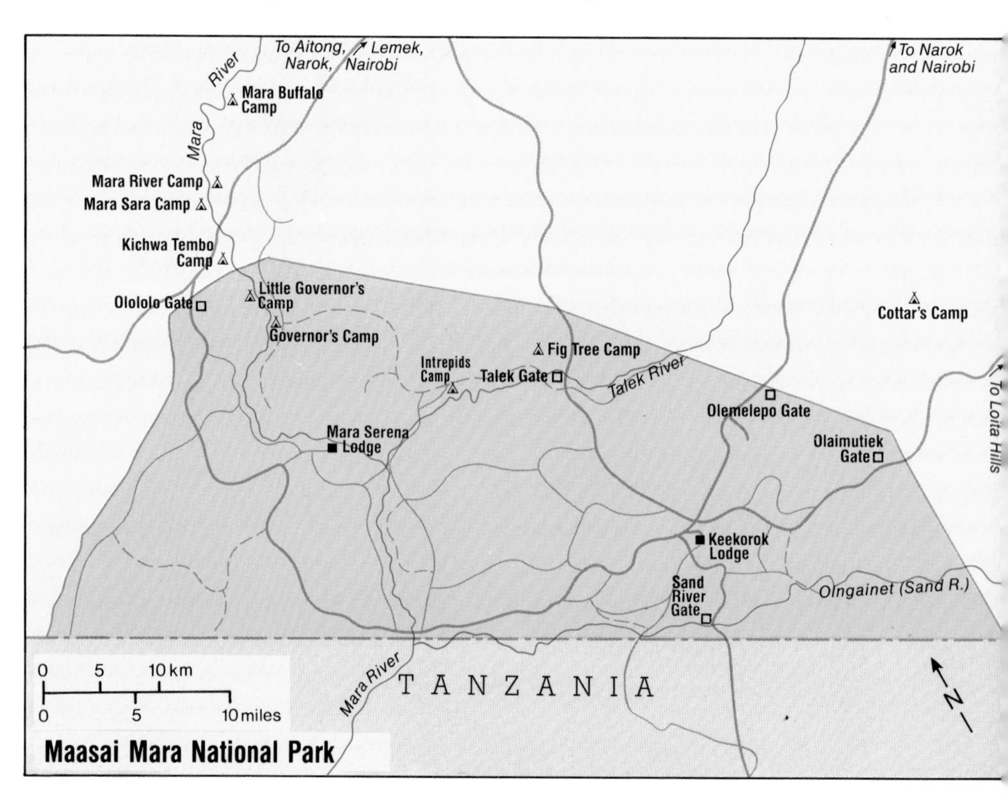

ENDANGERED SPECIES

Three of the more endangered species in East Africa today are the black rhino, the elephant and the wild or hunting dog.

Rhino: At the start of the 20th century the myopic rhino was well represented in East Africa, but lucrative trade for various parts of its body has taken a devastating toll. Africa lost about 50 percent of its rhinos in the 1970s and the early 1980s. Since then there has been a further 40 percent decline in population. Kenya had 20,000 rhinos in 1970 but by 1987 only about 500 remained.

Lions and hyaenas are natural enemies of young rhinos, but man is the biggest predator of adult rhinos. For hundreds of years man has used rhino horn for medicinal purposes. Chinese, Japanese and Korean pharmacists throughout Asia continue to sell popular rhino products for colds and flu and to reduce fevers. In some countries the skin, dung, intestines and bone are also used. Horn, however, is the most widely used rhino product and the tip of a fresh, dark horn is most sought after.

In 1985 the President of Kenya launched a special "Save the Rhino" programme. Anti-poaching operations were stepped up and remnant rhino populations identified.

Elephant: In 1960 many of East Africa's elephants were shot in order to protect crops. Today, as the value of ivory increases on international markets, elephants continue to be slaughtered for their tusks.

East Africa has lost 145,000 elephants in the last 10 years and there are now only 109,000 left.

Wild dog: East Africa's wild dogs are endangered because they have lost a sizeable amount of their habitat to agriculture. They are also susceptible to epidemic disease and have been persecuted by man. They breed well in captivity, but now only about 10,000 remain in the wild.

Sorting out ivory recovered from poachers.

NAIROBI NATIONAL PARK

Nairobi National Park is unique for its location, barely six kilometres (four miles) from the centre of a capital city. Created in 1946, it was the first park in Kenya and stretches a modest 120 square kilometres (48 square miles) south from Nairobi to the Mbagathi-Athi River system. Completely wild but tranquil animals can be seen grazing or hunting against the Nairobi skyline, while jet aircraft from all over Africa, Europe and the Far East make their final approach to Jomo Kenyatta International Airport. To the west, at the very edge of the Great Rift Valley, the gently saw-toothed Ngong Hills rise to an altitude of 2,458 metres (8,070 feet). On the northern horizon, beyond the city, one can see the Aberdares mountain range and to the east, on a clear morning, the peak of Mount Kenya. Being so close to the city, the park attracts many visitors on day trips and earns enough money to help subsidize some of the more remote protected areas.

City living: Almost 37 kilometres (23 miles) of fencing along the park's western boundary prevents wildlife from straying into human settlements and the rapidly-expanding and not very attractive industrial area. To the south, migratory wildebeest, zebras and Coke's hartebeest (kongoni) enter and leave the park more or less freely through the **Kitengela** portion of the of the northern **Athi-Kapiti plains**. Their movements depend on the availability of grazing and water: the animals are mostly out of the park during the short (November) and long (April) rains. As the pastures dry out, the animals pull back north into the park, until their density appears to rival that of the Serengeti herds. Absolute numbers are much less, however, and there are now fewer wildebeest and zebra than there used to be in the whole ecosystem which once stretched as far as Thika and the whale-shaped hill, **Ol Doinyo Sa-**

Male ostrich heads for the big city.

buk, which can be seen in the distance to the north-east.

Developments of farm buildings, rural residences and fences along the south-western edge of the park have raised fears that this free movement of animals could be affected which would make Nairobi's famous park no more then a big zoo. It is hoped that land use development policies will take into account the future of the park and its role in the annual movements of the migratory grazers, as well as its enormous potential as a money-earner for surrounding citizens.

Plant power: Though small, the park has a good variety of habitats. It slopes from about 1,740 metres (5,712 feet) in the western forest down to approximately 1,500 metres (4,925 feet) in the south-eastern plains. Forest in the west occupies almost 6 percent of the land and receives a rainfall of 700 mm to 1,100 mm (27 inches to 43 inches) a year, compared to the much drier south-eastern tip. Tree species in the forest include crotons, Kenya olive, yellow-flowered, long-seeded markhaemia, and Cape chestnut, with its lavender bunches of blossoms. Forested areas are home to at least a dozen black rhinos, some of which have been transported from less friendly neighbourhoods, buffalos and giraffe. Smaller wildlife such as dikdik, suni, duiker and bushbuck can also be seen here.

Bisected by valleys, the plains have less rainfall—about 500 mm to 700 mm (19 inches to 27 inches). They are covered with one of the most characteristic grassland types of the region, composed of *Themeda* (red oat grass) and *Setaria*, spotted with acacias, desert dates (*Balanites*) and the occasional arrow-poison tree (*Akocantha*) which are dark green and often stunted looking from years of harvesting branches for rendering the bark into poison. Visitors are allowed out of their cars at **Observation Hill**, a spectacular spot on the forest edge overlooking the central plain. From here herds of wildebeest, gazelle and zebra can be easily observed.

Cheetah waits in the high grass.

In the open grassland just under the hill, a well-studied male ostrich, "Pointy-head", has held his territory for years, and he or his successor can usually be seen entertaining delighted onlookers by dancing to female ostriches or chasing off rival males.

The southern part of the park is criss-crossed by ridges, valleys and plains. Cliffs fall for about 100 metres (330 feet) to the valley floor. In the dry season, wildlife gather to drink at the isolated pools left in the gorges. The gorge cliffs, some of which are quite spectacular and attractive to rock-climbers, have vegetation ranging from cacti and bushes to grasses and moss. They are lined with acacia trees browsed by giraffe. Buffalo and rhino can be found here, together with lion and leopard. The latter feed on the numerous hyraxes which inhabit the cliff faces in, among other places, **Hyrax Gorge**.

The southern and eastern ends of the park consist of acacia wooded grassland bound by the Nairobi-Mombasa road and the **Mbagathi River**. This modest but perennial river becomes the Athi, which runs into the Galana, which becomes the Sabaki and eventually runs into the Indian Ocean just a little north of Malindi.

Trailing the animals: Visitors can park their cars and follow a 1.5-kilometre **nature trail** along the banks of the river which leads to a series of hippo- and crocodile-inhabited pools. Hippos usually give themselves away by their loud snorting. During the day animals stay submerged in the water and only come out at night to graze. Hippo trails can be seen etched into the bank. A few crocodiles can also be seen swimming about or sunning on banks, as well as impalas and hundreds of vervet monkeys. Another, slightly longer trail meanders through stands of yellow-barked fever acacias (*Acacia xanthophloea*) and their greenish-barked cousins, *Acacia kirkii*.

Birds are abundant at the edge of water, including kingfishers, darters, storks, herons, saddlebill storks and ibises. Hammerkops and beautiful

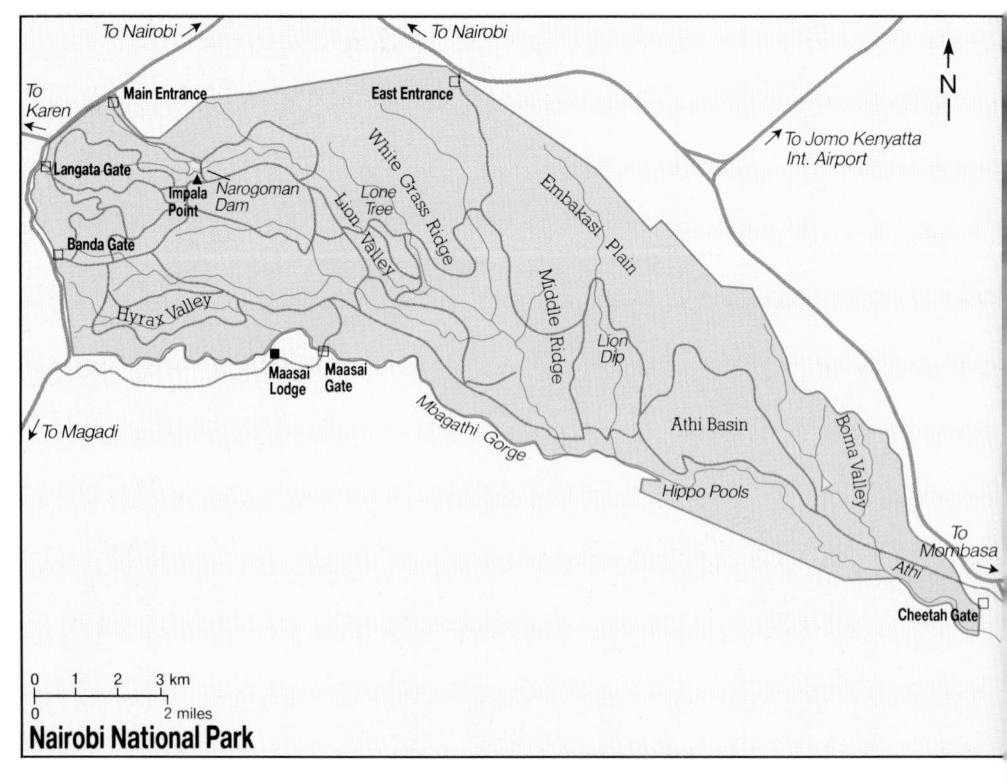

crowned cranes are also common. The open plain features birds like ostriches, marabou storks, cattle egrets, secretary birds and vultures.

The best time of year to visit is during the dry seasons, in February-March and again in August-September, in order to catch the return of the antelope herds that disperse southwards towards the Athi plains during the rains and return to the park to gather around the permanent water holes. Big cats and impalas are permanent residents; wildebeest, kongoni, zebra and to a lesser extent giraffe are migratory. There is a strong chance of seeing lion, cheetah, buffalo and rhino.

Nairobi National Park is an excellent introduction to the art and sport of wildlife viewing. Tours lasting four or five hours are available from most tour companies and can be taken either in the morning or afternoon. They can be arranged through any of the big hotels in Nairobi. Roads in the park are murram, but are kept in very good condition, so even small rental vehicles can carry you to an away-from-it-all experience in the morning, and back to the more intense experience of bargaining for local handicrafts in the Nairobi market before lunch.

OL DOINYO SABUK

Ol Doinyo Sabuk is a minute reserve of only 18 square kilometres (seven square miles) which used to be the residence of the American, Sir Northrop MacMillan, who is buried on his reputedly haunted homesite at the foot of the hill. The park is located just 15 kilometres (nine miles) northeast of Nairobi, near **Thika**. The hill, also known as *Kilima Mbogo* ("Buffalo Hill" in *Swahili*), is a granitic kopje rising above the surrounding plain. It is almost entirely forested except for a small patch at the top. The area is inhabited by buffalo and bushbuck, but they are hard to see because of the dense undergrowth. Admission is free. There is no accommodation available, but it makes a very interesting and pleasant day trip from Nairobi.

High plains drifter: lone topi.

THE ABERDARES

One of the three mountain parks, Aberdare covers 770 square kilometres (308 square miles) stretching from the 3,996-metre (13,120-foot) peak at **Ol Doinyo Lesatima** in the north to the nearly as high **Nyandarra Peaks** in the south. The eastern wall of the **Great Rift Valley** is the western boundary, and in the east there is an area known as the **Salient**. Two famous wildlife viewing lodges, **Treetops** and **The Ark**, are located in the eastern Salient. The park consists of high mountain rain forest, open moorland with hagenia woodlands and impenetrable bamboo forest.

Bamboo tunnels: Bamboo forests are found on some of the mountain slopes and extinct volcanoes and are criss-crossed by wildlife tracks that part the towering bamboo stalks. It is a fascinating experience to walk along these bamboo "tunnels". Extreme caution has to be exercised as they are regularly used by buffalo and elephant. Upon meeting wildlife, the only option is to clamber into the impenetrable bamboo and let the animals pass! Bamboo is a species of gigantic grass. Some stands of bamboo are apparently dead, as this particular type of bamboo to dies off after flowering every 30 years or so.

The moorland above the high treeline is covered in tussock grass with towering giant heather and typical alpine plants, some of which also occur in the Alps and Rocky Mountains. The view is phenomenal. **Karura** and **Guru waterfalls** are especially spectacular. Numerous rivers, such as the Chania, are faithfully restocked with brown and rainbow trout each year by the Kenya Fisheries Department—a most welcome relic of the colonial era. High fishing camps are operated on a self-help basis. The more exclusive camps are located outside the national park on the private reaches of fly fishing clubs. Fishing can also be arranged from some of the mountain lodges.

Far left, leopard gets his licks in.

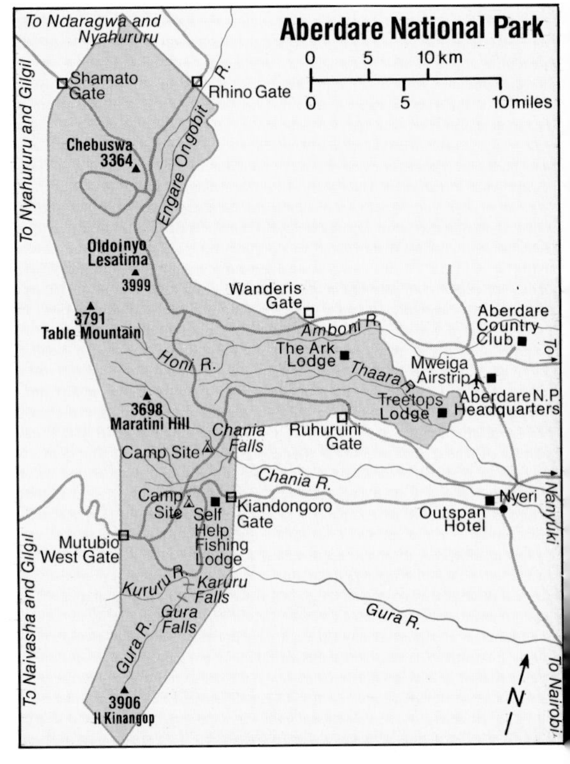

Not too tame: There were once a number of campsites within the park, but these had to be closed due to the lions that made themselves a nuisance. In this particular park larger wildlife is notorious for being aggressive, probably due to frequent unpleasant encounters with the high density human population in the surrounding area. Along a 37-kilometre (23-mile) stretch of park boundary above **Nyeri**, there is an experimental elephant proof ditch and electric fence to protect the maize shambas and people outside the park.

Some unusual and rare animals can be seen, such as bongo, a forest dwelling antelope that feeds at night and in the early morning. They are sleek and chestnut coloured, with about a dozen narrow, white, vertical stripes. Giant forest hogs can be recognised by their coarse, black hair and shy, retiring manner. *Asterix the Gaul* fans will be reminded of the boars beloved of *Obelisk*. Bushbuck are common—males are dark russet, with a white band across the chest—and hunchbacked red duiker occasionally creep out of the undergrowth to graze in open forest lanes.

An unusual phenomenon of the high forest parks is the occurrence of melanistic (all black) cats. There are numerous and well-authenticated reports of sightings of black leopards and serval cats, although these are usually fleeting glimpses as the animal dashes across the road.

The bronze-naped pigeon, dusky turtle and tambourine doves and the iridescent turaco are a marvellous sight as they clamber around in the upper branches of forest trees. Trumpeter and crowned hornbills can be vociferous in the early mornings and their heavy undulating flight styles across forest clearings is unmistakable. There are numerous forest dwelling barbets and at least 12 species of sunbird.

Both Treetops (where Elizabeth II learned she had become Queen of England) and The Ark lodges are famous for their rewarding wildlife viewing, especially by night when their water holes and salt-licks are floodlit.

Elephant-watching from The Ark.

MOUNT KENYA

The 3,350 metre (11,000 feet) contour line around Mount Kenya is the boundary of **Mount Kenya National Park**. Nowhere else on earth is there perpetual snow actually on the equator. Mount Kenya National Park offers glaciers, 30 small mountain lakes and some of the world's greatest ice climbing, all within a day's drive of Nairobi and the hot upland plains.

Large troops of black and white colobus monkeys are common. Sykes monkeys are also found in abundance. Melanistic leopards and serval cats exist here although they are rarely seen. Comical giant forest hogs are common and the more elusive bongo can sometimes be seen emerging from the forests in the evenings.

There are about 150 species of birds. Green ibis and crowned and silvery-cheeked hornbills are found by streams. Mountain chats and white-starred bush robins live in the bamboo forests. There are many varieties of mountain warbler and Jackson's francolin call loudly in the evening before nesting in the bamboo clumps. The almost totally green parrot you might see is actually called the "red-fronted" parrot! It has a small patch of red on the forehead and a patch on the shoulders and wing edges. There are also many iridescent sunbirds, the main species being emerald coloured malachite sunbirds and tufted malachite sunbirds, which are duller, with tufts on their shoulders.

The **Mountain Lodge** at 2,193 metres (7,200 feet) boasts the most consistent record of wildlife seen nightly. There is a bunker by the floodlit waterhole, so that the drinking animals can be viewed virtually eye to eye. The light is sufficient for satisfactory photographs. Safaris can be arranged deeper into the mountain forest, and fishing enthusiasts can try their luck at trout fishing up the valley, with rods hired from the lodge.

Far left, national flag flies on Mount Kenya.

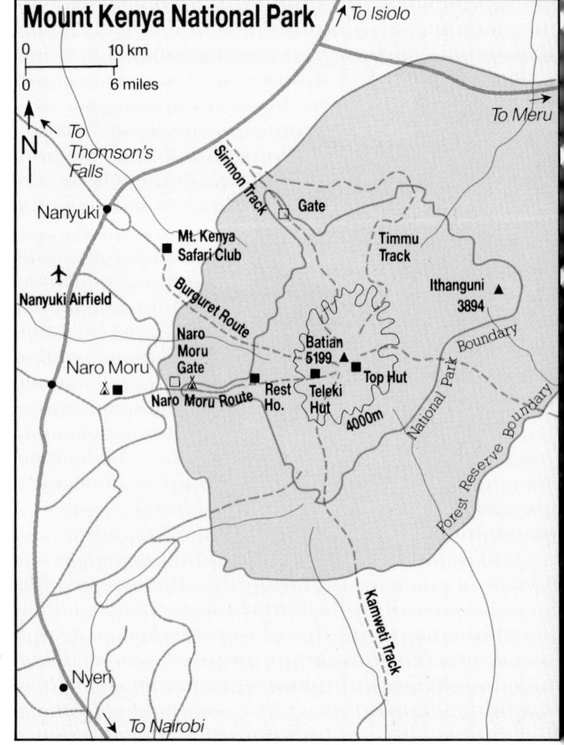

WILDLIFE PHOTOGRAPHY

East Africa must be one of the destinations serious outdoor photographers have on their shortlist of places to visit. Nowhere else in the world is it possible to find such variety and concentrations of free-roaming creatures, not to mention the scenery.

Wildlife photography lagged behind wildlife painting as a form of illustration mainly because early cameras were too big to carry in the field. A turn-of-the-century picture shows an earnest photographer pursuing a rhino with a camera the size of a microwave oven. It weighed 7,168 grams (16 pounds). Exposure took minutes—too long to capture an unrestrained creature on film.

By the middle of this century wildlife photography had changed dramatically. Faster film, telephoto lenses and motordrives gave photographers additional flexibility.

Low angle: remote-control camera and subject.

Safari-goers will find that the most suitable cameras are 35 mm SLR (single lens reflex) models with interchangeable lenses.

Lenses are a wildlife photographer's most important accessories. A 600 mm telephoto lens is the practical, upper size limit for work in the field. But it cannot be handheld and tripods are very cumbersome when used in vehicles. Zoom lenses—in the 80 mm to 200 mm range—are more versatile and offer better value for money.

Telephoto lenses are prone to camera shake since they magnify image movement. Accurate focusing is also essential with long focal lenses. Shoulder supports help to reduce camera shake; so do sand or bean bags. These look like small pillows filled with dried beans or sand. Lay the bag down on the roof of your open safari vehicle and mould the lens on to it. Many safari companies provide these bags.

Recent years have seen the introduction of very fast colour film. The sharpness and grain of these films are perfectly acceptable even if pictures are blown up to A4 size. Films in Africa, when available, are expensive so visitors should bring more film than they expect to shoot.

On the equator, the best time to take pictures is before 10 a.m. and after 3.30 p.m. When it is overcast, midday hours can be acceptable.

During dry times of the year dust can be a serious hazard to your equipment. It is particularly important to protect the camera from wind-driven dust when loading and unloading film.

The majority of safari pictures are taken from the roofhatch of a vehicle, but taking all your pictures from there creates stereotyped images. Shooting from a lower angle out of a side window often results in more dramatic shots.

Timing, as they say, is everything. Most experienced wildlife photographers will tell you that when it comes to taking memorable pictures it all boils down to being in the right place at the right time. For that reason even old hands can be surpassed by a relative newcomer to the wildlife game.

Meru, Kora and Rahole

Meru National Park achieved world recognition with Joy Adamson's *Born Free*, the story of Elsa the lioness that was rehabilitated to the wild. The similar tale of Pippa, her cheetah, was told in *The Spotted Sphinx*. After her release, Pippa eventually gave birth to two litters of fine cubs. Despite being one of the major national parks in Kenya and a very beautiful one, it is off the mainstream circuit for the majority of visitors. Still, it is strongly recommended you make an effort to visit this park.

The park covers an area of 800 square kilometres (320 square miles), lying to the west of Mount Kenya in the semi-arid area of the country. It straddles the equator and ranges from an altitude of 1,000 metres (3,300 feet) in the foothills of the **Nyambeni Range** (the northern boundary) to less than 300 metres (990 feet) on the Tana River in the south.

The main tourist roads are in the western part with a few roads in the remote east. The eastern park boundary is bordered by the Bisanadi National Reserve and the Kora National Reserve. To the north of Kora is the Rahole National Reserve, which means that altogether there is an area of 4,670 square kilometres (1,868 square miles) of wildlife sanctuary.

Vegetation is mainly bushland with combretum bush prevailing in the north and commiphora in the south. The north-east is dominated by grassland with borassus palms and acacia woodland. There is plenty of water, the main perennial river being the Tana, which is the longest in Kenya. Many other small streams occur in the park. Most are bordered by riverine forest. Some valleys are partially flooded during the rainy season, providing a swampy grassland habitat favoured by buffalo and waterbuck.

Reviving stock: Animal life is now plentiful, but game had virtually vanished by 1959, when the local council of

Pale chanting goshawk stops for directions.

the *Wameru* tribe seized the initiative from the colonial government and designated the area for conservation. Large numbers of buffalo can usually be found around the swamps and river. Big herds of elephant used to be seen quite often in the swamp area near the **Meru Mulika Lodge**. The Tana River provides a sanctuary for hippopotamus and crocodile. Black rhinos used to be abundant in the park, but sadly they too have been heavily poached, as was the small protected herd of white rhinos. In 1988 the five remaining animals were killed by poachers, aided, it is alleged, by a disgruntled park ranger who used to protect them. The white rhinos had been introduced from South Africa in the hope that they would breed to establish a viable herd. The name "white" does not refer to the colour of the animals but is a misinterpretation of the Africaans word *weit* which means "wide" and refers to their broad mouths. White rhinos are mainly grazers whereas black rhinos are browsers.

Leopards have also been the focus of a re-population bid. Over the years, many have been brought in from other parts of Kenya.

Meru supports a range of species more usually found in northern protected areas, such as Grevy's zebra (with narrower stripes than the more common Burchell's zebra), beisa oryx and reticulated giraffe (rust-red coloured with distinctive thin white lines creating a "crazy paving" effect). Dikdik, gerenuk (which supposedly do not need water and survive on dew) and the big cats are abundant, but sometimes difficult to see because of the tall grass cover and thick bush. Eland and kongoni prefer the wetter grassland areas. Lesser kudu—either alone or in pairs—can be found in thickets or in valley bottoms in the evening.

Birdwatchers should look out for the relatively uncommon palm-nut vulture, which feeds on a mixture of palm nuts and carrion. In addition, the palm swift can be seen building its nest on the underside of palm fronds. Pel's fishing owl and the rare Peter's finfoot live near

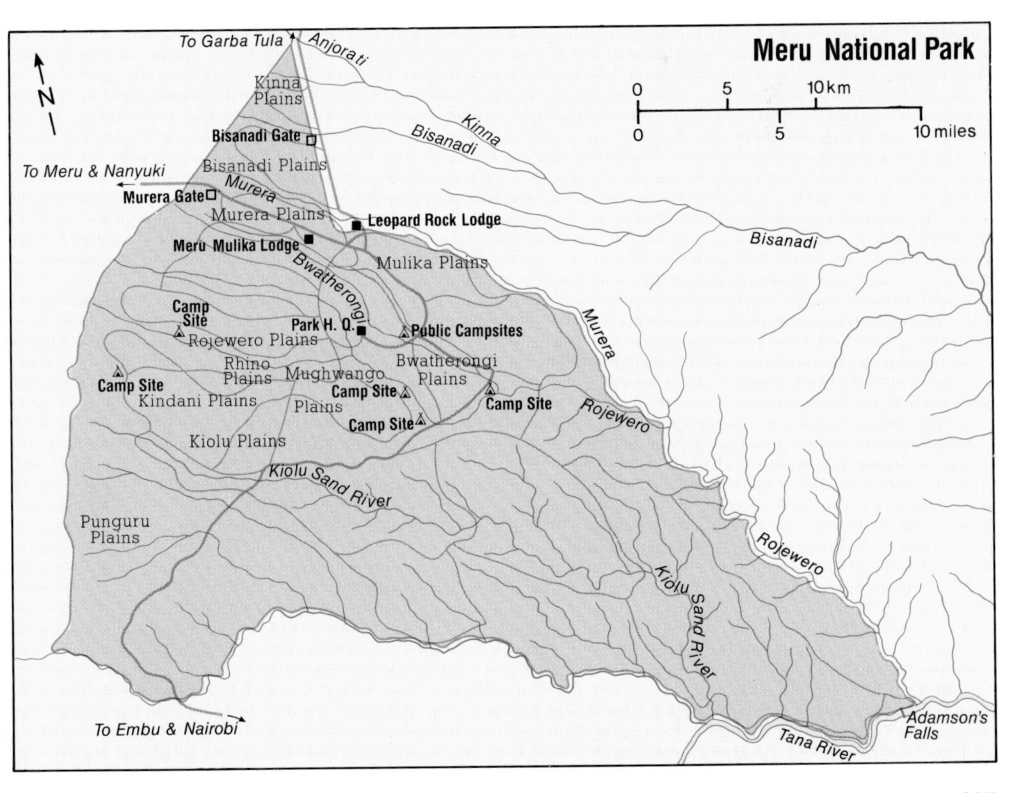

the Tana River. Peter's finfoot resemble long-necked slender ducks or small cormorants. They are very secretive and are usually seen swimming under overhanging trees close to the bank. Helmeted and vulturine (with slender, striped necks) guinea fowl are common.

Wilderness area: One section of the park has been designated a wilderness area, in which there are no roads. This area can only be reached by four-wheel drive vehicles, accompanied by an experienced ranger. Similarly, the 600 square kilometres (240 square miles) of the **Bisanadi National Reserve**, which adjoins Meru National Park, is undeveloped for tourism. Access is difficult even with four-wheel drive vehicles. The habitat, flora and fauna are similar to Meru, with more spectacular elongated rocky outcrops.

Two routes lead from Nairobi to Meru National Park: one around Mount Kenya, through **Nanyuki** and the other one through **Embu**. Both roads go to **Meru** town from where it is 78 kilometres (48 miles) to the park. If you go via Nanyuki, you can enter the park from the west using the **Murera Gate**.

Accommodation at Meru is available at the well-appointed Meru Mulika Lodge which has numerous thatched huts in an attractive setting. Below the lodge, large herds of elephant wander across **Mulika Swamp**. Some visitors prefer to fly in to avoid the slow and winding road.

Not far away, on the banks of **Murera River**, is **Leopard Rock Self-help Lodge** which has 10 *bandas* (rustic cabins with bathroom). You should bring your own food. There are also several other campsites which are marked on most maps.

KORA NATIONAL RESERVE

Bordering on the middle reaches of the Tana River, the 1,790 square kilometres (700 square miles) of the **Kora National Reserve** were made famous by George Adamson and Tony Fitzjohn who engaged for years in the dangerous business of re-introducing captive lions and leopards to the wild. Tragically,

Despite protection, these white rhinos were poached in 1988.

Adamson was ambushed and killed by bandits in August 1988. The remote Kora region is adjoined to Rahole National Reserve, and is composed of riverine woodland along the Tana River and miles of bushland in the interior. It is also renowned for rocky outcrops with their own unique habitats and fauna. Wildlife will not be seen in great numbers, but there are occasional sightings of lion, lesser kudu, elephant and waterbuck. The river is beautiful in this section and is well stocked with hippos and crocodiles.

Kora was the site of a major ecological survey in 1983 carried out by the National Museums of Kenya and the Royal Geographical Society with support from the US National Aeronautic and Space Administration and the United Nations Environment Programme. The results gave valuable insights and management information for a wild part of Africa increasingly encroached by Somali pastoralists.

Kora is about 130 kilometres (80 miles) from the township of **Garissa**. There are no tourist facilities in what is essentially an area set aside for scientific research. Visits can, on occasion, be arranged to **Adamson's Camp** or the riverside research station which is now a ranger post.

RAHOLE NATIONAL RESERVE

North across the Tana River from Kora is the **Rahole National Reserve**. This reserve of 1,270 square kilometres (508 square miles) was developed to illustrate the potential for local wildlife to co-exist with tribes that live in the area. However, the experiment seems to have failed as poaching is rife and settlements abound. Until it can be developed to ensure a tourism industry with revenues for local pastoralists, the area will remain undistinguished and unremarkable, although there is plenty of splendid scenery.

There are no tourist facilities except at Meru, 40 kilometres (25 miles) upstream. However, in the dry season Rahole is really only accessible from the east, via the Garsen-Garissa road.

The late George Adamson taking a walk in remote Kora.

SAMBURU AND BUFFALO SPRINGS

Samburu and Buffalo Springs reserves lie in what used to be called the Northern Frontier District, a vast area of semi-desert and desert that stretches north from Mount Kenya to Sudan and Ethiopia. It is a stark, rugged landscape where nomads, who have changed little over centuries, still move their herds across the ecosystem chasing the ephemeral growth of grass. It is the emptiness and wildness that makes a visit to these reserves such an unforgettable experience.

Two into one: The two reserves, Samburu to the north of the **Ewaso Ngiro** (brown water) **River** and Buffalo Springs to the south are usually treated as one unit, by tour companies as well as wildlife. A bridge across the Ewaso Ngiro River a little way upstream of **Samburu Lodge** connects the reserves.

The major, central part of both reserves is dry, open, thorn bush country, which only becomes green during the rains. The river which originates on the **Laikipia Plateau**, fed by the runoff from the Aberdares and Mount Kenya, is a permanent source of water for animals and is lined by acacias, tamarind and doum palms.

A variety of animals can be found, including diminishing numbers of elephant and numerous buffalo and waterbuck that feed on the vegetation around the river and in the adjourning swamps. Impala herds, with one male guarding up to 50 females and young, graze along the riverine vegetation. Grevy's zebra, beisa oryx, reticulated giraffe and gerenuk are only found in this sort of dry semi-arid country. Grevy's zebra used to be poached for its fine, narrow striped skin by local poachers or exported to European zoos by expatriate entrepreneurs, and they are only slowly re-establishing a viable population. Oryx are very shy, relatively scarce animals with beautifully marked heads and long, straight horns. Dikdik are far more

Termite mounds and weaver nests are common Samburu sights.

common and particularly like the rocky hills and dry, acacia woodland to be found here.

Dinner guests: Crocodiles sun themselves on the banks of rivers. Lion, cheetah and leopard are also fairly easy to see, thanks to the sparse grass cover. If you fail to see a leopard on one of your game drives, you can always watch them in the evening from the verandah of your lodge as they come to eat the bait hung from nearby trees.

Smaller mammals include ground squirrels which are common around the lodges and dwarf mongooses are frequently seen scampering across the open ground looking for food.

Birds are abundant, including the blue-legged, northern Somali race of ostrich, which is particularly prominent during the breeding season. Numerous flocks of helmeted and vulturine guinea fowls can be seen especially in the afternoon as they go to the river to drink. Martial eagles, one of the largest of the eagles, are often seen perching on a vantage point scanning for movements in the grass indicating potential prey. Other birds of prey, such as bateleur and pygmy falcon are also common. Along the banks of the rivers, kingfishers and Layard's black weavers are found. The rare bright green and red chested Narina's trogan, a bird related to the parrot, is also found in the riverine woodland.

From Nairobi to the reserves, it is approximately 300 kilometres (186 miles) on tarmac up to **Isiolo**, then on dirt road for another 53 kilometres (33 miles). The most convenient entrance to the reserves is the **Gare Mara Gate**, 20 kilometres (12.5 miles) north of Isiolo, through the Buffalo Spring Reserve. Another entrance, three kilometres before reaching **Archer's Post**, is called **Buffalo Springs Gate**. The road directly into Samburu Reserve, reached from the township of Archer's Post, is in bad condition, but the journey is made more interesting by the several Samburu *manyattas* (enclosed villages) passed on the way.

The luxurious Samburu Lodge is on the north side of the river by the western

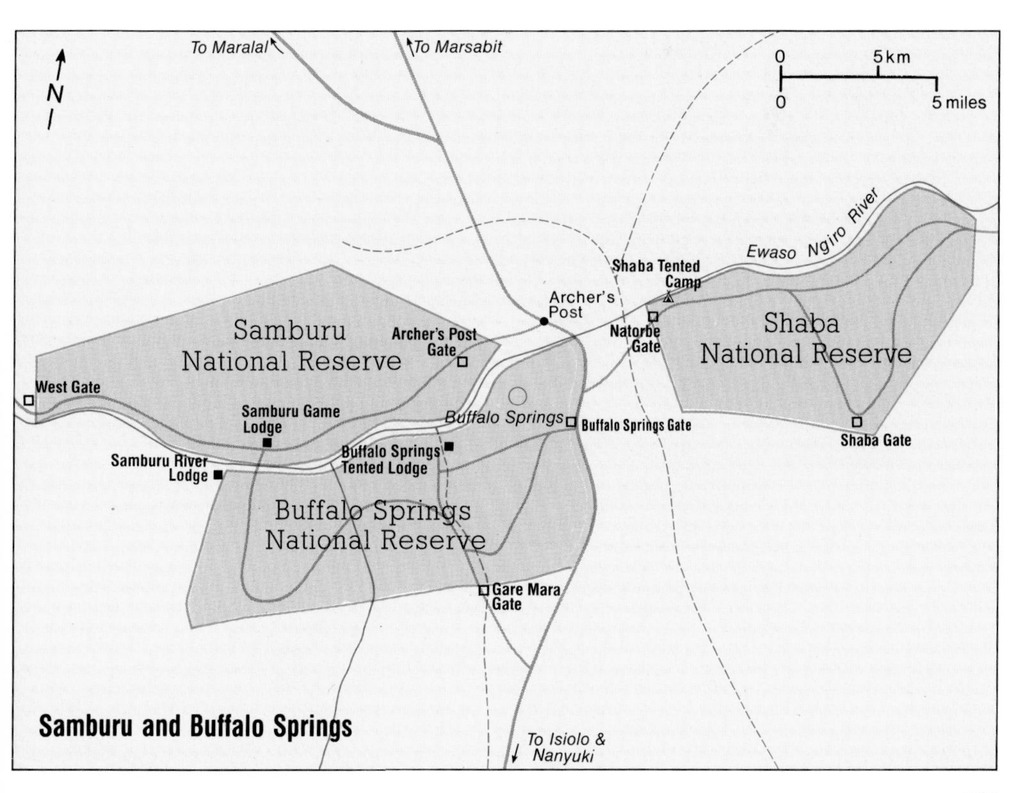

Samburu and Buffalo Springs

boundary. You can watch crocodiles being fed on leftovers, and there is a platform where goat carcasses are hung from trees to attract the big cats, especially leopards. A tarmac airstrip is located nearby for tourists who do not wish to endure the long drive.

The equally well-appointed **Buffalo Springs Tent Lodge** is on the south bank, not far from the eastern gate. A short distance away are the clear pools of Buffalo Springs. The story goes that during World War II an Italian bomb, dropped by a plane from occupied Somalia, missed Isiolo township and formed the pool.

SHABA NATIONAL RESERVE

Shaba National Reserve lies in the northern district to the south of the Uaso Nyiro River, covering 240 square kilometres (96 square miles). It takes its name from **Mount Shaba**, a copper-coloured sandstone hill which lies partially in the reserve which is famous for its lava flows that oozed down from the **Nyambeni Hills** only 5,000 years ago. The western side of the reserve is bushed grassland savannah, dotted with thorn bushes, gradually becoming acacia woodland nearer Mount Shaba. Beyond the mountain the vegetation becomes grassland plains. A series of springs bubble up in the river in the north-eastern side of the reserve. One spring, **Penny's Drop**, was named after Joy Adamson's leopard Penny, which she released back to the wild in Shaba Reserve. It was in her Shaba camp that Joy Adamson died in 1980.

Although heavy poaching in Shaba has made animals very shy, you might be lucky enough to see elephant, lion, cheetah, leopard, waterbuck, as well as all the animals specially adapted to the dry region: beisa oryx, gerenuk, Grevy's zebra, reticulated giraffe and Somali ostrich.

The reserve is about 300 kilometres (186 miles) from Nairobi. Access is by a turn-off two kilometres (just under one mile) short of Archer's Post through the **Natorbe Gate**, seven kilometres (4.3 miles) from the main road.

Shy dikdik move in pairs.

MARSABIT

Marsabit National Park and Reserve is situated in what used to be known as Kenya's "Northern Frontier District". Marsabit itself is an improbable forested volcanic mountain which rises like an oasis out of the dry black lava-strewn surrounding semi-desert.

Climatic oddity: The park encompasses the mountain and is only some 20 square kilometres (eight square miles) in extent. The much older surrounding reserve covers approximately 2,100 square kilometres (840 square miles). The eastern slopes of **Mount Marsabit** are dry and barren, whereas the western slope is covered in perpetual mist and enjoys heavy rainfall which has given rise to dense rainforest. The reason for this climatic oddity is that when hot air blown off the surrounding desert rises and cools over the mountain, clouds are formed and rain occurs.

The mountain microclimate allows for the growth of lush, tropical, evergreen forest in which elephant and greater kudu can be found. Herds of buffalo—some claim a smaller "mountain" variety—are also common.

Ahmed the elephant: Lake Paradise, at the bottom of one of the mountain's craters, hosts a variety of waterbirds. Some of the biggest elephant tuskers have lived here, including the famous Ahmed, now sadly dead. Ahmed was renowned for his splendid downsweeping tusks and was protected by presidential decree. In the late 1970s, just a couple of years before his death, he was assigned a ranger to watch over him and to keep his location known for visitors. From his teeth, it was clear that he was about 65 (old for an elephant), had an abscess on one side of his jaw, and was obliged to chew on the other side. He was on his last set of molars, and their diminishing and smoothing surfaces were clearly incapable of processing the necessary amount of vegetation. So he had the added distinction of

Teeth to be proud of.

being one of the last great tuskers to die a natural death: debilitation caused by undernourishment. His position has been filled by another big tusker known as Mohammed II, who can be seen in the crater swamp in front of the **Marsabit Lodge**.

Marsabit National Park is 560 kilometres (347 miles) north of Nairobi. The asphalt road finishes at Isiolo which means that 270 kilometres (167 miles) has to be driven on corrugated dirt road. In order to drive to Marsabit one must obtain permits from Provincial Headquarters at Isiolo. From here, it is recommended to drive in convoy and to carry petrol, water and supplies for the journey. Various charter companies at Wilson Airport in Nairobi will arrange flights to the reserve.

Accommodation in the park is provided by the Marsabit Lodge at the edge of the forest, overlooking the swamp and lake in the **Sokortre Dika** crater. The lodge is comfortable and food is provided as well as lunch boxes for game drives. Nights are cold so a sweater is needed for when you are not sitting around the lodge fireplace. Rooms face the crater lake where greater kudu, reticulated giraffe and, if you are lucky, the elusive Lammergeyer can be seen. Campsites are available—the most beautiful is the one located at Lake Paradise.

LOSAI NATIONAL RESERVE

Located to the south-west of Marsabit, across the **Kaisut Desert**, the 1,735-square-kilometre (694-square-mile) **Losai National Reserve** is an area of impenetrable mountain forest on the north-eastern edge of the central highlands. The Great North road runs through the eastern section, but it is a wild area and is generally inaccessible, except by four-wheel drive vehicle in the dry season. For the more intrepid visitor, a walk up the seasonally-dry bed of the **Milgis** is an incomparable experience. It is like a broad avenue, bordered by some of the most magnificent umbrella acacias (*Acacia tortilis*) in Kenya.

Sand dune north of Marsabit.

RIFT VALLEY LAKES

LAKE NAKURU

Lake Nakuru National Park was created in 1961 as a bird sanctuary. Originally, it comprised only the lake and its immediate surroundings, including the escarpment at its western side known as **Baboon Cliffs**. It was expanded in 1974 with help from the World Wildlife Fund, and now includes an extensive area of savannah to the south. Today the total area of the park is about 200 square kilometres (80 square miles). The name Nakuru is derived from the *Maasai* word *en-akuro*, meaning "swirling dust".

Pretty in pink: The park is famous for its concentrations of both greater and lesser flamingoes, which cover the lake in a layer of shocking pink. The numbers of flamingoes fluctuates in accordance with the availability of blue-green algae known as *Spirulinga.* The algae is sensitive to the salt concentration of the water so its "blooming" is related to the amount of rainfall which in turn determines the amount of water available to dilute or concentrate the salt solution. The lake level varies considerably. In the late 1950s, for example, it dried out completely and the resulting dustbowl made life unbearable in the busy farming town of **Nakuru** on the lakeside. When conditions are right, there can be around two million flamingoes milling about the shallows—the spectacle is truly awe-inspiring. When disturbed, the pink clouds reeling through the sky are an amazing sight.

For many years it was a mystery where flamingoes went to nest in reliable numbers since they do not nest on the lake. The well-known ornithologist, Leslie Brown, spent many years trying to establish their nesting grounds. He even took flying lessons, and eventually found their regular nesting site on another Rift Valley water body, **Lake Natron**, just over the border in Tanza-

Massed flamingoes on Lake Bogoria.

nia. His fascinating book *The Mystery of the Flamingoes* is essential reading for any birdwatcher visiting Kenya.

Birds of paradise: The park is considered an ornithological paradise. Over 400 varieties of birds can be seen altogether, although not at the same time since many are migrant visitors from the northern hemisphere. During the European and East Asian winters the park becomes an important feeding ground for migrant waders. Among them are little stints, curlew sandpipers, marsh sandpipers and greenshanks. Great numbers of pelicans can be seen at the southern and eastern shores. The numbers of these birds has increased considerably since the alkaline and high temperature-tolerant *Tilapia grahami* fish were introduced to the lake in the early 1960s. Pelicans feed by working as a team, herding the fish towards each other, and then dipping into the water in unison to increase their catch.

Verreaux's eagles can be seen using the updrafts around Baboon Cliffs to search for prey along the cliffs. Other birds of prey commonly seen are long crested eagles, Augur buzzards, harrier eagles, fish eagles, gabar goshawks and harrier hawks. The acacia woodlands harbour a number of birds, including red-chested cuckoos (the bird which chants: "it-will-rain, it-will-rain" before the rains), African hoopoes and grey-headed kingfishers.

The park has now been fenced in to make a rhino sanctuary. Rhinos have been moved here from elsewhere in Kenya and they seem to be thriving in this environment. Other species of mammal include lion, leopard and hyaena. It is the best place in Kenya to see Bohor reedbuck and Defassa waterbuck. A herd of Rothschild giraffe was introduced in 1977.

Lake Nakuru is 150 kilometres (93 miles) from Nairobi on a good tarmac road. Part of the way is along the edge of the **Rift Valley Escarpment** and the views are spectacular.

Lion Hill Camp is perched on higher ground by the eastern boundary overlooking the lake. It is also adjacent to

Flamingoes perform a courtship dance.

Kenya's finest euphorbia forest with grotesque, giant cactus-like trees. On cold evenings a fire is lit by the bar. **Lake Nakuru Lodge** used to be part of Lord Delamare's estate and, apart from the main manor house, there are new *bandas*. Safari vehicles are available for hire and an airstrip is close by. There are also two well-maintained campsites with good water supplies.

LAKE NAIVASHA

Lake Naivasha is one of the Rift Valley's cleanest freshwater lakes, renowned equally for its great variety of birdlife, its scenic beauty, and its role as a retreat for the zany white settlers of "Happy Valley" fame. The infamous, pink painted **Gin Palace** still perches preposterously overlooking the southeastern shoreline, fringed with papyrus and secluded lagoons with splendid blue water lilies. The lake is little more than an hour's drive from Nairobi, using the Trans-Africa Highway.

At present the lake covers about 150 square kilometres (60 square miles) since recent rains have been good. The water level has varied markedly from year to year, having almost dried out in the 1890s. The hectarage and the fortunes of the lakeside vegetable farmers flutuate widely.

The ecology of the lake has been changed considerably by human intervention. Sport fishing was introduced in the 1920s and later, species like the American red swamp crayfish and black bass were introduced both for commercial and sport fishing. Various aquatic plants were also introduced, the most prominent being water hyacynth, which forms thick carpets of vegetation and can become a serious problem to waterways. The South American coypu, an aquatic rodent, is also present in the park, having escaped from a fur farm. The area surrounding the lake is extensively irrigated to grow fruits and vegetables. Controversy about the detrimental effects of these introductions and the irrigation schemes has been going on for years. No final conclusion has been reached.

White pelicans on a fishing expedition.

Crescent Island: A little crater lake at the bottom of a small volcano at the western side of Lake Naivasha and the wildlife sanctuary on **Crescent Island** can be visited. Crescent Island is a peninsula or island, depending on the lake level, joined to the mainland by a causeway. In 1988, the lake level dropped sufficiently to allow Crescent Island to become part of the mainland.

On the eastern end of the island sailing is possible at a private boating club. Sunday regattas are an incongruous sight on the bottom of the Rift Valley and the view of the Rift Valley walls from the lake is an altogether exhilarating experience.

Among the resident birds are fish eagles, ospreys, lily-trotters, black crakes and a variety of herons. Hippo also live in the lake. A number of mammals can be seen grazing in the surrounding lake environs, such as zebra, impala, buffalo, giraffe, Kongoni and, at night, hippos.

Lake Naivasha is 80 kilometres (50 miles) from Nairobi on the main Nairobi-Kisumu road. The old road, tracking an ancient elephant trail, snakes down the eastern Rift Valley Wall Escarpment and is to be avoided except by the bravest: it has been relegated to the lorry transport category. The newer road skirts along the top of the escarpment from **Limuru**, only dropping down into the Rift just south of Naivasha. From either, there is a magnificent view of Lake Naivasha and the extinct volcanoes, Suswa and Longonot, in the valley bottom.

Accommodation is available at **Lake Naivasha Hotel** or **Safariland Lodge**. The resplendent traditional Kenya Sunday Lunch at Lake Naivasha Hotel is recommended, even if you are not staying there. The view over the lake, from the well-manicured lawn, in the shade of yellow-barked fever trees, will not be soon forgotten.

There are a series of campsites on the southern side of the lake, probably the best known being **Fisherman's Camp**, where *bandas* can be rented or you can pitch your own tent. *Bandas* are also

Mount Longonot overlooks Lake Naivasha.

available at **YMCA Camp** and at **Top Camp**. If you have camping gear, Safariland Lodge also has a campsite.

Going to Hell: Hell's Gate National Park, a dramatically beautiful slice through the volcanic ridge south of Lake Naivasha, has only been recently created. It lies some 13 kilometres (eight miles) south-east of Naivasha and is about 68 square kilometres (27 square miles) in area.

The park is an impressive gorge with towering cliffs. Close to the entrance is **Fisher's Tower**, a lone 25-metre- (82-foot-) high rock. Powerful geysers, which gave the park its name, have been harnessed with foreign aid to generate electricity. The geothermal electricity project has been carefully executed so that it does not affect the beauty of the park.

Among the birds to be seen are a colony of Ruppell's vulture and a pair of resident lammergeyers that breed on the cliffs. The lammergeyers have developed the habit of scavenging bones, flying them to considerable heights and dropping them onto rocks to crack open and reveal the bone marrow. There are several "dropping points" in Hell's Gate. There are also Verreaux's eagles, the largest of the East African eagles, invariably seen soaring in pairs, and many other notable birds of prey. Secretary birds have taken up residence in a low acacia tree near the track which cuts up through the gorge. Mammals found in the park include Thomson's gazelle, antelope, zebra, hyrax, cheetah and leopard.

Camping is the only available accommodation. It might be wise to enter into a private arrangement with a local Maasai warrior to guard your vehicle for the night.

LONGONOT NATIONAL RESERVE

Longonot National Reserve encompasses an extinct volcanic crater, protruding from the floor of the Great Rift Valley and towering impressively over the southern side of Lake Naivasha. It is very visible from the Nairobi-Naivasha road. In all, Longonot Reserve covers

Hippos abound in lake waters.

152 square kilometres (61 square miles). **Mount Longonot** can be climbed on foot from various access points.

Plains wildlife can be seen on the lower slopes, and various other animals, including leopards, are occasionally found in the crater. As might be expected from its variable altitude and vegetation, it is a haven for birds. On the middle-to-lower slopes, stands of leleshwa grow like weeds and are relatively impervious to the frequent grass fires which rage up from the plains in the dry season. The wood of leleshwa is extremely hard and good for carving in the manner of briar: the fragrant, light green and slightly hairy leaves are used for personal hygiene by trekking *Maasai* who tuck a bunch of them under each arm.

LAKE BOGORIA

Lake Bogoria National Reserve was gazetted to protect the herds of greater kudu which live mainly on the western slopes of the **Laikipia Escarpment**, which towers over the lake to the east. The reserve covers approximately 110 square kilometres (44 square kilometres) and includes the shallow soda lake which attracts huge flocks of flamingoes. Bogoria was formerly known as Lake Hannington during the colonial era. It was named after the missionary bishop who was murdered in Uganda.

Hot rocks: Hot springs and spectacular spluttering steam jets are one of the main attractions and are located approximately three-quarters of the way down the western lake shore heading south. The water in the springs is boiling hot and spouts up from subterranean aquifers surrounded by magma heated rock. Extreme caution should be exercised in walking around the springs. Some of the apparently solid ground is merely a crust on top of extremely hot mud. The sides of the larger springs are treacherously slippery, and more than one unfortunate soul has died after complications from scalding. If you get a bit hungry there is often a helpful fellow who, for a few shillings, will arrange to

Secretary Birds live on a diet of snakes and other reptiles.

have fresh maize boiled to a turn on the end of a pole in the springs.

The entire lake is fed by hot, sulphur rich springs which make swimming impossible but have created an environment of particular interest for the geomorphologist, ornithologist, or those interested in the special adaptations of fish. Even the layman will marvel at the primaeval scenery.

Lake Bogoria (formerly Lake Hannington) is 64 kilometres (40 miles) north of Nakuru and can be reached either from the south by taking a turning to the right about 38 kilometres (24 miles) from Nakuru or by continuing along the tarmac road until near to **Marigat** and then turning right at the sign post for **Loboi Gate**. The approach from the south gives a spectacular view of the lake which can suddenly be seen after a bend in the road. However, only four-wheel drive vehicles should be taken and even then progress is very slow, particularly along the lava flows near the entrance to the reserve. The northern road is good and can be negotiated by any saloon car up to the hot springs.

The only accommodation is **Acacia**, **Fig Tree** and **Riverside campsites** at the southern end of the lake. No facilities are available. All necessary water has to be brought along since the water in the lake is not drinkable. There is also a campsite just outside the northern entrance to the reserve, where water is available.

LAKE BARINGO

Lake Baringo lies 15 kilometres (nine miles) north of the little town of Marigat. It is a freshwater lake, twice the size of Lake Naivasha. It is home to great numbers of birds and hippos, which can be seen in the evenings grazing at the shoreside. Lake Baringo does not enjoy any particular protected status, except for the respect of the surrounding population for the value of its fish and the employment generated by tourism.

The lake has noticeably more sediment than other rift valley lakes, which

Green surroundings of Lake Naivasha.

can be seen from its brownish colour viewed from the shoreline or detected in the light it reflects to earth resources satellites. It has probably always been like this although much development aid is currently being spent to forestall erosion. The lake level fluctuates from year to year, sometimes by several metres, depending on rainfall in the surrounding hills.

The colony of nesting goliath herons on **Gibraltar Island** attracts many ornithologists. It is also possible to see Verreaux's eagles, bristle-crowned starlings and Hemprich's hornbill on the escarpment on the western side of the lake. Many other species of birds live in the acacia woodland bordering the lake, including west Nile red bishops and silverbirds.

First class accommodation is available either at the **Lake Baringo Club** or at **Island Camp**, on **Ol Kokwa Island**. The Lake Baringo Club offers a number of local excursions, including guided birdwatch walks, boat trips, water-skiing, and, in contrast, camel rides. Island Camp is a luxury tented lodge with swimming pool and water sport facilities. Camping facilities, as well as *bandas* are available at **Robert's Camp** on the lakeshore. Showers and toilets are provided.

LAKE MAGADI

Soda lake: The shallow, highly alkaline **Lake Magadi** is the southernmost of the Rift Valley lakes in Kenya. It lies in a semi-desert region where temperatures soar to over 40 C (104 F) during midday. A vigorous commercial enterprise which extracts sodium and calcium salts is supported by evaporation pans excavated into the lake, which is almost poisonously rich in salts because it has no external drainage. All of the rainfall runoff from the surrounding countryside brings with it dissolved minerals and ends up in Magadi (which, not surprizingly, means "soda" in *Maa*, the Maasai language). Since there is no outlet to the lake, the searing heat and fierce sun evaporates much of the water, leaving a concentrated salt solution. This is spread out over some 100 square kilometres (40 square miles) and the lake bed appears to be one enormous sheet of white.

A number of hot springs can be found around the periphery of lake, the most accessible being to the south. Water birds are abundant, most notably the chestnut-banded sand plover which, in Kenya, can be found only at this lake. Flamingoes are usually very prominent. The southern shore is one of the sites in Kenya of annual gatherings of ornithologists who, during the northern hemisphere winter, set up skeins of fine "mist nets" to capture and ring-band some of the hundreds of thousands of birds which have migrated from the southern reaches of Europe to spend winter in East Africa. As you watch birds being extracted by expert hands from the tangles of a mist net, it is astounding to realise that their last port of call was quite likely near the southern steppes of Russia.

The trip to Lake Magadi is stark, beautiful and, thanks to the Magadi Soda Company, tarmacked all the way from the northern edge of the Ngong Hills. The road drops through a series of spectacular step faults down the eastern wall of the Rift Valley, from 2,100 metres (6,894 feet) at the Ngong Hills to 609 metres (2,000 feet) at the lake. Wildlife such as gerenuk and giraffe and fringe-eared oryx can be seen on the way.

Site for sore eyes: On the way to Magadi, it is worth stopping off for a look at **Olookisaili**, one of Kenya's most important archaeological excavation sites, on the shores of an ancient Pleistocene lake. There, *Homo habilis* hunted a fauna much richer than today's, and *Homo sapiens* may spend a rustic night in the self-help *bandas*. Arrangements must be made beforehand at the National Museums of Kenya office in Nairobi.

At Magadi itself, there is no accommodation available, unless you are fortunate enough to be invited to the Soda Company's Club. However, camping is allowed although no facilities are provided. It is essential to bring your own water and other provisions.

Right, soda shores of Lake Magadi.

Lake Turkana

Lake Turkana, known from Count Teleiki's first exclamation as the Jade Sea, is so far off the beaten track that it gives the impression that time has stood still. The surrounding scrub desert enhances the colour of the lake which looks even more impressive because there is no contrasting greenery. It is set in the midst of volcanic formations and dry rivers known as "luggas" and is the northern-most lake in the Kenyan part of the Great Rift Valley, where temperatures can reach up to 50 C (122 F). From here, our earliest ancestors probably started organised hunting, harvesting and society. Today, the ethnic peoples inhabiting this region are nomadic pastoralists who seem hardly affected by modern life.

The largest population of Nile crocodiles in Kenya lives on the shores of the lake and a population of the largest freshwater fish in East Africa, the Nile perch, inhabits the lake and can grow to over 100 kilograms (220 pounds), much to the joy of sport fishermen.

The area is rich in birdlife. In (northern hemisphere) spring, black-tailed godwits and spotted redshanks can be seen in full breeding plumage. European migrants use this area as a stopover on their way north. Birds of prey are also abundant.

Ferguson's Gulf is 64 kilometres (40 miles) north of **Lodwar**. It presents some of the best fishing opportunities in the country. Accommodation can be found at **Lake Turkana Fishing Lodge**, which is reached by taking a boat from the lake shore across Ferguson's Gulf. The lodge is comfortable, with basic facilities. It even has a swimming pool filled with lakewater, which has an unmistakably soapy feel. Nothing harmful, except crocodiles, could survive in that! The food is good and the menu consists mainly of lake fish.

The gulf itself was quite dry during 1989 but it will certainly fill again as the

Central Island and its crater lakes.

lake level rises with the next good period of rains. As you drive to the jetty from the airstrip or the end of the tarmac road from Nairobi, the hulk of a beached Norwegian fishing boat can be seen, a monument to a development aid scheme gone wrong. The *Turkana* are adaptable people and are able to survive in such a harsh environment by embracing fishing as well as herding. They also enjoy a deserved reputation as fearless watchmen.

The **Cherangani Hills** to the south and west of Ferguson's Gulf are one of the great unknown trekking areas in the world. The view from the ridge tops is magnificent yet has only been seen by a handful of hearty walkers.

Central Island, in the middle of Lake Turkana, can be reached by boat from Ferguson's Gulf. The island's main crater lake is a nesting point for an extremely large number of crocodiles and many water birds. Once on the island, the rest of the trip has to be undertaken on foot, which can be very strenuous in the heat of the day. Until recently, the island was populated more by immigrant *Luo* fishermen from Lake Victoria than by crocodiles.

South Island is the birdwatcher's and Nile perch fisherman's paradise of Lake Turkana. Care must be exercised if you are camping rough since the ever-present large population of crocodiles can be dangerous. The airstrip on the island is a twisted, rock-strewn horror and a challenge for every bush pilot in Kenya!

Adjacent to the eastern shores of Lake Turkana, **Sibiloi National Park** covers 2,500 square kilometres (1,000 square miles) of barren, semi-arid bushland. Richard Leakey has found many important fossils of early man and animals in this ancient location and a fascinating archaeological museum has been opened at **Koobi Fora**. Grevy's zebra and beisa oryx can be seen. There is no water here so come prepared if venturing into this area by car. Access is mostly by plane, unless you have at least a week and access to a sturdy four-wheel drive vehicle.

Far right, another regal sunset.

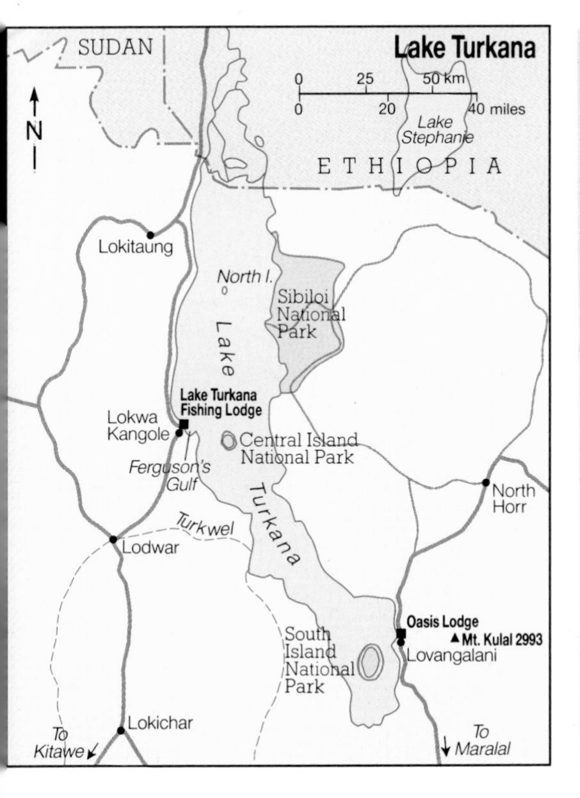

Mount Elgon National Park is famous for its forest which has gigantic podocarpus trees and impressive stands of juniper and elgon olive. The mountain, like most in Kenya, is a long-extinct volcano. The park covers approximately 170 square kilometres (68 square miles). It lies between 2,500 and 4,300 metres (8,200 and 14,100 feet) up the side of the mountain, the summit of which is in Uganda. Despite its small size, the park has a variety of habitats, from savannah to woodlands and mountain forest to alpine moorlands.

Salt caves: In the side of the mountain are vast caves, the most famous of which is **Kitum**, into which elephants have walked for thousands of years. There is good reason to believe that elephants dug the caves themselves since they enter them every two or three days to get salt. There, in complete darkness, nearly 200 metres (660 feet) into the side of the mountain, they use their tusks to gouge mineral salts out of the cave walls. This remarkable population has learned to move about in the dark *en masse*, with mothers guiding their young past dangerous holes and crevasses using their trunks. Like nearly all other elephants in East Africa, they have been heavily poached in the last few years and they are now under serious threat of complete extinction, despite the valiant efforts of Kenya's Wildlife Conservation and Management Department and the Anti-Poaching Unit.

Colobus monkey, leopard, golden cat, black-fronted duiker, elephant and buffalo live in the rain forest and on the moorland. The park also has a wealth of afro-alpine flowers.

The entrance to the park is 27 kilometres (17 miles) from **Kitale** along the Endebess road. Although there is no accommodation in the park, one can stay at **Mount Elgon Lodge** near the entrance. The lodge is a converted farm

Comfortable Mount Elgon Lodge.

house with a beautiful view. Alternatively, one can stay at **Kitale Hotel** or at **Lokitela Guest House**, run by the Mills family on their farm (19 kilometres/12 miles) west of Kitale.

It is also possible to climb Mount Elgon. For more information contact the Mountain Club of Kenya.

KAKAMEGA FOREST

Kakamega Forest National Reserve was gazetted in 1985, but the status of the park which was scheduled for gazetting two years earlier is still unclear. The 97 square kilometres (39 square miles) comprise the most eastern area of the Congo Central African rain forest, to the east of Kakamega, and is known for its birdlife, mammals and vegetation. Many of the birds and mammals that occur here cannot be found anywhere else in Kenya. Characteristic species include grey parrots, great blue turacos, Ross's turacos, red-chested owlets and African broadbills, wattle-eyes and many others. The Angolan black and white colobus may also be found, although in smaller numbers.

LAMBWE VALLEY

Lambwe Valley National Park covers 120 square kilometres (48 square miles) of tall grasslands mixed with woodlands and acacia thickets in **South Nyanza** near **Homa Bay**, Lake Victoria. The reserve was created to preserve the lovely roan antelope, a small population of which are resident. Other mammals include oribi and Jackson's hartebeest. Giraffes, zebra and ostrich have been introduced from Samburu to help reduce the dense vegetation. Birds are abundant in this area and can be easily observed.

Lambwe is in one of the most highly populated areas of Kenya, and has long been a centre of controversy. Tsetse flies abound and it is considered a reservoir for human and animal sleeping sickness. The park was subjected to large amounts of DDT spraying until the practice was stopped by conservationist movements. Nonetheless, the will of the people prevailed and the park was de-gazetted in 1989.

SAIWA SWAMP

Saiwa Swamp National Park is the smallest in Kenya, covering an area of only 1.9 square kilometres (0.76 square miles). It encloses the fringing belt of rain forest and the swamps fed by the **Saiwa River**.

Saving sitatungas: Saiwa was created specially to protect the sitatunga antelopes (about 80 to 100 in number) which live there. They spend most of the day half-submerged and hidden amid the floating vegetation. The hooves of this antelope are enormously elongated to about 18 centimetres (seven inches) which enable the animal to support its weight on the marshy ground. Camouflaged platforms have been built to allow visitors to see these rare animals. Another rare species is the Brazza's monkey, of which some 30 of the 200 left in Kenya live in the park. Other animals include spotted-necked otter, nocturnal potto and giant forest squirrel. The park lies 15 kilometres (nine miles) north of Kitale to the east of the main road to Lake Turkana.

Tusk marks in a salt cave.

COASTAL PARKS

Shimba Hills National Park is situated 56 kilometres (35 miles) south of **Mombasa** and consists of an area of 310 square kilometres (124 square miles), encompassing a line of gently rolling hills (rising up to 450 metres/ 1,500 feet) which run more or less parallel to the Kenya coastline. For sheer tranquility and lovely coastal views it is unbeatable. There are areas of open glades alternating with impressive stands of coastal rain forest. Herds of roan and spectacular sable antelope (occurring nowhere else in the country) can be found grazing here.

Fowl town: Lone bull buffalo can be seen and there is a resident herd which spends most of midday sheltering under the shade of one or two trees, normally in the area of the landing strip. There are signs of elephant everywhere and they can be seen going to drink in the mornings and evenings. They sometimes even wander into the outskirts of **Kwale** town. However, most of the time the herd are deep in the seemingly impenetrable forest in the middle of the park. Forest loving red duiker, bushbuck, waterbuck and grey duiker are also found. Lions and leopards are also present, although not in abundance. Birdlife, however, is profuse. Keep an eye out for the spur fowl—*kwale* in *Swahili*—from which the town takes its name.

An easy drive from any of the beach resorts south of Mombasa, the Shimba Hills offer a cool change from the humidity of the coastal strip. There are stunning views of the ocean andTanzania's Usambara and Pare mountains.

There is a new lodge just outside the park gate and self-help thatched *bandas* are located on the edge of an escarpment. Bring food and bedding for these. Otherwise, there's plenty of beachfront accommodation between the townships of **Tiwi** and **Diani**, a 21-kilometre (13-mile) stretch south of Mombasa.

This sable antelope has perfectly proportioned horns.

MALINDI-WATAMU

These parks are situated on the coast by **Malindi**, and cover an area of beaches from the high tide line and coral reefs out to the edge of the continental shelf. The coral gardens abound with mostly unspoilt underwater wildlife. Snorkeling and scuba diving are the best way to see the myriad aquatic flora and fauna. The fish are habituated to people and will eat out of your hand.

There is perennial concern that the Malindi reefs are under threat from siltation from the effluent of the **Sabalki River** just north of Malindi town. The coral polyps are unable to feed and breath under the layer of accumulated silt. There are numerous luxury hotels at Malindi and **Watamu**, and these give easy access on a daily basis.

MIDA CREEK

Mida Creek is located on the north coast of Kenya, a few kilometres south of **Gedi**, an early Islamic coastal city that was abandoned in the 17th century for no apparent reason. The ruins are well kept and worth visiting.

Mudflats and mangroves: The creek is a series of tidal mudflats surrounded by mangrove trees. It is known for the numbers of waders that stop there between March and May on their way north during the annual migration. Birds seen include sanderlings, turnstones, curlew sandpipers, greenshanks, little stints, wimbrels, Terek sandpipers and grey plovers. Ospreys, several species of terns and the rare, non-migrant crab plover also inhabit the creek. In the mangroves flocks of the brilliant carmine bee-eaters flash in the sunlight.

Snorkeling is possible for the more daring. When the tide is right you can enter the creek at park headquarters and float down and back up the dark, murky water with the current. Look out for the sea carp weighing up to to 182 kilograms (400 pounds) that lurk in the caves.

The creek is reached via the Mombasa-Malindi road. Turn right 35 kilometres (22 miles) after the Kilifi ferry.

A slow day in Shimba.

KISITE-MPUNGE

Established in 1978 as Kenya's second marine reserve, this park encompasses the outer and inner **Mpunguti Islands**, and the sand bar of **Kisite**, south of **Wasini** at **Shimoni**. The water is crystal clear and ideal for snorkeling. The only access is by boat from Shimoni which takes 90 minutes. *Shimo* in *Swahili* means hole—named after the large cave that was used as a holding tank for slaves before shipment. There is excellent deep sea fishing here.

ARABUKU-SOKOKE

The **Arabuku-Sokoke Forest Reserve** runs parallel to the coastline between **Kilifi** and Gedi for a stretch of about 360 square kilometres (144 square miles). Tightly surrounded by farmland, it is the last remaining extensive patch of brachestygia woodland and lowland coastal rain forest of azeleas left in Kenya.

Rare mammals: The forest is home to a number of interesting and rare mammals including Zanzibar duiker, Ader's duiker, the bristle-tailed and the yellow-rumped elephant shrew. All are difficult to see because of their shy nature and small size. Elephant shrews, with their elephantine snouts, are the largest members of this peculiar African family, measuring 50 centimetres (20 inches) long.

A number of rare and common local birds are also found in this forest. These include the Sokoke scops owl, thick-billed cuckoo, Retz helmet shrike, Amani sunbird, African pitta (rare), Fisher's turaco, Sokoke pipit and southern-banded harrier eagles. Clarke's weavers occur in flocks of 100 or more but their nests have never been found. Another memorable sight are the thousands of butterflies that drink along the pools near the forest tracks.

Take the Mombasa-Malindi road to reach the forest. Some 20 kilometres (12 miles) after crossing the gorge with the Kilifi ferry, the forest can be seen on the left. There are several access routes, most of which require four-wheel drive vehicles.

The popular coastal town of Malindi.

TO THE EAST

Remote and undeveloped, **Arawale National Reserve** consists of a 533-square-kilometre (213-square-mile) triangle adjacent to the eastern banks of the **Tana River** in the north-eastern part of Kenya. It was created in 1974 specifically for the protection of Hunter's hartebeest or hirole, a rarely-seen antelope with lyre-shaped horns similar to the impala's and a white stripe across the forehead between the eyes. Small herds comprising a territorial males accompanied by his females, tend to favour the open grassland patches between the more dense bushland. Grevy's zebra and lesser kudu are also found, as are herds of buffalo and rapidly diminishing numbers of elephants.

Hunter's hartebeest is only found in this part of Kenya and although potentially threatened because of range restriction and small numbers, they seem to be holding their own. They are under little threat, since there is not much competetion with man for living space in this remote part of the world and there has been no official hunting allowed in Kenya since the blanket ban on hunting wildlife imposed in 1977. Although the range of Hunter's hartebeest probably extends eastward into southern Somalia, the species has never been reliably recorded on the western banks of the Tana. There are no accommodation or camping facilities in Arawale.

TANA RIVER PRIMATE RESERVE

Halfway between the Arawale Reserve and the mouth of the Tana River at the Indian Ocean, some 130 kilometers (81 miles) north of Malindi, lies the **Tana River Primate Reserve**. This is a small protected area of riverine forest covering 169 square kilometres (68 square miles) which was gazetted in 1976 specifically to protect the world's only population of one of the four subspecies of crested mangabey monkey, the endangered Tana mangabey (*Cer-*

Rich colours of Ross's turaco.

cocebus galeritus galeritus). There are two subspecies which live in the high forests of Zaire, about which almost nothing is known, and another endangered subspecies in the Uzungwa mountains of Tanzania, the Sanje crested mangabey, which was only discovered by scientists in 1981.

The Tana mangabey is at serious risk and, despite protection in the reserve, its population has dropped from around 2,000 in the early 1970s to between 800 and 1,000 today. The major threats to its narrow belt of riverine habitat include clearing for agriculture, felling of large trees for making traditional dugout canoes, annual burning of the flood plain grasses which is inexorably eroding away the forest edge, and natural changes to the river course in this relatively flat, flood plain area. The Tana mangabey is therefore included in the most threatened category of the 1973 International Convention on Trade in Endangered Species of Wild Fauna and Flora (CITES), which makes it difficult, but not impossible to export this rare beast. However, habitat changes due to the inevitable expansion of human populations rather than trade are likely to bring about its demise in the near future.

There is also a small group of threatened red colobus monkey as well as red river hog and numerous hippos, crocodiles and other riverine animals. Birdlife is prolific.

Visitors must either camp or attempt to book at **Baomo Lodge** (often closed) to the north of the reserve.

BONI, DODORI AND KIUNGA

Boni National Reserve is situated on Kenya's eastern border with Somalia, with one corner just touching the Indian Ocean, in 1,339 square kilometres (535 square miles) of undeveloped wilderness. It is the only area in Kenya with coastal lowland groundwater forest. It was set aside partly as a buffer, and partly as a refuge for large populations of coastal topi and elephants so that the latter would journey down through neighbouring **Dodori National Reserve** to the seashore, often wading out at low tide to "Elephant Island".

Dodori National Reserve is 877 square kilometres (350 square miles) in area, and is situated just north of **Lamu**. It almost joins on to Boni National Reserve to the north and **Kiunga Marine National Reserve** adjacent to the coast. Dodori was established in 1976 to provide protection of major breeding grounds for the local topi population.

Elephants are still found in small numbers as are lesser kudu. Birdlife is rich, with many waterbirds, birds of prey and nothern migrants. Dugongs, or seacows—the original mermaids of seafaring lore—used to breed along the coast, but their numbers are rapidly declining due to encounters with local fishermen.

With 250 square kilometres (100 square miles) of islands, beaches and coral reefs in north-east Kenya by the Somali border, the Kiunga Marine National Reserve is a prime sea-bird breeding ground, most notably for Hemprich's gull which nests between July and October.

Right, deserted beach on the north coast.

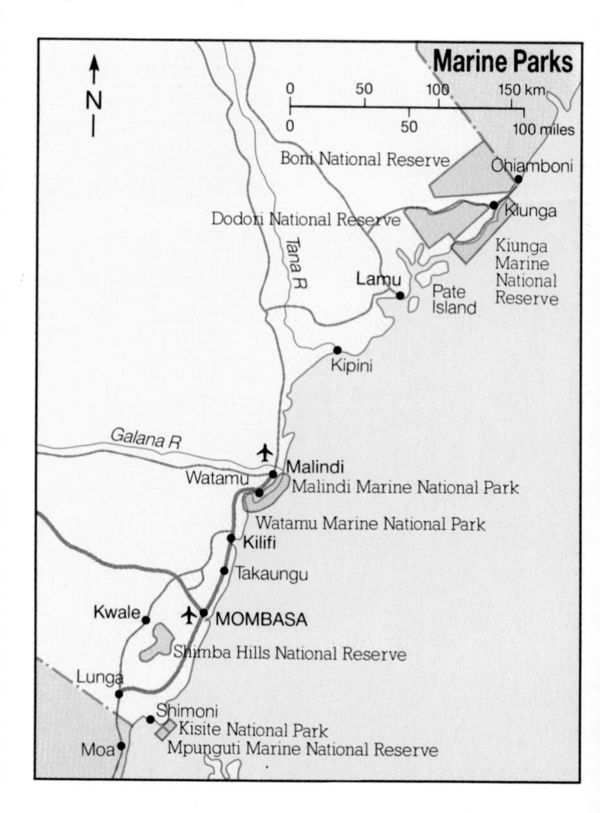

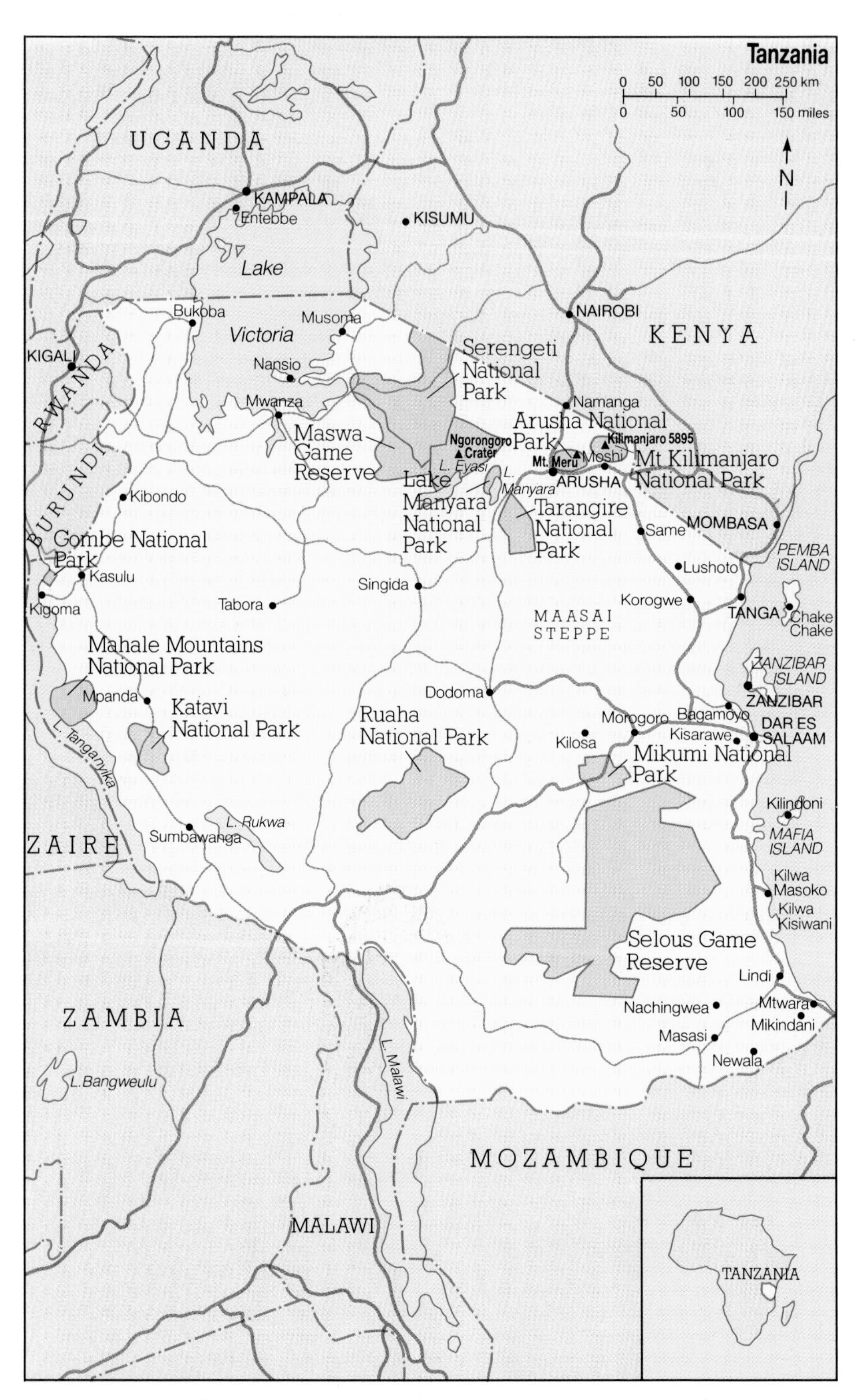
Tanzania
0 50 100 150 200 250 km
0 50 100 150 miles
N
UGANDA
KAMPALA
Entebbe
KISUMU
Lake
Victoria
Bukoba
Musoma
Nansio
Mwanza
KIGALI
RWANDA
BURUNDI
NAIROBI
KENYA
Serengeti National Park
Namanga
Arusha National Park
Maswa Game Reserve
Ngorongoro Crater
Kilimanjaro 5895
Mt. Meru
Moshi
ARUSHA
Mt Kilimanjaro National Park
L. Eyasi
L. Manyara
Lake Manyara National Park
Tarangire National Park
Kibondo
Gombe National Park
Kasulu
Kigoma
Tabora
Singida
Same
MOMBASA
PEMBA ISLAND
Lushoto
Korogwe
TANGA
Chake Chake
MAASAI STEPPE
Mahale Mountains National Park
Mpanda
Katavi National Park
ZANZIBAR ISLAND
ZANZIBAR
Dodoma
Ruaha National Park
Morogoro
Bagamoyo
DAR ES SALAAM
Kisarawe
Kilosa
Mikumi National Park
L. Tanganyika
Kilindoni
MAFIA ISLAND
L. Rukwa
Sumbawanga
ZAIRE
Kilwa Masoko
Kilwa Kisiwani
Selous Game Reserve
Lindi
Nachingwea
Mtwara
Mikindani
Masasi
Newala
ZAMBIA
L. Bangweulu
L. Malawi
MOZAMBIQUE
MALAWI
TANZANIA

TANZANIA

Situated just south of the equator, Tanzania stretches from the Indian Ocean westward across more than 1,000 kilometres (620 miles) to Lake Tanganyika, which marks the boundary with Zaire and the Congo forests. Lake Victoria indents Tanzania's northern frontier and Lake Malawi the southern. Some 800 kilometres (496 miles) of unspoiled coastline strewn with islands contrast with the vast interior where Africa's highest mountain, Kilimanjaro, stands along with many other spectacular mountains amid enormous stretches of woodland, bush, savannah and marsh. Cleaving much of the country is the Great Rift Valley. The combination of scenery, magnificent wildlife and luxurious lodges makes the safari experience in Tanzania an exquisite one.

Tanzania is a republic made up of Tanganyika, the mainland, and Zanzibar Island. The total area of the nation is about 380,000 square kilometres (147,820 square miles) and the population in 1988 was about 22 million. People are concentrated in the coastal areas, around Lake Victoria and on the fertile slopes of the volcanoes in the north. There are over one million people in the capital city, Dar es Salaam ("Peaceful Harbour"), but only a few hundred thousand in any of the secondary towns: the majority live in rural areas.

Tanzania's history as a nation started in 1961 when Tanganyika, the mainland, became the first colony in East Africa to receive independence from Britain. Zanzibar became independent in 1963 and the two nations joined to become the United Republic of Tanzania in 1964. Julius Nyerere, from the Lake Victoria area, was the first president until he retired in 1985 to become the chairman of the political party, Chama cha Mapinduzi. His successor is Ali Hassan Mwinyi, from Zanzibar.

Tanzania's unrecorded history covers a span of millions of years. Some visible traces remain at prehistoric and stone age sites all over the country. At Laetoli, are the ancient footprints of human ancestors preserved in ash on the slopes of volcanoes making up the Ngorongoro Highlands. The stones and bones of our ancestors have been left at Olduvai Gorge and fascinating rock paintings at Kondoa and other sites testify to the artistic skills and abilities of ancient hunters and gatherers.

<u>Preceding pages:</u> baobab trees can have trunks of up to 10 metres (33 feet) in diameter. These fill up the landscape in Tarangire National Park.

SERENGETI

With its wide open spaces, bright blue skies and creature-filled landscape, the fabled **Serengeti** typifies everyone's version of "dream Africa". It is one of the last places in the world where vast numbers of large animals can be seen in their natural habitat.

Serengeti was established in 1951. It is Tanzania's largest and oldest national park. Its rolling plains, bordered with ranges of hills and sprinkled with rocky outcrops, spread out over almost 14,500 square kilometres (5,790 square miles). Reflecting the varied terrain is the park's annual rainfall, which ranges from about 500 mm (19.5 inches) in the south-west to about 1,200 mm (46.8 inches) in the north and west.

In spite of its fame, Serengeti has not been ruined by tourism. Roads into the park can be rough going and local tracks are not much smoother. Accommodation is comfortable but not luxurious, there are no balloon rides, night safaris, foot safaris or fashionable camps. Visitors and staff are expected to adapt to the environment, instead of changing the environment to cater to human whim. It is this lack of human influence which gives the park its charm.

Serengeti can be visited in either dry or wet season. During the rains from November to May, the annual wildebeest migration takes place. When it is wettest and food is plentiful wildebeest spread far out on the plains, only retreating to the bush and woodland when the savannah dries out. When all the green grass is eaten they gather along the western edge of the plains then trickle north and west searching for food until they reach the Maasai Mara in the north where there is more pasture and permanent water. When the rains begin again in November, they stream back southwards to seek the fresh, nutritious grass that grows on the volcanic soils in the lee of the **Ngorongoro Highlands**.

During the dry months, from June to

Far left, this cheetah cub resembles a fur ball.

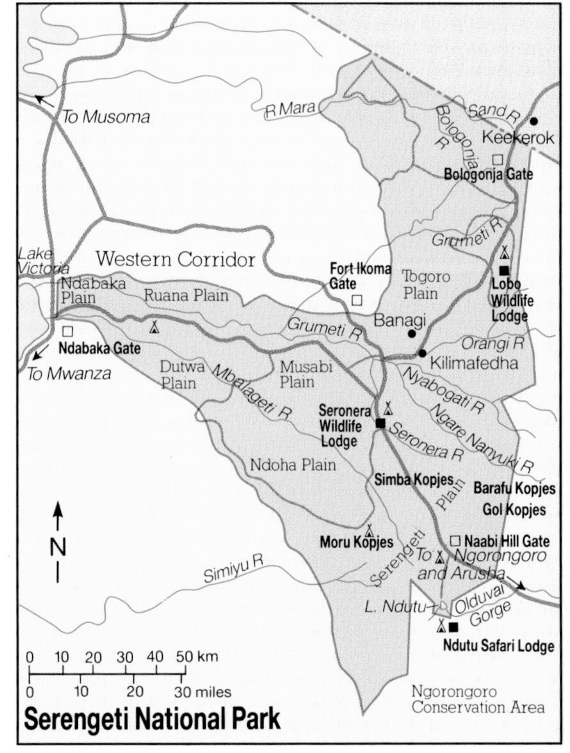

September, the wildebeest along with several hundred thousand zebra are massed on the short grass plains in the south of the park.

In February, most female wildebeest calve, so visitors may see a birth and the period of imprinting immediately after. This is the short time when mother and calf get to know one another intimately enough to survive most separations. The calf can run as fast as its mother within half an hour of being born, even before it has its first drink of milk. Some of the other ungulates giving birth at this time include zebras, gazelles and eland, although their birth season is somnewhat more extended.

But life and death are inseparable: vulnerable new-born calves attract predators, which in turn are accompanied by scavengers that clean up the carcasses left scattered on the plains. Only the strong survive.

The Serengeti is "dream Africa".

Gazelles, particularly Thomson's, are the first to take advantage of the new grass. Zebras, elands, topi and hartebeest move from the woodland edge out to the plains.

Some buffalo are tempted out along the watercourses. It is a myth that predators come out to follow their prey. Most predators are resident in a well-defined range and those that move out on to the plains most likely live adjacent all year round. So you won't find vast numbers of lions, hyaenas, jackals, cheetahs or wild dogs congregating with the vast herds. Hyaena, jackal or wild dog dens on the plains are special attractions and well worth a look; you might find these predators hunting or tending their young.

The rain in the plains: The plains are incredibly beautiful during the rains, with dramatic skies full of storm clouds. Birds arrive from Asia and Europe to take advantage of the many insects that are present then. The bird migration can be even more impressive than the wildebeests'. Thousands of European storks cover the plains, kestrels and harriers hover or swoop after rodents and dung beetles, caspian plovers from the Russian steppe fly in tight forma-

tion, European swallows, swifts and terns pursue flying insects and yellow wagtails catch grasshoppers.

When the plains dry up, their character changes completely. The unending green becomes a vast expanse of short, brown grass. The lack of trees becomes more obvious. *Kopjes* stick out more and seem more interesting because much life still clings to their water-filled crevices. Most of the large animals depart, leaving only a few hardy warthogs, Grant's gazelles, topi and hartebeest. Small creatures go underground; the plains look bare but lizards, mice and beetles are still hidden under rocks, in holes and grass clumps.

During the dry season animals retreat to places where there is permanent water. If you follow the many tracks around the rivers in the centre of the park at **Seronera**, or northwards to **Lobo**, or westwards to **Kirawira**, you will always find many animals along the rivers, in open grasslands and woodlands. The main residents are impalas, topis, buffalos, giraffes, waterbucks, reedbucks, bushbucks, dikdiks, warthogs, baboons and vervet monkeys. The smaller or shyer animals include mongooses, civets, genets, servals, caracals, leopards, bat-eared foxes, bushbabies, oribi, duikers, pangolins, aardvarks, aardwolves and many birds.

Serengeti has a few oddities. In the western corridor, which follows the **Grumeti River** to **Lake Victoria**, there are black and white colobus monkeys, more like those of Uganda and eastern Kenya than those found at Mount Kilimanjaro and Meru. On the open grasslands by the Grumeti are wide-ranging groups of orange-haired patas monkeys. There might still be a remnant group of roan antelope around **Banagi Hill**. Their presence here so far out of their normal range gives credence to the idea that in former times the vegetation of Serengeti was like that of southern Tanzania, which is composed of miombo woodland.

Near the end of the Grumeti River where it begins to enter the ancient

Seronera Lodge is designed to complement its surroundings.

floodplains of Lake Victoria there are Nile crocodiles up to four metres (13 feet) long. Small and medium crocodiles travel as far upstream as Seronera where they inhabit some of the pools in the **Seronera River**.

Studying Serengeti: Father and son Bernard and Michael Grzimek, were the first to make systematic studies of the Serengeti ecosystem. They flew their famous zebra striped airplane from Frankfurt to Tanzania in the 1950s and pioneered aerial census methods over the plains and woodlands. Thanks to their research there are records of the low numbers of wildebeest at that time. The Grzimeks also studied the grasslands and produced a best-selling book and subsequent movie, *Serengeti Shall Not Die*, which did much to make people aware that Serengeti existed. The book was later translated into 24 languages.

By the early 1960s Serengeti was well known enough to attract funding and cooperation in setting up a complete and well-equipped research facility near Seronera. In the mid-1960s, scientists at the Serengeti Research Institute (SRI) started work in earnest. An ecological monitoring programme began which systematically counted animals, measured rainfall and studied burning. Studies were started on lions, hyaenas, wildebeest, zebras, giraffes, impalas, buffalos, elephants, gazelles, topis, birds, vegetation and soils.

Martial eagle is poised for take-off.

Most of this first generation of scientists left by the mid-1970s and were followed by others who continued certain species studies or tackled new subjects such as mongooses. Much of what we now know about East African wildlife comes from this research on Serengeti animals. Scientific vigour, which declined in the 1970s, has now revived and SRI is again producing much important information on wildlife ecology. One of their most important findings was to recognise the crucial need to monitor rainfall, fire, vegetation and animal numbers. We are learning that the balance of nature only works like a see-saw in motion, always changing, never static. But the complexities of the ecosystem have only been glimpsed. We cannot hope to "manage" any wildlife area successfully without understanding how life there manages itself without human interference.

Serengeti is a unique living laboratory that continues to play a vital role in promoting understanding of wild animals and their environment.

The Musoma road leading from Arusha across the Rift Valley and over the shoulder of Ngorongoro carries on across the Serengeti plain through the park's centre at Seronera and leaves the park at the west by **Ikoma Gate**. This is the major route to and through Serengeti but there are two other gates: **Bologonja Gate** in the north at the border between Kenya and Tanzania and **Ndabaka Gate** in the west which almost touches on the main road between Musoma and Mwanza around the shore of Lake Victoria. The main road is passable all year round but other roads in the park are not maintained regularly and their condition should be checked before setting out.

TZ
9031

Ngorongoro Conservation Area

Ngorongoro Crater has been called the eighth wonder of the natural world. It has been designated a World Heritage Site and a Biosphere Reserve and the crater deserves every bit of its fame—yet it represents only about 3 percent of the entire **Ngorongoro Conservation Area** (NCA), which covers 8,280 square kilometres (3,196 square miles), stretching from **Lake Eyasi** in the Rift Valley north to the Serengeti Plains. The region contains at least seven extinct volcanoes and is probably one of the most varied terrains in East Africa. Altitude in the park varies from Lake Eyasi at 1,000 metres (3,280 feet) to the peak of **Lolmalasin** mountain at 3,640 metres (11,940 feet).

The Ngorongoro Highlands on their southern and eastern slopes are clothed in forest which captures water for all the farmland to the south and also supports Manyara's special ground water forest at the base of the rift. The massif of old volcanoes surrounding the Ngorongoro **caldera** includes several craters and mountains of beauty and importance.

The plains to the north are critical to the migration of wildebeest. Note that at least half of the Serengeti Plains are in Ngorongoro Conservation Area and another plain, the **Salei**, is protected only by the NCA. Annual rainfall ranges from around 300 mm (11.7 inches) in the dry regions such as Lake Eyasi to over 1,000 mm (39 inches) up in the highlands.

Archaeological sites: The NCA was set up in the late 1950s in response to complaints by the *Maasai* that they were being pushed out of their traditional grazing areas by the formation of Serengeti National Park, which at that time included Ngorongoro. The conservation area was set up as a unique test of the idea that humans who lead a pastoral life-style, utilizing livestock only for sustenance, should be able to co-exist with wildlife. The area was also to cater to tourists, protect the forest to the south and guard unique archaeological sites such as **Olduvai Gorge** and **Laetoli**. The Laetoli site has footprints of our hominid ancestors that are dated firmly at over 3.5 million years old.

At Olduvai many fossils and artifacts dating from the last two million years have been found embedded in rich deposits containing remains of contemporary animals and plants. A visit to the small and excellent museum at Olduvai gives one a chance to fathom the beginnings of mankind, within an area where the first humans are believed to have evolved.

There are other archaeological sites round the conservation area, some even in the crater itself. Humans have been living here for millions of years so it is fitting that the conservation area enables local tribes to live with the wildlife that has also evolved here.

Wildlife in the crater: Ngorongoro is not only a fascinating lesson in land use and the modern problems of how to reconcile human desires with nature, it is also one of the most picturesque set-

Left, encounter on the crater floor. Right, an ostrich egg for breakfast.

tings for viewing wildlife. All animals are so used to cars that you can get close enough both to study their behaviour and to take excellent photographs.

All the usual grassland dwelling animals live in the crater: wildebeest, zebras, Thomson's gazelles, Grant's gazelles, elands, hartebeest, buffalos, warthogs, ostriches, kori bustards, crowned cranes, and predators such as bat-eared foxes, jackals, hyaenas, lions and, rarely seen, cheetahs.

In the forest and swamps are some elephants, all males, perhaps because there is not enough food for the normally much larger female and calf groups. Also near the forests are a few waterbuck and bushbuck. Around the marshes and streams are reedbuck and serval cat, and birds—masses of ducks, geese, herons, ibis, plovers, widow and whydah birds. Flamingoes often provide a pink fringe to the little soda lake in the crater's centre, and you can drive close enough to see and photograph them well. Hippos can be seen at two exceptionally pretty sites: the pool at **Mandusi swamp** and the lake at **Ngoitokitok springs** where visitors can eat their picnic lunch in the shade of fig trees where starlings, weavers and fish eagles perch. You should, however, beware of the African kites which swoop with incredible accuracy to steal food.

There is another picnic site in the **Lerai Forest** (a *Maasai* word for yellow-barked acacia trees) where there is running water and vervet monkeys and large numbers of baboons. Leopards live along the **Munge River** and forested slopes all round, yet they are seldom seen. Rhinos also take refuge in the forest but are often in the open on the soda flats or along watercourses. Some mountain reedbuck live in the brush on the inner crater wall. Impala and giraffe can be seen only on the outer slopes, never in the crater, because there is not enough of their food to support a breeding population. Male elephants are only temporary feeders in the crater and leave from time to time to join the female groups which range the forest

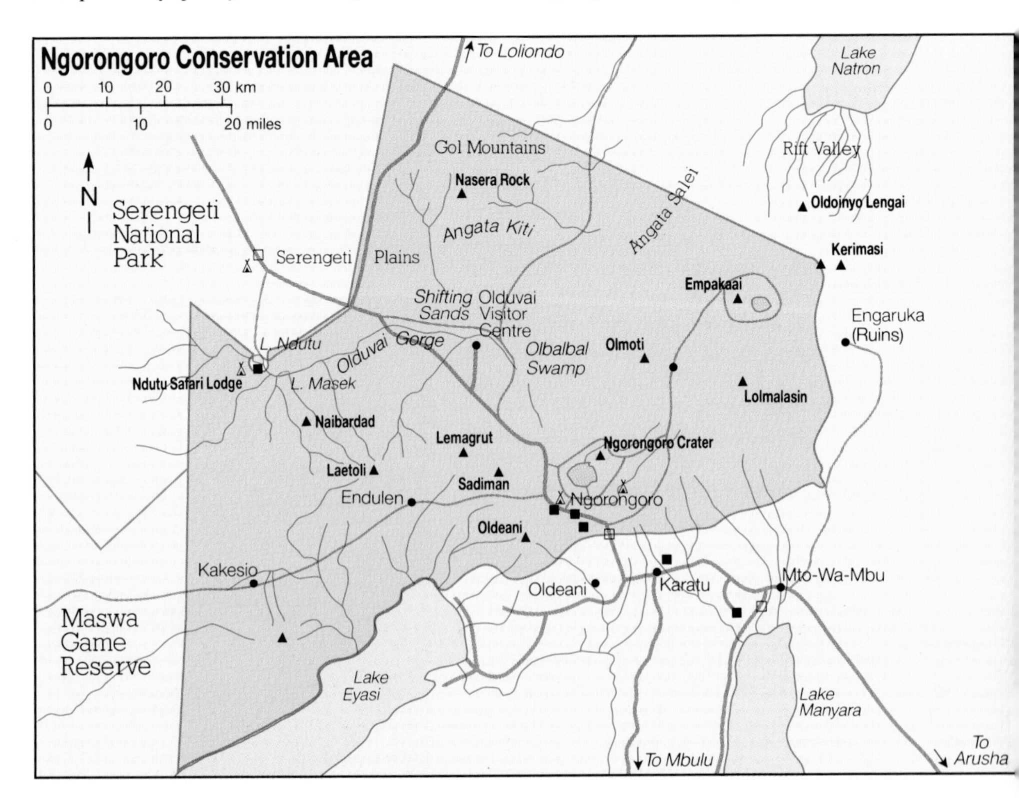

outside.

It is impossible to visit the crater without noticing all the wildflowers and the beautiful mantle of forest along the south. Even the dry, thorny northern side has picturesque plants. If you can reach it the remoteness and pristine beauty of **Empaakai Crater** and the lovely walk from **Olmoti's bowl** to climb down among mountain wildflowers is worth the effort. From here you can see the stark cone of the still active volcano **Oldoinyo Lengai** at the north-east corner of the conservation area.

Granite cliffs: The granite outcrops and the ancient crystalline **Gol Mountains** with their great gorges are reached by crossing bumpy plains strewn with windswept sand dunes. Here, sparkling pink granite cliffs provide ancient nesting sites for griffon vultures. Right at the northern bases of the great volcanoes lie the archaeological sites of Olduvai and Laetoli. In the western end of Olduvai Gorge, thick acacia woodland surrounds **Ndutu**, with its precious lakes providing water for the animals there and also those from the surrounding savannah.

The Serengeti and the Salei plains are the location of the annual migrations of wildebeest, zebras, gazelles, elands, ostriches, and other birds. The vast green sward dotted with kopjes is a fitting scene for the births and deaths of great numbers of animals.

So Ngorongoro means more than just the famous crater. With its streams and lakes, forests, open grasslands, bushy slopes and wildlife, NCA is a microcosm of East Africa and its historical significance for humankind makes it unique among game parks.

The main road between Arusha and Lake Victoria cuts right through the conservation area, running around the west rim of Ngorongoro Crater. It is gravel surfaced and seldom muddy enough to make the crater inaccessible; however, all roads in the conservation area vary greatly in their condition, especially during the heavy rains of March to May. Go with a good driver and check first at park headquarters.

Bat-eared foxes play peek-a-boo.

KILIMANJARO

Kilimanjaro is a huge mountain but the park that protects its upper reaches is really quite small, only 756 square kilometres (292 square miles). It was established in 1973 and ranges in altitude from 1,000 metres (3,280 feet) to 5,895 metres (19,340 feet) at Kibo peak, the highest point in Africa. The lower slopes are protected by being both a forest and game reserve. The park may be expanded in the future but now has several corridors or rights of way through the forest. The alternative routes to the peaks follow these lush forest corridors.

The peaks: Kilimanjaro is composed of three extinct volcanoes: **Kibo**, **Mawenzi** (5,145 metres/16,890 feet) and **Shira** (4,002 metres/13,140 feet). Kibo, being highest, is the peak that most visitors wish to climb. The top is breathtaking, not only because it is a long, steep climb at very high altitude but also because the views from Kibo are stunning. From the rim you can survey (weather permitting!) a vast surround of plains and other Rift Valley mountains (most notably Mount Meru to the west). Looking inward one sees the ash cone and, around the interior, sparkling glacier ice carved by wind and rain and melted by the sun into fantastic shapes. The exhilaration of having successfully reached the top adds to the sense of being on Africa's highest point.

Most people reach Kibo by way of a well-planned route along the eastern slopes. There are chalets and huts along the way, providing food and rest at each night's stop. From the park gate at **Marangu** there are two trails; one directly up a fairly open ridge; the other through forest with monkeys and birds—Hartlaub's turaco is especially easy to see. The trails join well before reaching the first huts at **Mandara** at 2,700 metres (8,856 feet).

On the second day one can visit

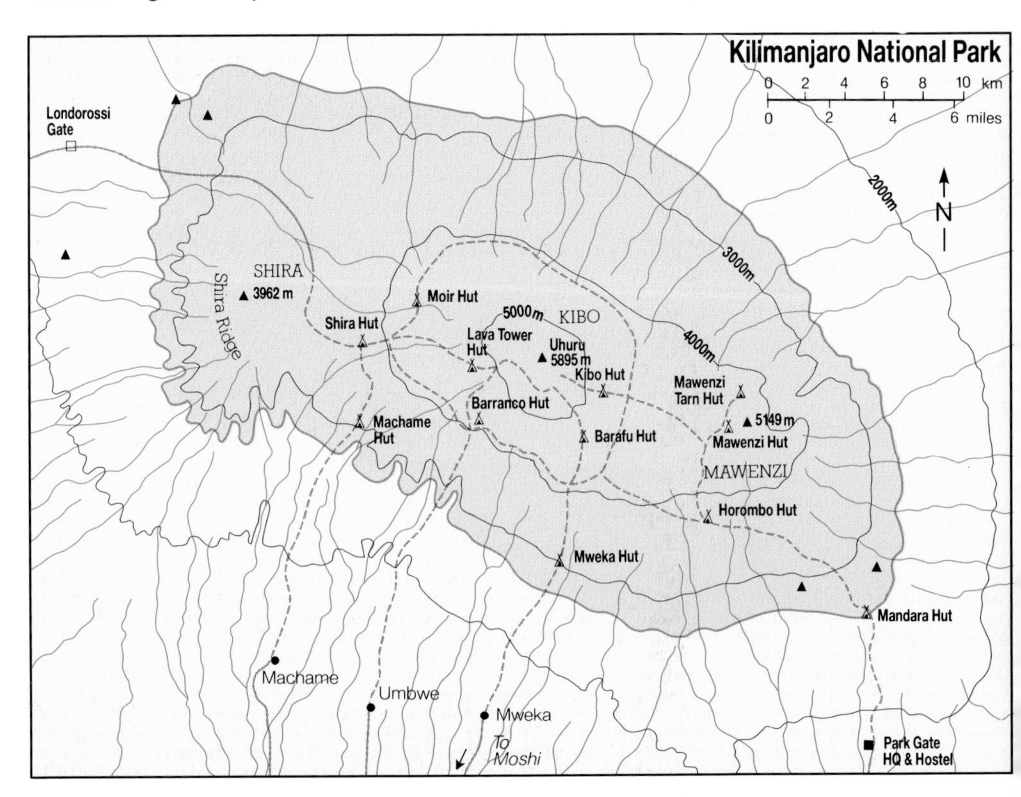

Maundi Crater, one of the many little "parasitic cones" that deck the flanks of the mountain. Here you can see the huge creamy protea flower in its native habitat. Also along the way are great hagenia trees with clusters of dark red flowers, bright green hypericum bushes with large yellow flowers, and the dramatic red or yellow spikes of poker flowers. A common bushy tree that might look familiar to visitors from northern climes is the giant heather with delicate little white flowers.

The trail continues across valleys and streams and emerges finally on to the upper moorlands where one is greeted by the strange forms of Senecio and lobelia trees. The high moor and grassland is decked with beautiful grasses, flowers and birds. The air is crisp, the views splendid; the altitude here is not yet high enough to spoil your enjoyment of this delightful walk.

The second set of huts at **Horombo** 3,807 metres (12,500 feet) are located on the slope above a particularly pretty and interesting valley. From here there are trails to less-visited areas of Kilimanjaro, such as the circuit route, Mawenzi and **Mawenzi Tarn**. Since the number of bunks is double that of the other hut complexes it is worth staying an extra day to explore this striking and beautiful area lying just below the "altitude sickness" zone.

Once above the heath and moorland, the climb ceases to be a major wildlife experience. Even so, there are still many fascinating sites: compact everlasting plants with pretty, hardy blooms, intrepid spiders hunting among rocks that bear lichens perhaps hundreds of years old. And there are the transients across **The Saddle**: tracks of eland, leopard or African hunting dogs and lovely butterflies blown high on the wind, or croaking white-naped ravens soaring above your head.

After crossing the rocky, bare Saddle between the flattened peak of Kibo and the jagged peak of Mawenzi, there is a rise that leads to **Kibo Hut** at 4,721 metres (15,500 feet). Here climbers usually spend the third night—short

Icefield remnants on Mount Kilimanjaro.

though it is! Ordinarily, you are roused just after midnight in order to get to the top of Kibo at dawn while the scree is still frozen and there is time to get all the way back down to Horombo Hut.

The long hike, usually in the dark, means one foot ahead of the other up numerous switchbacks to Kibo's summit—there are really two: **Gillman's Point** at 5,681 metres (18,650 feet) is a few hundred metres lower and a couple of kilometres closer than **Uhuru Peak**, Kibo's highest point at 5,895 metres (19,340 feet). There's not much wildlife up here, but early explorers found a leopard frozen on the rim and others were pursued by a pack of hunting dogs. Most people reach either of the two peaks of Kibo then head back down, stopping at Horombo overnight, bypassing Mandara and reaching the Marangu Gate by afternoon. Thus the total hike takes about five days.

Mawenzi, a ragged cone of hard but crumbling lava can be seen across The Saddle from Kibo. It tempts only the hard-core rock climbers and some have met their death there. There are several climbing routes up Mawenzi and four major alternative routes to the top of Kibo: **Mweka, Umbwe, Machame** and Shira**.** Each route has its own distinctive character.

Shira peak, on the west side of Kibo, is hidden by the bulk of the mountain. It is much eroded, hardly more than a shallow crater with some higher edges and forms a broad plateau. Shira is by far the most beautiful side of Kilimanjaro and has the advantage of being accessible by four-wheel drive vehicle, along a track which climbs from wheat farms and plantation forest through natural forest to moorland. From the plateau formed by the old Shira volcano one can climb further on to the flanks of Kibo from the west and join the circuit trails, or continue on up to the peak by way of the **Great Western Breach** or **Barranco**.

Shira plateau has wonderful and plentiful plant life and some large mammals as well; common duikers, small herds of eland and even the odd

Left and far left, *Helichrysum* flowers are a familiar sight.

stray lion or leopard. The scenery is magnificent and currently unspoiled by the clutter of accommodation or the litter of myriad climbers.

Plant life: Kilimanjaro is one of the best places in the world to see how plant life changes with altitude. Lush rain forest gives way to highland desert further up the mountain. Besides freezing and desiccation caused by the increase in altitude, daily fluctuations in temperature and intense radiation mean that some days plants have ice crystals instead of water at their roots with hot sun on the leaves. As the soil unfreezes, the roots can be disturbed by soil movements and the leaves can dry out.

Only about 50 kinds of hardy plants survive above 4,000 metres (13,120 feet) on Kilimanjaro. Besides the lichens and mosses that cling to rocks and soil there are several flowering plants. Tussock grasses survive by growing in a clump, the old leaves forming a cushion that insulates the roots and retains moisture. Leaves are long and thin to reduce evaporation.

Giant senecios and lobelias survive by insulating their trunks with their dead leaves. They also produce a sort of "anti-freeze" fluid in their leaves and have tiny, protected flowers.

Easy access: Kilimanjaro National Park is one of the most accessible parks in Tanzania. It has a good tarmac road from **Moshi**, the town at the mountain's foot, to Marangu, the main entry gate. Moshi itself lies on the main road and railway from Dar es Salaam to Arusha and is 56 kilometres (34.7 miles) east of Kilimanjaro International Airport. Roads go to all points except the northern side (which borders on Kenya and is currently closed to climbers).

Climbing the mountain by routes other than the main Marangu route requires prior clearance from the park headquarters at Marangu.

The best weather for climbing is January, February or September and the mountain is best avoided in April and May. Rainfall varies between 1,500 mm (58.5 inches) at the park boundary to 100 mm (3.9 inches) at the top.

Trekkers take a break at the base of Kibo.

Lake Manyara

Lake Manyara National Park is spectacularly set on a narrow band of lakeshore along the western wall of the Great Rift Valley. The park covers just 330 square kilometres (132 square miles), two thirds of which is taken up by the lake. Altitude varies between 960 metres (3,149 feet) at the lake to 2,000 metres (6,560 feet) at the top of the escarpment. Rainfall is variable, ranging from 250 mm to 1200 mm (9.75 inches to 46.8 inches) yearly, but the springs and streams emerging from the base of the rift wall water a forest that could not otherwise grow in such a dry area.

The approach to Manyara is dramatic because the rift wall is so clearly defined and can be seen running north and south into the hazy distances. The road to Lake Victoria passes the north gate and park headquarters of Manyara. If permission is gained beforehand one can enter and exit by the southern gate which is on a seasonal track that roughly follows the base of the rift, joining the main Arusha-Dodoma road at **Magagu**. This track should be avoided during the wet season.

Driving across the valley, one can see giraffes and often a variety of other plains dwellers such as wildebeest, zebra and ostrich, even before reaching the park.

Mto-wa-Mbu village (the name means "mosquito river") spreads out below the **Simba River**, just a couple of kilometres from the Manyara park gate. The village has a thriving local market with a colourful mix of peoples—it was claimed that over 100 different languages could be heard here.

Bubbling brooks: At the park entrance the water that has travelled so far from the Ngorongoro Highlands, underground through lava rock, emerges in abundance. The bubbling brooks and clear streams water a mature forest composed largely of mahogany trees,

Lovely Lake Manyara.

fig trees, fat sausage trees, crotons with heart-shaped leaves, and many others typical of riversides and upland forests.

The cool, shady forest is a welcome and beautiful respite from the normally hot dry glare of the Rift Valley floor. Other primates find the forest a good place too; there are many troops of baboons and vervets in the more open patches of trees. Blue monkeys are a special treat here because they are habituated to cars and people, and the trees are not too thick or tall. Elephants occasionally loom out of the forest where they shelter and feed. More abundant but more rarely seen are the shy, solitary or nocturnal animals such as bushbuck, rhinos, aardvarks, pangolins, civets, leopards, and wild cats. Waterbucks in Manyara are the common type with a white "bull's eye" ring on their rumps.

Big termite mounds dot the forest floor and above, silvery-cheeked hornbill and crowned hornbill call. At the forest edge and in more open areas, ground hornbills are often seen feeding in family groups. Grasslands stretch long the lake flats, providing food for wildebeest, zebra, gazelle, ostrich, buffalo and warthogs. Giraffe browse among the thorn bushes and on the lake perimeters. Sometimes young bulls neck fight. The older, almost black giraffe with huge knobs on their heads are the breeding bulls.

The Mosquito River cuts through forest and across the grassy plain to enter the lake at the north end. It emerges from a thick stand of yellow-barked acacias which are well browsed by giraffes and it offers a place for hippos and birds to rest and bathe.

Life at the lake: The birdlife at Mto-wa-Mbu is stunning in its abundance and diversity. Most of the birds have been feeding in the lake and they come to the freshwater river to drink and to wash the sticky soda from their feathers. Pelicans, storks and cormorants are the main bathers along with crowds of Egyptian geese, spur-winged geese, strange sounding whistling ducks, plus terns, gulls, thickknees and others.

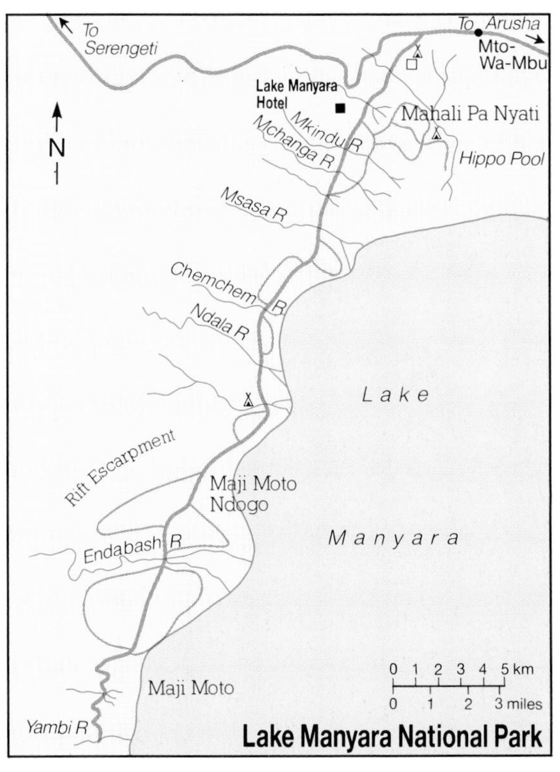

Far right, yellow-billed stork goes fishing.

The entry of the river into the lake is hidden by sedges and rushes but this upper end of the lake is often fringed pink by flamingoes in their thousands. Flamingoes are forever on the move, searching for the right food in lakes up and down the Rift Valley. Unless the lake is particularly wet or dry, flamingoes gather there in vast numbers, turning the lake pink all along its length. Although it is difficult to get close, the sight, sound and smell of a million flamingoes is unforgettable.

A large herd of buffalo lives on the flats, where they become prey for the local lions. The lions can sometimes be seen feeding on a kill by the swamp where the river enters the lake.

The rest of this park lies to the south. Look out for the "wild mango" tree which grows on the fringe of the forest. It is a bushy tree with long, shiny green leaves and masses of white waxy flowers that look like gardenia and fill the air with a magical fragrance of jasmine.

Leaving the forest and lakeshore you enter a more shrubby habitat and eventually emerge into acacia woodland. The dominant trees are the familiar flat-topped *Acacia tortilis.* In places there are thick stands of bushes that hide animals from view but in the many open areas are long views out to the lake shore where ostriches parade, elephants bathe in the river or lake shores and impala wander with baboons through the low trees.

Packs of banded mongooses and dwarf mongooses gather around termite mounds and dikdik pairs hide in bushes. Lions sometimes rest on the broad limbs of acacia trees, and you might spot the occasional leopard. Most abundant of all are the tsetse flies which keep out livestock and protect the area from human encroachment.

Birds are plentiful, especially during the migrations of Eurasian species from October to April. Then large numbers of European bee-eaters and rollers join the local species, as well as various cuckoos, buzzards, hawks, falcons and eagles.

Where the cliffs get more rocky and

Family of cormorants.

you can see the white streaks of the rock hyrax "toilets", you might see the Verreaux's eagle. It is the largest East African eagle, mostly black with white scapulars and rump. They often soar in pairs, patrolling the cliffs in search of hyraxes, their main prey. Although Verreaux's eagles travel the length of Manyara, their main haunts are around the **Endabash** area in the south.

The Endabash River: Wander along some of the tracks towards the escarpment or down along the **Endabash River**. Where the river enters the lake, there is a broad pasture with many grazing animals spread across it. Along the river the mix of trees and bush forms an unusually rich composite of riverine forest. The Endabash cascades down a steep granite cliff behind the ranger post. Sometimes the river crossing that leads across Endabash to the south is difficult because of high water so a ranger must take you there. But if you can cross, the visit to **Maji Moto** ("hot springs") less than five kilometres (three miles) further on is well worth the trip. These springs come out from the base of a granite cliff on which grow Terminalia trees with reddish pods that look like flowers, and some Euphorbias, lots of mixed brush and big gardenia bushes. A stunted old baobab marks the lower side of the point where you walk down to the pool to test the water. Be careful of slippery stones: the water is definitely HOT, about 60 C (140 F).

If you head further south, there is another large grassy plain just past Maji Moto. A long waterfall falls from the rift wall at the edge of the plain, then you enter a strange forest made up almost entirely of *Acacia robusta*. This southern extension to the park was farmed until the 1960s and you may still find signs of human settlement, although the area has reverted to native vegetation.

After the ranger post and jacaranda trees, you know you have left the park because suddenly there are coffee bushes and tractors and barns—a reminder that Manyara is only a tiny strip of the wild, between human habitation, the lake and rift wall.

Poolside at the Lake Manyara Hotel.

SELOUS

Selous Game Reserve is the largest in Africa, covering some 51,200 square kilometres (19,763 square miles). Its immense size and remoteness from populated areas make it particularly attractive both to wildlife and adventurers who can explore on foot or in a boat.

The varied terrain is undulating with rocky outcrops cut by many, usually dry, watercourses. Altitude varies between 110 metres (361 feet) and 1,250 metres (4,100 feet).

It is best to visit Selous between July and March, out of the rainy season. An average of 600 mm to 1,000 mm (24 inches to 39 inches) of rain falls yearly, most of it from March to June.

The pristine landscape is dominated by the great **Rufiji River** system, the largest in East Africa. Three huge sand rivers, the **Great Ruaha**, **Kilombero** and **Luwegu** flow through the reserve and meet before being channelled through the spectacular **Stiegler's Gorge**. Named after a German explorer who was killed by an elephant there in 1907, the gorge can be traversed by a swaying footbridge over the raging waters.

Selous was originally established as a hunting preserve by the Germans just after the turn of the 20th century. It is still visited by occasional hunters who pay dearly for the privilege.

Selous was named in 1922 after the naturalist, elephant-hunter and explorer, Frederick Courtney Selous who, as captain of a British unit, was killed in action during the advance against the Germans at the end of the First World War I. His grave is north along a tributary of the **Behobeho River.**

Over 1,700 species of plants have been identified in Selous, along with hundreds of bird species. Since the area is relatively well watered, there is an abundance of trees and bush, providing excellent shelter for the wildlife.

Large and lesser game: Selous is

Cooling down in the Ruaha River.

most famous for its huge herds of elephants, although their numbers have been reduced by poachers. Huge buffalo herds come to drink at the river in dry season. A boat trip on the Rufiji will reveal hippos and crocodiles in abundance, and you may also see waterbuck, the southern reedbuck and bushbuck. The beautiful sable antelope with its long curved horns and the slightly larger roan antelope and greater kudu like to hide in the tall grass while lesser kudu are more often seen among the rocky hills and dry bush. Impalas and Lichtenstein's hartebeest are common as are zebras and the southern race of wildebeest, which is marked with a distinctive white chevron on its long face. Rhinos have been almost exterminated by poachers but small numbers may still exist.

Selous is of great scientific importance because it has a history of research and a wildlife station in the reserve. At **Kingupira** in the east, a Miombo Research Centre was established in 1966. Monitoring has shown, sadly, that the estimated 100,000 elephant population in 1977 dropped to about 50,000 in 1987.

The easiest way to reach Selous is by air to one of the airstrips (each of the four major camps/lodges has its own strip). The Tanzam railway from Dar es Salaam runs along the northern boundary of the reserve. The views are superb. There are several stations four to six hours from Dar es Salaam where you can disembark and enter Selous, provided transport or a foot safari has been organized in advance. The major station is at **Fuga**.

Getting to Selous by road takes about eight hours or more from Dar es Salaam. If arriving from the east, you need to plan to come in the dry season and with a four-wheel drive vehicle. The main, and much better but longer road goes west, although access to the reserve is only via tracks leading off south from **Morogoro** or **Mikumi**. Once inside Selous you can travel about only by four-wheel drive, on foot, with an armed escort or in a boat along the river.

Starting off on a foot safari.

RUAHA

Ruaha is Tanzania's second largest national park, covering 10,200 square kilometres (3,937 square miles) of undulating plateau with some mountains, rocky hills and two extensive river valleys on the east and western borders. Together with two important game reserves, **Rungwa** and **Kizigio**, that buffer Ruaha's northern boundary, the total protected area of some 25,600 square kilometres (9,886 square miles) makes for a very impressive wildlife area indeed. Ruaha was originally created in 1964 from half of Rungwa.

The park lies on a combination of ancient granite and more recent sediments brought about by the extensive rifting through Tanzania. On average only 600 mm (24 inches) of rain per year falls on poor soils on the plateau and richer valley soils. In the east, where rainfall is least, this amount just barely supports a bushland and trees that are adapted to dry conditions, including spiny commiphora and acacia species, and the splendid baobabs.

The mountains in the far west of the park catch more rain and so are covered by an evergreen upland forest that is yet to be fully explored and appreciated. Rainfall increases up to 800 mm (31 inches) towards the west and south, supporting a miombo woodland that covers half of Ruaha.

Miombo is composed predominantly of *Brachystegia* trees of which there are about 15 different species in Tanzania. Together with many other plants these trees create a rather special woodland. Some trees such as "mninga" (*Pterocarpus angolensis*) are valuable for timber. Miombo trees are all about 10 metres (33 feet) in height and there is a sparse understory. The woodland has a gentle, rather pretty but monotonous appearance during most the year. But just before the start of the rainy season, after the tall grass catches fire, the forest is transformed into a breathtaking,

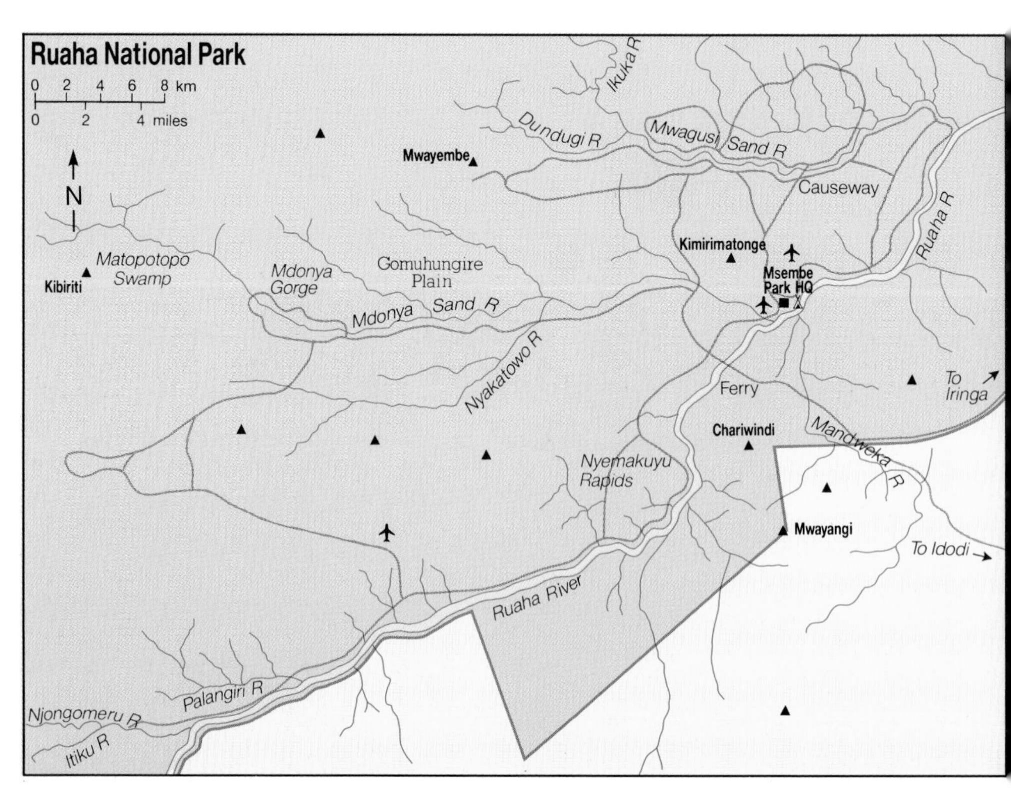

multi-hued tapestry of pale green, yellow, russet, purple, red, crimson and copper.

The Ruaha River: The park is bound by two long rivers but the **Ruaha River** in the south is the major attraction. Here, water flows over a broad sandy bed, roaring and wide in wet season, a trickle in dry season. The river itself and its fringing vegetation provide water, food and homes for many creatures, including human visitors. Most of the passable tracks and accommodation are along the river banks or in the drier areas. Visitors should persevere and be adventurous enough to explore the more remote and wilder miombo and western mountains. You will be rewarded with some of the country's most spectacular scenery.

More than 1,600 plant species and approximately 400 bird species have been recorded here. The vegetation, especially the distinctive flowers found in the miombo during the rainy season, attracts many insects and birds. The tsetse fly, with its irritating bite, is so abundant that it has dissuaded both people and livestock from coming into the area.

Some larger mammals are spread over a wide area of the park, but only in very small concentrations. These include both greater and lesser kudu, the reddish Lichtenstein's hartebeest and sable and roan antelopes. Elephants were once so abundant that there was fear of them destroying the huge *Acacia albida* trees so characteristic of the river banks. Now, as in most other parks in East Africa, elephant numbers have been drastically reduced by poaching for ivory.

Crocodiles basking on the sandbanks and bathing hippos enliven the river and you might even see a clawless otter catching fish. Reedbuck, waterbuck and buffalo frequent the river edges along with yellow baboons and vervet monkeys.

In grassland and bushland there are giraffes, eland, impala, zebra, warthog, Grant's gazelle, dikdik, mongoose, ostrich, the elusive ratel civet, wild cat,

Arid section of Ruaha.

porcupines and other small creatures. Predators one is likely to see include lion, cheetah, leopard, jackal and the African hunting dog.

From October to November and March to April thousands of birds flock to Ruaha on their annual migration from Europe and Asia. Together with resident hornbills, green wood-hoopoes, kingfishers, plovers, bee-eaters, sunbirds and water birds, skimmers can be seen dragging their specialized lower bills along the water in search of a meal. Nearby, hammerkops can be seen making repairs to their gigantic nests or swooping down on unsuspecting frogs.

Research difficulties: Scientists have sporadically studied some of the vegetation, baboons and elephants in Ruaha but, although there is scope to study the rarer large animals such as sable, roan, lesser and greater kudu, it is unlikely that extensive wildlife research will occur here. The logistical difficulties, especially the park's remoteness and seasonal isolation make such work difficult. This means that interested visitors can still contribute substantially to information about Ruaha.

Road access to Ruaha is via the historic highland town of **Iringa**, which is 112 kilometres (70 miles) from the park. From the park boundary it is a further eight kilometres (five miles) to the Ruaha River. Here you check in at a guard post before driving to a wooden ferry which is pulled by hand across the river. This is an exciting way to enter the park when the river is low but during the rainy season access is difficult or impossible.

Other roads and entry points do exist but they should be carefully checked up on before a visit to determine feasibility. The best time to visit the park is during the dry season, from June to November.

There is an airstrip for light aircraft at park headquarters at **Msembe** beside the Ruaha River. Flying time from Nairobi is about three and a half hours. Contact Tanzania National Parks (TANAPA) on arrival in Tanzania for more details.

Impalas on the move.

MIKUMI

Mikumi National Park, established in 1964, is Tanzania's third largest park and the one most popular with residents because of its accessibility and proximity to populated areas. The grassy flood plain surrounded by tiers of hills and misty mountains rising to 2,000 metres (5,660 feet) in the east gives Mikumi a particularly attractive landscape covering 3,230 square kilometres (1,274 square miles). Rainfall varies between 600 mm and 800 mm (24 inches to 31 inches): the wettest season is between November and May with a drier period between January and March.

Many distinctive trees grow here, including the tall Sterculia (*Sterculia appendiculata*) with long, smooth, pale yellow trunks, the well-shaped afzelias (*Afzelia quanzensis*), chunky sausage trees (*Kigelia aethiopum*), fat baobabs (*Adansonia digitata*) and the tree which gives the park its name, the Borassus palm (*Borassus flabellifer*) with a swelling high on the trunk.

Wonderful views: The **Mkata River** area and flood plain at an altitude of 500 metres (1,640 feet) is an open area well covered by roads. Animals in the tourist area are mostly habituated to cars and thus easy to watch and photograph. Further afield in the hills, animals are shyer and the heavier vegetation makes glimpses of them more exciting. The southern hills are covered with *Brachystegia* woodland and cut by watercourses. There are hot springs and wonderful views.

North of the main road is the area most frequented by visitors and one can find a large variety of animals in a relatively short time. There are herds of elephants, pools crowded with hippos, wallowing buffalos, swamps with reedbucks and waterbirds, and plains dotted with yellow baboons, warthogs, wildebeest, zebras, elands and Lichtenstein's hartebeest. In bushier spots you may see greater kudu. Lions and leopards are

Bath time for these buffalos.

often seen but packs of African hunting dogs in open areas provide a special treat as they are often very tame.

In the hills impala are widespread and elephants are often encountered. To the south of the park sable antelope are sometimes seen. Bush and woodland birds are abundant and include many species of hornbills, sunbirds, cuckoos, shrikes and birds of prey.

Animal research: Mikumi is a very important educational park with a large hostel for university students and visiting school groups who are shown how to do animal counts, measure tree growth, gauge the extent of fires and study the distribution of grasses, among other tasks. Mikumi also has an Animal Behaviour Research Unit whose scientists have been studying the behaviour of yellow baboons since 1974.

One finding in particular illustrates the importance of ecology on social behaviour. The optimal time for giving birth for baboons in Mikumi is March to July at the end of the rainy season. For some months afterwards there is enough food to keep a certain number of nursing mothers and their babies healthy. But during the critical weaning period at six months of age, food is not so plentiful. Baboons have developed a way to decide which females will bear babies so that only the "right" number of young are born.

Adult females in the troop form coalitions. Two or more females threaten and attack other females, some of whom become disturbed at not having babies that season, although they always have another chance later on. In this way the more dominant females have fewer competitors for the food that their own children will need during weaning.

Mikumi is 300 kilometres (186 miles) from Dar es Salaam along the road from the coast to Zambia via Morogoro and Iringa. This road cuts through the park for about 50 kilometres (31 miles). There is a railway station at **Mikumi village**, about 22 kilometres (13 miles) from the park entrance. Light aircraft can land on the airstrip at park headquarters.

Thorny perch for this red-billed hornbill.

TARANGIRE

Tarangire National Park was established in 1970 and covers 2,600 square kilometres (1,000 square miles) of gently undulating plains with two large *Mbunga* (flat pans) that are seasonal swamps in the south. A river cuts through numerous rocky hills rising from 1,000 metres (3,280 feet) to 1,675 metres (5,495 feet) at the top of **Kirogwa**.

Tarangire is best seen during the dry season when there are great concentrations of animals. Rainfall averages 800 mm (31.2 inches) per year and falls mostly between November and May, when the southern roads and unbridged river crossings become impassable.

However, during the dry season, Tarangire is the main refuge for wildlife from the surrounding areas on the floor of the Rift Valley. These animals move about an ecosystem ranging across 20,000 square kilometres (8,000 square miles) from Lake Natron in the north to Maasai Steppe in the south, including Lake Manyara.

River of life: The importance of the Tarangire River to the animals in this ecosystem cannot be over-estimated: wildebeest—the rift valley race with lighter coloured, wider horns—zebras, elands, elephants, Coke's hartebeest, buffalos and the elegant fringe-eared oryx flock to its waters and valleys in their thousands. These migrants join the resident waterbuck, impalas, warthogs, dikdiks, giraffes, lesser kudus and there are even a few rhinos, plus predators such as lions, leopards and cheetahs.

There are pools and open water in the river all year round, but during dry season much of the water travels underneath the sandy riverbed. To get at the water, elephants dig holes which other animals also use for drinking.

The high bluff at the end of a row of tents at **Tarangire Lodge** is a particularly good spot for game viewing. At dusk local baboons gather at a bend in

Plants adapt to the dry landscape.

the river to groom and fight. Near dark they climb into the tall doum palms to be safe from leopards. This scene also occurs at the large grove of palms just after crossing the **Engelhard Bridge**, opposite the campsites.

From your camp or lodge you can hear the barks of baboons, especially strident when a leopard is near (recognisable by its deep sawing roar). Listen too for lion roars and for elephants trumpeting.

At the permanent water spots in the river are numerous birds, such as kingfishers, herons—especially goliath heron—ducks, geese, hammerkops and more. Tarangire has a wealth of birdlife at all times and it is one of the very best spots to see flocks of green wood hoopoes, African hoopoes, green and yellow "brown" parrots, Fischer's and yellow-collared lovebirds, bare-faced and white-bellied go-away birds, Tanzanian endemic ashy starling, a large number of different kinds of doves, pigeons, mousebirds, cuckoos, swifts, swallows, hornbills, and many others.

Swamps and baobabs: The large swamps in the south which dry up gradually after the rains stop form huge pastures for elephants with seepages that provide water for many birds. Many pythons gather here. These huge, very beautifully patterned, non-poisonous snakes—they kill their prey by squeezing it to death—leave the marshes when dry and take refuge from predators by going up into the crowns of the flat-topped acacia. They wind themselves into a coil that looks like a nest from the ground. During the long dry season, pythons "aestivate", not eating but conserving their resources until rats, birds and small animals are plentiful in the swamps again. Sometimes they congregate, either for habitat or to seek mates.

In north Tarangire, the **Lemiyon** and **Matete** areas have rolling landscapes with monumental baobabs. These giant "upside down" trees are often hollow or punctured with holes where elephants have ripped off the bark or poked through the fibrous trunk. Poachers

Far left, a huge baobab.

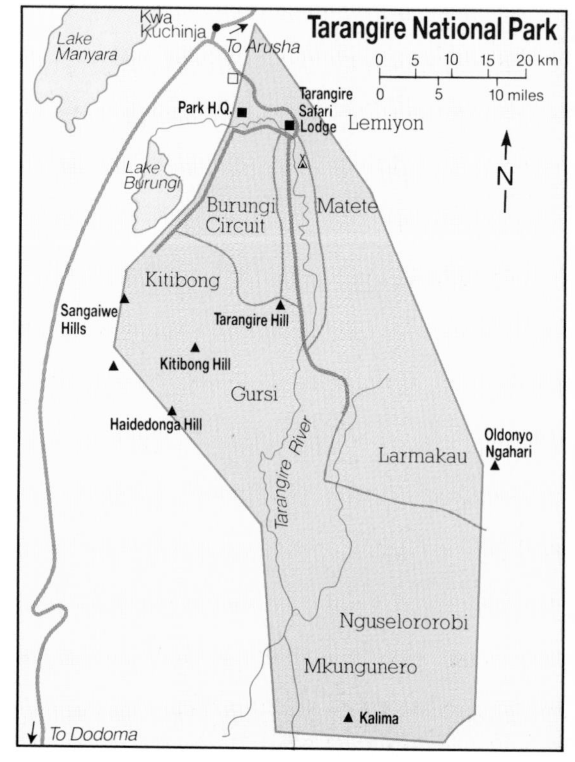

sometimes use these ready-made caves as hideouts. More usually, barn owls as well as any number of bats (there are at least 10 species in Tarangire) nest inside the baobabs.

Many birds use baobabs as nest sites during the rainy season. The large communal nests of the red-billed buffalo weavers are especially conspicuous. If you see a hornbill visiting a certain site on a baobab look for its nest-hole too; the male walls up his mate inside, feeding her through a narrow slit, until the eggs hatch. Then she emerges to help feed the nestlings, who remain sealed in until big enough to fly. Baobab flowers are pollinated by bats and the pithy pods are deliciously astringent. Fruits fall off in dry season and are eagerly snapped up by a variety of animals.

Other trees include stout sausage trees along the river. Huge fruits dangle down from thick ropey stems and are eaten by baboons, rhinos and elephants. When sausage trees bloom the nectar-filled blossoms open only at night, intended solely for the fertilizing touch of certain fruit bats. In the early morning the trees are literally attacked by groups of vervet monkeys and baboons who dash up to suck any remaining nectar. In the very early morning hordes of insects rush to feed on the maroon blooms.

Common flat-topped acacia can be found in most areas, along with multi-branched *Commiphora* and many short, leafy *Combretum* and bushy *Dalbergia* trees—better known as African blackwood or ebony—whose very dense, hard wood is used for carvings. These trees are usually damaged or stunted from elephant attacks. Except for *combretums*, all these and most other trees have spines or poisonous latex to deter animals.

Tarangire is a good park to see how plants adapt to an arid area with meagre seasonal rain.

Tarangire is easy to reach: a tarmac road heads south-west from Arusha, and in about two hours (112 kilometres or 70 miles) you will reach the park entrance at **Minjingu**.

Topi have distinctive black markings.

ARUSHA

Arusha National Park (ANP) is close to major human cultivation and settlements. It is small—only 137 square kilometres (53 square miles)—yet has varied terrain with a distinct feeling of wildness. There are fresh to alkaline lakes separated by hummocks in the **Momela** area; a small crater, **Ngurdoto**, with steep inner walls: and the extinct volcano, **Mount Meru**, with a breached crater and many valleys on its flanks leading down to farm land on the lower slopes.

Altitude ranges from 1,500 metres (4,920 feet) at Momela to 4,566 metres (14,990 feet) at the top of Mount Meru. Most rain falls between November and May, although there is often a less rainy period from December to February when skies clear to offer magnificent views of Mount Meru and Kilimanjaro. In the dry season views tend to be a little more hazy.

Ngurdoto, Momela and Mount Meru are easily accessible. Since there are no lions in the park, a number of observation points have been prepared where you can get out of your vehicle. This feeling of freedom, with time to look about on foot, greatly contributes to the enchantment of a visit here.

There are two main entrances: **Ngurdoto Gate** lies at the base of Ngurdoto Crater: **Momela Gate** is situated where the main Ngare-Nanyuki track crosses a high ridge connecting Mount Meru with the rest of the park. **Ngurdoto Crater** is always a good entry point to the park. There is a small museum at the gate surrounded by forest trees. Here you can get out of your vehicle and scan the treetops with your binoculars. Look into the ascending branches of the open-crowned African olive, one of the most valuable timber trees in Africa. The deep, staccato roars you might hear are made by the local troop of black and white colobus monkeys. The population in ANP is particularly striking,

The extinct crater of Mount Meru.

being a breed which lives only in highland areas such as here and on Mount Kilimanjaro. The hair of their cape and tail is exceptionally long and white and streams behind them as they leap amongst the trees. There are troops of colobus throughout the park's forests. You may also meet the blue monkey, mostly grey and black with a long slender tail.

The huge silvery-cheeked hornbill is usually seen in pairs, has an ivory coloured casque above its beak and a call sounding like maniacal laughter.

The crater: Surrounding Ngurdoto Crater are very beautiful forests containing huge strangler figtrees with long roots descending from on high. Wild mangos, medium sized trees with shiny, dark green leaves and clusters of white waxy flowers and inedible fruit, are also common.

The crater itself is about 2.4 kilometres (1.5 miles) across. The inside has been designated a private sanctuary for wildlife, and no humans are allowed onto the crater floor.

From any viewpoint on the rim of Ngurdoto you look out over a wide bowl; the bottom is usually marshy and grades into grassy swards around the edges before disappearing into lush forest on the inner slopes. Palm trees lean out of places where water seeps on the walls and stand out among the lacy-leafed *Albizia*, dark *Bersama,* gnarled *Nuxia* and tall *Olea* trees on the rim.

At **Leitong**, the highest point at 1,853 metres (6,077 feet) on the north side of the crater, you can see unusual lobelia plants as well as a splendid view across the crater. In the sky eagles, ravens, buzzards and hawks ride the up-draughts. If it is clear you can see to Kilimanjaro.

Returning through lower forest on the outside slopes of Ngurdoto look for the shy red duiker. This rust-red antelope is larger than a dikdik but might be confused with a female bushbuck, which is also reddish with white spots on its throat and sides.

In the lower forest, trees grow in decorative groves and clumps, with a

Camping inside the crater.

very short green sward spreading below to catch the dappled light. The most common trees are crotons, recognizable by the silvery undersides of their pale green leaves. *Croton macrostachyus* has heart-shaped leaves that turn orange when old. It is very widespread in open areas all over the park; *Croton megalocarpus* is taller, with narrower leaves and is even more abundant.

The road between Ngurdoto Crater and the Momela lakes passes by several side-roads leading to viewpoints. They all offer different perspectives of the ponds, marshes, valleys and lakes. Hides and observation points near the Momela lakes look out on the surrounding terrain which was formed when Mount Meru blew its top and muddy, rocky avalanches spread debris all the way to the foot of neighbouring Kilimanjaro. From a high point, such as **Boma la Megi**, you can see the widespread array of hills and lakes in and beyond the park.

The Momela lakes: In this area lakes range from fresh to alkaline water, most of it coming from underground streams. They sometimes appear to be different colours, as a result of algal growths. Each lake attracts a distinctive clientele: the pinker, lesser flamingo chooses lakes rich in blue-green algae, its only food. The paler, greater flamingo mainly eats crustacea which feed on algae and, because it is less numerous and taller, can exploit different and deeper lakes. Pelicans visit from time to time. Large varieties of ducks, geese and waders, both residents and migrants, can also be seen.

Leaving the lakes there is a broad undulating area of grass and bush. In patches of trees and thick bush embellishing the hills, look carefully for herds of elephants or buffalo, solitary bushbuck or eland and the tiny dikdik—if you see one you know its mate is somewhere near. The most commonly seen animals throughout this area between Momela lakes and Mount Meru are giraffes, which are very tame.

To explore Mount Meru, one must cross the **Ngare Nanyuki River** and

Typical forest foliage.

ascend the steep mountain track in a four-wheel drive vehicle. This extinct volcano is covered with treasures: the lower slopes have marshy, flower-filled glades, hidden waterfalls, clear streams, butterfly- and bushbuck-flecked clearings and a blossoming, bushy forest with leaping colobus monkeys.

Beyond the drive-through fig tree arch, the upper forest has enormous junipers and podos with olive pigeons, red-fronted parrots or Hartlaub's turacos in the uppermost branches. Above the **Kitoto** viewpoint this forest gives way to giant heather. In the flat floor of Meru's crater is a heath dotted with lilies, red hot pokers and other flowers. There is also a dead forest covered in pale green lichen. Many high altitude plants grow in this vast amphitheatre with slopes of bare scree, steep cliffs of crumbling rock and a jagged summit.

There are mountain huts and guides for those who want to make the steep climb up the ridge of Meru's cone to the peak at 4,566 metres (14,990 feet). The walk up the mountain from its base is not only relatively easy but thoroughly rewarding.

Conservation and conflict: Arusha National Park, established in 1960, is named after the large town spread over its south-west flank, about 35 kilometres (22 miles) away. The town in turn is named after the *Warusha* people who live mostly on the west and driest side of the mountain. They are said to be *Maasai* people who settled around Mount Meru in order to grow crops instead of leading a totally pastoral way of life, dependent on livestock. Another large group of people, called the *Meru* who are relatives to the *Wachagga*, live on Mount Kilimanjaro.

Wildlife enthusiasts use Arusha as the gateway to game parks in the region, including Serengeti, Mount Kilimanjaro, Tarangire and Ngorongoro Conservation Area.

Arusha is a showcase for nature.

From many places in the park one can see evidence of the squeeze along the boundaries by people and farms. In addition to cultivated plots, a broad band of mostly pine forest stretches over many kilometres of the southern and western flanks. The trees were planted many years ago to provide timber for the area. Along the borders the native vegetation abruptly changes where humans have modified the area for their own uses.

In the pristine forest on Mount Meru people secretly cut down trees to provide wood for homes and tourist lodges and even to make carvings for tourists to buy. Trophy poachers prowl the slopes in search of the last rhinos and elephants. In Ngurdoto Crater, meat poachers have cleared out most of the larger wildlife, and around the Momela lakes people illegally fish, reducing food for birds and scaring away wildlife. Although they suffer from a lack of funds, staff and equipment, the park authorities continue to fight the endless battle to protect wildlife.

Arusha demonstrates the considerable dangers of a relatively self-contained ecological unit being pressed on all sides by the demands and self-interests of humanity.

Gombe, Mahale and Other Parks

Gombe and Mahale national parks are both located on the eastern shore of Lake Tanganyika. The primary purpose of both is to protect endangered populations of chimpanzees (*Pan troglodytes*). The two parks are, however, very different in ease of access, size, terrain and ecology.

Gombe National Park was established in 1968 and covers an area of 52 square kilometres (20 square miles). From the lake shore to the ridge tops ranges between 770 and 1,440 metres (2,525 and 4,723 feet) with an average annual rainfall of 1,000 mm (39 inches). This park can be reached in 40 minutes by speedboat from Kigoma. Alternative transport by water taxi takes between four and six hours.

Even so, Gombe National Park is relatively easier to get to than Mahale and has numerous other primates in addition to chimps: olive baboons, red colobus, red-tailed monkeys, blue monkeys and vervet monkeys. Another common primate has very important effects on this little park: *Homo sapiens.*

Human beings live on all sides of Gombe except for the lake. Even here, many fishermen come to spread out their catch of small *dagaa* fish (a bit like white bait) to dry on the shore. Poaching and encroaching are major problems.

A boat ride along the impressively big Lake Tanganyika is a memorable part of a visit to Gombe. But there is so much more: lush vegetation, beautiful birds, butterflies and other insects, wildlife, helpful rangers and researchers and, above all, the thrill of meeting chimpanzees face to face in their natural habitat. You can swim in the lake and go snorkeling to look for the many aquarium fish that come from this lake.

Mahale: Mahale Mountains National Park is much larger and more remote than Gombe, with a more complex landscape, many more animals and better potential for long-term survival

Wizened youngster at Gombe.

as an ecosystem.

It was established in 1986 and covers an area of 1,200 square kilometres (480 square miles). Altitude ranges from 770 metres (2,525 feet) along the lake shore to 2,460 metres (8,070 feet) on the mountain tops. There are many steep valleys with miombo woodland plains in the south. Average rainfall is 1,000 mm (39 inches) per year. Access to Mahale is by boat from Kigoma which can take any time from 10 to 24 hours. It may be wise to check the state of the airstrip first.

The Mahale Mountains rise from the lakeshore to misty heights, their steep slopes swathed in thick jungle-like vegetation. The mountains provide both a habitat for many kinds of animal life and also help to protect the park from human encroachment. The lake shore confines the western side of the park and the tsetse fly-infested miombo woodland to the south helps buffer inroads from that side.

People live and farm along Mahale's northern boundary (there is a village where you can stop to get supplies of fresh food), but it is somewhat isolated here due to the **Malagarasi River.** It enters Lake Tanganyika to the north of Mahale and has a vast swamp behind its wide mouth, across which no road has yet been built. Local people use the lake as their road, with a variety of different size boats. Transport inside Mahale is strictly by foot!

Mahale is a remnant of the forests that once stretched across all of equatorial Africa and harbours a wide variety of animals: bushbucks, waterbucks, bushpigs, buffalos, elephants, leopards, the same wide variety of primates as at Gombe and many different kinds of forest birds. In the south are giraffes, roan and sable antelopes, kudus, elands, lions and a host of miombo birds. Clawless otters feed off the fish in the lake.

Championing chimps: Dr Jane Goodall and her team in Gombe, together with Drs Itani, Nishida and their colleagues in Mahale, have been studying chimpanzee behaviour since the

Old elephant eyes.

early 1960s. These normally shy apes have become completely habituated to human observers. The studies have revolutionised our understanding of our closest animal relative.

The chimps live in large, loosely-knit communities where family bonds are very important throughout life. The community's range is defended by a coalition of related males.

Mostly vegetarian, chimps also eat substantial amounts of ants and termites, using stems and grasses as simple tools to extract the insects from their nests. They also frequently cooperate to hunt and eat larger prey such as monkeys and bushpigs.

Chimps are excitable and emotional; between the males of a community there is often a vigorous struggle for status, and between communities gang warfare has been seen, sometimes resulting in death. They are capable of infanticide, and of dying of grief at the loss of a mother. Yet on the whole they are peaceful, likeable, more handsome than their threadbare cousins in zoos, and more like ourselves than we may care to admit.

OTHER NOTABLE AREAS

Eastern Arc mountains: These are three mountain ranges in Eastern Tanzania, close enough to the Indian Ocean to receive high rainfall. Lush forests have clothed these ancient crystalline mountains ever since jungle vegetation million years ago, by climatic change and by the formation of the Great Rift Valley. And like isolated islands, each has developed its own endemic flora and fauna. Out of 2,000 species of plants so far identified in these mountains, some 25 to 30 percent are found nowhere else in the world. There are also many rare and unique reptiles, amphibians and insects.

The **Usambara Mountains** in the north have patches of protected forest where one can walk in virgin areas and see wild African violets (*Saintpaulia ionantha*), as well as orchids, wild coffee bushes and buttressed trees. Though mammals are rare, there are many small

Giraffes on a moonlit stroll.

creatures such as millipedes, frogs and chameleons that are unique to the Usambaras. It has been said that these forests constitute the richest biological community in Africa.

The **Uluguru Mountains**, inland from the central coast, are particularly beautiful. Rising from the **Morogoro Plain** in a compact and rugged bunch, their steep slopes are covered with forest and full of interesting plants and birds.

The **Uzungwa Mountains** are the source of the Kilombero River which flows through the Selous Game Reserve. There is a proposal to make about 1,200 square kilometres (480 square miles) of the mountains into a national park which would protect the forested habitat of much unique wildlife, including an unusual red colobus monkey (*Colobus badius gordonorum*). In the Uzungwa-Kilombero area a new species of weaver bird, a new cisticola, and the rufous-winged sunbird were only discovered in the 1980s. Even a new subspecies of monkey, the Sanje crested mangabey was only described in 1981.

Around Lake Rukwa: Rukwa Valley, Katavi National Park and **Uwanda Game Reserve**, located in south-west Tanzania, comprise a large area surrounding **Lake Rukwa**. Seasonal flooding and drying up of the lake creates a wide shore of grass that attracts large numbers of grazing animals, most notably topi, buffalos, elands and zebras. Elephants, hippos, Defassa waterbuck, Lichtenstein's hartebeest, southern reedbuck, impala, roan, greater kudu and giraffe are also evident. Perhaps the least common species is the puku (*Kobus vardoni*), whose closest relative is the Uganda kob (*Adenota kob*). The puku is a reddish brown colour, rather stocky and shaggy with thick lyrate horns. Around the lake and in the valley over 400 species of birds have been identified.

Occasionally Lake Rukwa dries up to the point where hippos, crocodiles and fish are confined to the few remaining muddy bogs. The coming of the first

A rare roan antelope.

rain storms, usually in November, causes the grasses and sedges on the perimeter of the flood plain to sprout and fill many ponds and pools that lure waterfowl. A great British biologist, Vesey Fitzgerald, wrote: "there can be no more fascinating scene in the whole of Africa than the lawns of new grass in the (Rukwa) valley at this season".

Northern Tanzania: Mkomazi and **Umba Game Reserves** along the northern border of Tanzania to the east of Kilimanjaro, cover 2,500 square kilometres (1,000 square miles) of varied habitats, including dry plains with many valleys, part of the Pare Mountains, and steppe stretching eastward towards the Indian Ocean. This stretch of land is an important buffer zone for animals in the vast Tsavo National Park across the invisible border between Tanzania and Kenya.

There is encroachment on all sides of the reserves by pastoralists and cultivators; sometimes livestock seems more common than wildlife. Increased use of the reserves by non-hunting visitors will lead to improvements in facilities and protection but meanwhile, the landscape remains incredibly beautiful. There is a fascinating variety of semi-arid plants, including many euphorbia with their poisonous latex and the knobbly-spined *Erythrina burtii* with striking red flowers. Numerous birds including thousands of migrating European swallows (*Hirundo rustica*) and golden-breasted starlings (*Cosmopsarus regius*) live or visit this area.

Rubondo Island: Rubondo Island National Park is a large island in Lake Victoria. The whole area covers 457 square kilometres (182 square miles), half of which is water. A flight over the southern part of this largest lake in Africa will show Rubondo as the only island still extensively covered with forest. The trees and bush are set off most attractively by a fringe of shore with beaches and papyrus swamps. Professor Bernard Grzimek of the Frankfurt Zoo, who did so much to encourage the protection of homes for wild animals, took a special interest in

Bristle-like mop of the crowned crane.

Rubondo and helped to get it gazetted as a park in 1977.

Because Rubondo is offshore and the human population is restricted, some of the more endangered species such as rhinos, elephants, roan antelopes, suni, black and white colobus and chimpanzees were introduced. They are now thriving and it is hoped that their descendants will repopulate areas where numbers are currently depleted.

Hippos, bushbuck, sitatunga, vervet monkeys, mongooses, and crocodiles are among the indigenous species. There are no large predators on the island (though poachers come in canoes to kill rhinos), so it will be of interest to see how the populations of native and introduced species will regulate themselves. Besides the mammals, Rubondo has a good variety of reptiles, amphibians and insects and there are numerous lake, swamp, plains and forest birds—you can probably see 100 different kinds of birds during a walk around part of the island.

Zanzibar: Zanzibar and its **Jozani Forest Reserve** is an off-the-beaten-track delight. Zanzibar Island, with its old Arab town and narrow streets, palaces, dank caves, churches and groves of coconuts, cinnamon, cloves, cocoa and pepper, all testify to Zanzibar's history as an important centre for trade, including the infamous slave trade. The beaches of Zanzibar are lovely and there are some good areas for snorkeling among the wonderful fishes and corals on the reef.

Jozani, on the southern part of the island, is a small reserve with pandanus palms and other remnants of a once-flourishing indigenous vegetation plus introduced plots of non-native trees. While the reserve is of interest botanically, the presence of duiker, such as the Zanzibar blue duiker (*Cephalophus monticola sundervalli*) and the dwarf red duiker (*Cephalophus callipygus adersi*) and the brightly coloured and very rare Zanzibar red colobus (*Colobus badius kirkii*) make Jozani especially interesting. Be sure to bring binoculars for watching the monkeys and birds.

Luminescent colours of the green-headed sunbird.

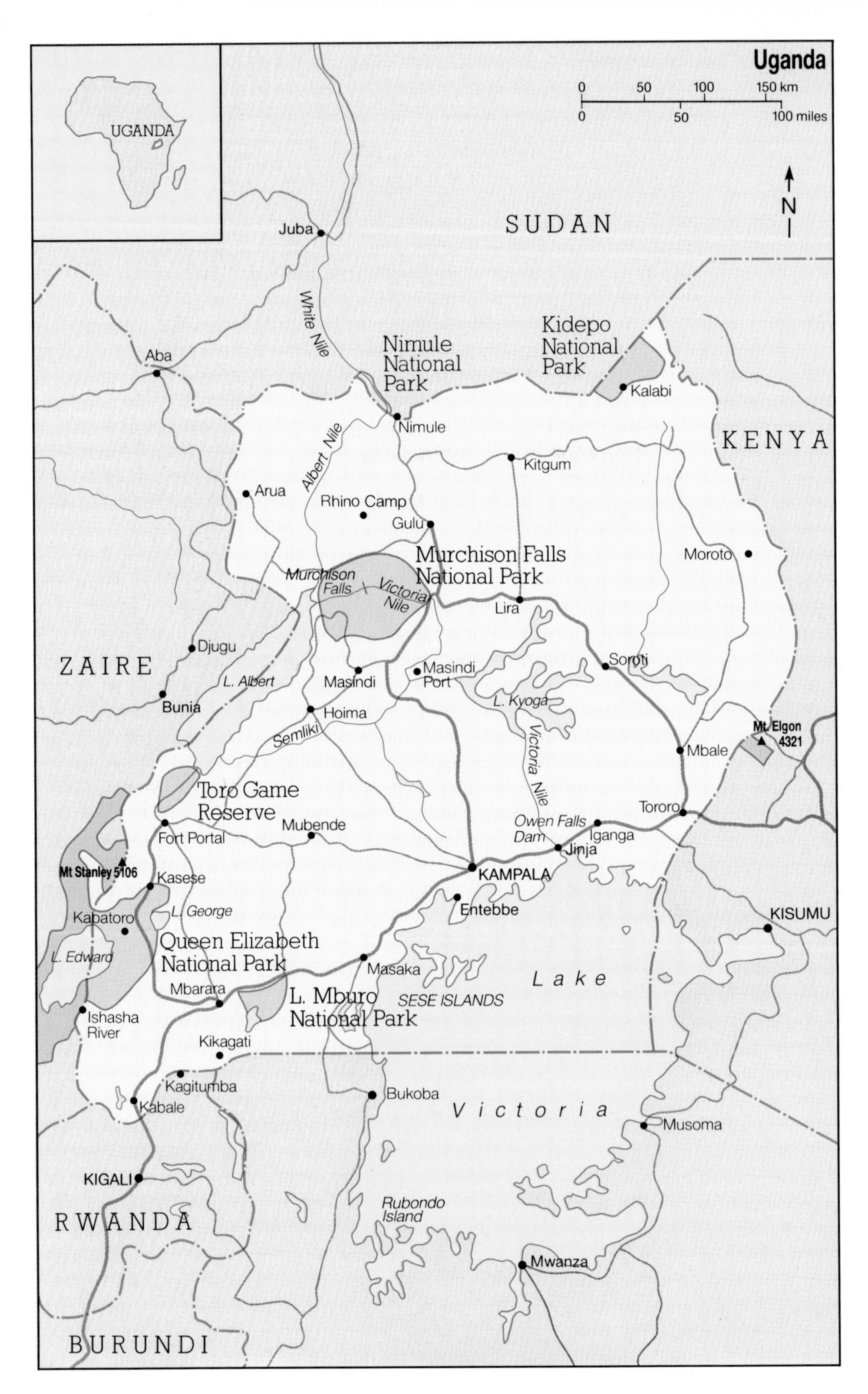
Uganda
0 50 100 150 km
0 50 100 miles
N
UGANDA
SUDAN
KENYA
ZAIRE
RWANDA
BURUNDI
Juba
White Nile
Aba
Nimule National Park
Kidepo National Park
Kalabi
Nimule
Albert Nile
Kitgum
Arua
Rhino Camp
Gulu
Moroto
Murchison Falls National Park
Murchison Falls
Victoria Nile
Lira
Djugu
Masindi
Masindi Port
Soroti
L. Albert
L. Kyoga
Bunia
Hoima
Semliki
Victoria Nile
Mt. Elgon 4321
Mbale
Toro Game Reserve
Tororo
Fort Portal
Mubende
Owen Falls Dam
Iganga
Jinja
Mt Stanley 5106
Kasese
KAMPALA
Entebbe
L. George
Kabatoro
KISUMU
Queen Elizabeth National Park
L. Edward
Masaka
Lake
Mbarara
L. Mburo National Park
SESE ISLANDS
Ishasha River
Kikagati
Kagitumba
Bukoba
Kabale
Victoria
Musoma
KIGALI
Rubondo Island
Mwanza

UGANDA

Uganda's position in East Africa is unique. It is at the centre of migratory routes for animals travelling between the north, south, east and west of the continent. It combines the best of these worlds, and enjoys a greater diversity of animal and plant species. Tropical forest, lakes, snow-capped mountains and endless plains are host to wildlife usually associated with West and Central Africa and Tanzania, Kenya and Somalia.

Unfortunately, almost two decades of instability and civil war have left deep scars in Uganda's once-spectacular national parks and game reserves, and the tourist flow has been reduced to a trickle. The regimes of Idi Amin (1971-79), Milton Obote (1980-85) and Tito Okello (August 1985 to January 1986) paved the way for rampant poaching, large-scale encroachment of land and the local population's negative, or at best indifferent attitude towards wildlife. At the same time the infrastructure suffered tremendously and efforts by well-meaning government officials, game wardens and rangers were futile.

However, since President Yoweri Kaguta Museveni took power in 1986 things have gradually improved. Serious efforts are being made to restore Uganda's old glory as a tourist resort. The Ministry of Tourism has been whipping tourist organisations into shape and the export of animals such as parrots and monkeys is now strictly controlled.

Uganda today has four national parks, 12 game reserves, 14 controlled hunting areas (hunting has been banned since 1980) and eight animal sanctuaries. Park names were changed under Amin's rule, but are now known by their original names. Maps still bear Amin's changes and visitors need to take note of the "double names" of each region. Under Amin, Murchison Falls became Kabalega Falls, Queen Elizabeth National Park became Ruwenzori National Park, Lake Albert was Lake Mobutu Sese Seko and Lake Edward changed to Lake Idi Amin Dada. As conditions gradually improve, some of the hitherto neglected wildlife areas, especialy the mountains and forests, will again become accessible to nature lovers.

Preceding pages: spectacular Murchison Falls plunges some 350 metres into the Nile River.

Queen Elizabeth National Park

Situated on the edge of the majestic Ruwenzori mountains on the border with Zaire, **Queen Elizabeth National Park** (QEP) is the jewel in Uganda's crown. The park, with its exceptional mixture of plains, lakes, mountains, craters and tropical forest spread over 1,980 square kilometres (792 square miles), is one of the most beautiful and versatile spots in Africa.

The park is surrounded by other conservation areas: the Ruwenzori mountains in the west, Virunga National Park of Zaire in the south, Kigezi Game Reserve in the south-west, the Rift Valley and Lake George in the west, and Kibale Forest Corridor and Game Reserve in the north.

The varied natural land features have endowed QEP with a rich variety of fauna. Elephants, hippo, waterbuck, bohor's reedbuck, bushbuck, Uganda kob, warthog, sitatunga, baboon, chimpanzee, lion, leopard and topi are just a few of the mammals that can be found.

Four hundred and ninety-two species of birds have already been identified, mainly along the 32-kilometre (20-mile) **Kazinga Channel**, which connects **Lake George** with **Lake Edward**. The same channel is the abode of hundreds of hippos. The **Mweya Lodge**, built on a bluff overlooking Lake Edward, has been recently restored.

Poaching problems: Tragically, heavy poaching during the 1970s turned QEP into something of a nightmare for conservationists. Elephant, buffalo and hippo were the main victims. The number of elephants dwindled from 3,884 in 1966 to an paltry 153 in 1980. These days, however, the numbers are slowly increasing and have almost reached the magic 1,000 mark. This is mainly due to improved security, more efficient anti-poaching units, and better cooperation with the local population.

Left, unique tree-climbing lions.

The position of hippo and buffalo, which also suffered from mass butchering, is somewhat less alarming. Many young animals survived and both species are capable of making a quick recovery, provided sufficient protection continues to be given.

The second problem facing QEP is encroachment of land, which indirectly gives rise to poaching. A large number of fishing villages have appeared over the last 20 years, resulting in overfishing of the lakes and overgrazing, with park land illegally taken over for agricultural use.

Natural beauty: Nevertheless, QEP's natural beauty will always remain. To drive along the **Kikorondo craters** route with the mountain massif of the Ruwenzori ons one side, and on the other the bright green plains, will be forever a blissful experience. The craters, numbering over 80, including seven lake craters, are mainly found in the north and east. They were formed approximately 20 million years ago during the mid-Pleistocene era. One of the biggest, **Lake Katwe**, is currently used for salt mining.

Just as spectacular is a trip on the Kazinga Channel, which is more like a lake and is fed by numerous streams. Here, spotting a hippo is as easy as spotting a star on a cloudless night. Bird lovers will have a field day watching the largest concentration of birds in Uganda. Large white- and pink-backed pelicans (*Pelecanus onocrotalus* and *P. rufescens*) are worth looking out for.

Vegetation in the park is dominated by bush thickets mixed with short grass and grassland. But savannah, swamps and semi-deciduous forest can also be found.

Nature has divided the area into three sections: north and south of the channel, with the south side split by the thick, nearly impenetrable **Maramagambo Forest**. Despite its relative proximity, it is no longer possible to reach the **Ishasha** side (south of the forest) from the Mweya side.

Although difficult to reach Ishasha in the south, this section of the park should definitely be included in a safari since it is widely described as the most unspoilt game area south of the Nile. Apart from one small village, there are few signs of human habitation.

The main attractions within these 70 square kilometres (28 square miles), which can only be reached by four-wheel drive vehicle from **Kabale** or by aircraft, include tree-climbing lions, the highest density of Ugandan kob in the world, the only place in QEP where topi roam, two pools literally jam-packed with hippos, and a gallery forest along the rivers. This is the habitat for monkeys, chimpanzees and giant forest hogs and numerous forest birds such as turacos and hornbills.

Branching out: Lions have chosen the fig tree for their uncommon behaviour. No good explanation for this curious habit has yet been given, although some zoologists claim that it may have something to do with avoiding flies that carry diseases. Others think that the lions climb to relax after feeding because it is cooler in the trees. Whatever the reason, it is indeed bizarre to see a huge, lazy lion resting on a tree branch with his belly bulging over. The only other place in the world where tree climbing lions can be found is Lake Manyara National Park in Tanzania.

Topi, which feature only on the Ishasha side of QEP, have been heavily poached by Tanzanian troops after they ousted Idi Amin in 1979. Conservationists claim that one year of Tanzanian soldiers on the rampage did more damage to the topi population than eight years of Amin.

At certain times of the year, when cloud cover is scarce, it is possible to see from Ishasha both the Ruwenzori mountains and the Virunga volcanoes in Rwanda.

At present, QEP—especially the northern side—is Uganda's main tourist attraction. It is virtually the only place in Uganda that needs advance booking during Christmas and Easter.

Both sides of the park have enormous potential for tourism, if only the roads to Mweya and Ishasha and between the various sectors can be improved. There is an airstrip for light aircraft at Mweya, while larger planes can land at Kasese.

Far left, inquisitive ground squirrel.

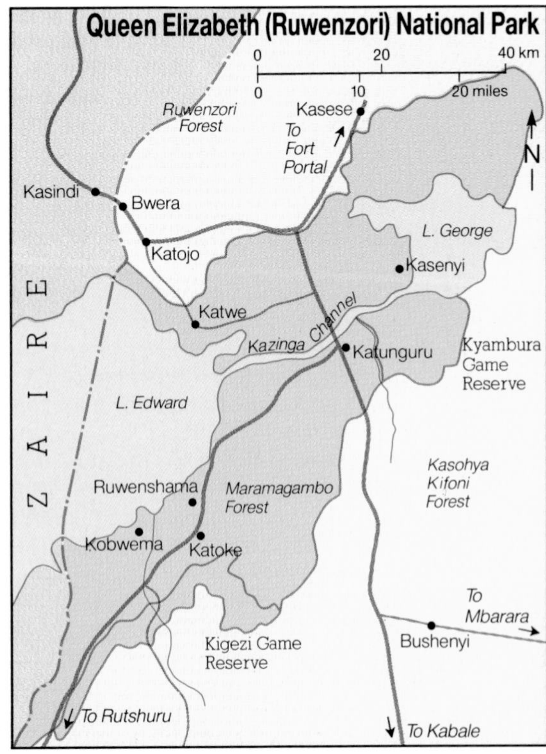

MURCHISON FALLS

Before the tragic effects of the Amin regime in the late 1970s, **Murchison Falls National Park** (MFNP) was the pride of the country and the main attraction for tourists. Its abundant game and dazzling scenery, together with a spectacular waterfall and the excellent fishing possibilities (especially for huge nile perch) had made it into a very appealing holiday resort. Two luxurious lodges, at **Chobe** and **Paraa**, offered all possible comforts to visitors.

The 3,900-square-kilometre (1,560-square-mile) park, the biggest in Uganda, is split into two sections by the **River Nile**, and is situated in the northwest of the country, close to where the Nile enters Lake Albert. The southern sector in **Masindi district** is bigger than in the north, but has fewer animals.

The park was gazetted in 1952 after sleeping sickness earlier this century made the area unsuitable for the grazing of livestock. It derived its name from the glorious **Murchison Falls** near the park headquarters at Paraa, where the Nile narrows to a mere two metres (6.6 feet) before it plunges 350 metres (1,150 feet) into the calmer waters below. Here is a completely different world where hippos, crocodiles and water birds quietly roam and elephant and buffalo gather to quench their thirst. This is a beautiful stretch of the river.

Loss of wildlife: The present state of the game in MFNP recounts the sad history of wildlife in Uganda from the mid-1970s onwards. The days when elephant were the biggest threat to vegetation were abruptly curtailed. The devastating extent of poaching in MFNP has been unequalled elsewhere.

In the early 1980s carcasses outnumbered living elephants by a shocking three to one. Numbers of elephants were reduced from 15,000 in the early 1970s to only 1,200 in 1980. Continuing conflicts, especially in the northern sector, have made recent countings impossible,

Storm clouds approach the park.

although game wardens are of the opinion—perhaps somewhat optimistically—that herds are once again on the increase.

Meanwhile, white and black rhinoceros are now completely extinct, and herds of buffalo, hippo and crocodile have been decimated in a similarly dramatic manner.

Other animals have fared somewhat better and Uganda kob, Jackson's hartebeest, oribi, bohor's reedbuck, duiker, warthog, Rothschild's giraffe, lion, leopard, hyaena and numerous birds can still be found in the park.

Like most parks in the country, MFNP offers a wide variation in vegetation, ranging from semi-deciduous closed forest with *Cynometra alexandri* in the northern sector, to swamp vegetation along the valley floors. Savannah and shrubby acacia woodland cover large sections of the park.

In the 1960s and early 1970s woodlands suffered much damage from burning and being trampled by large herds of elephants. The lack of mid-term growth of trees has also led to the vast expansion of grasslands.

The future for MFNP still looks relatively grim. The northern sector remains insecure because of pockets of anti-government rebels, making work or visits to this remotesection of the park virtually impossible, or at least very risky. Only the more adventurous will travel here.

Meanwhile, access to the southern part is hindered by badly maintained roads. **Paraa Lodge** was restocked in 1987 but looted again later that same year. Moreover, the ferry which connects the two sections of the park has been out of order for many years.

On the brighter side, construction of the main road to Murchison Falls, built with the assistance of the European Economic Community, was finished in early 1989, and has opened up the southern section to the public once again. The first visitors to the northern section were reported in January 1989. There are also airstrips for light aircraft near the lodges.

Far left, feeding time for this youngster.

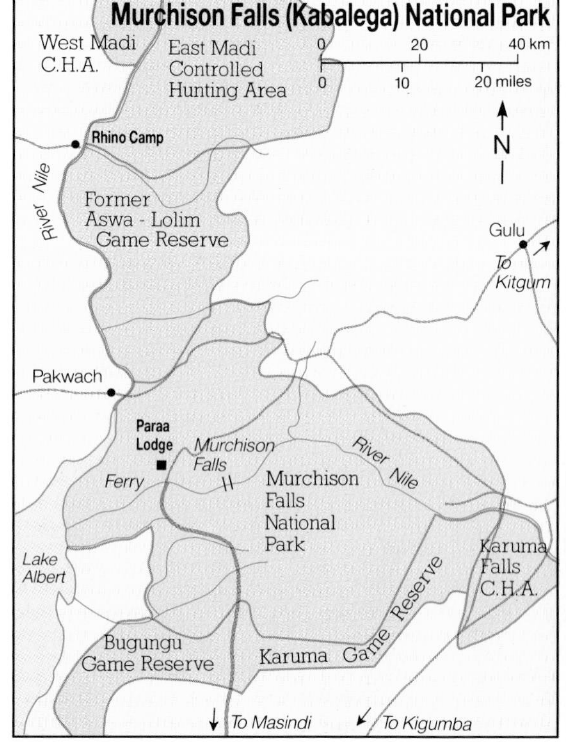

Kidepo Valley

Situated in an almost inaccessible part of Uganda—the north-east corner of **Karamoja** that borders **Sudan** and is close to the semi-desert **Turkana district** of Kenya—**Kidepo Valley National Park** (KVNP) may well be the most unspoilt park in the world.

The 1,400 square kilometres (560 square miles) of KVNP deserve more attention than they currently receive. Persistent insecurity in the east between 1987 and 1989 practically cut off the area from the rest of Uganda. Anyone interested in visiting Kidepo had to rely on aircraft, provided that clearance could be obtained, and even then, lack of transport facilities within the park, and a lodge without food or diesel for electricity, made a visit here one of the least likely options for tourists in Uganda.

Years of civil war turned KVNP into a temporary free-for-all where cattle rustlers and poachers from Uganda and Sudan gladly participated. The presence now of an army detachment and some 70 armed game rangers has greatly improved security in the park and poaching is controlled to a considerable extent. This should make the park safe for tourists.

This area was made into a game reserve in 1958; four years later it was gazetted as a national park. In 1967 conservationists welcomed an extension to the southern region with the addition of the **Narus River**, used by wild animals during dry season grazing. Kidepo means *to pick* in the local dialect and refers to the coconuts left to pick at a spot near the Narus River.

Wild Africa: High, freakish mountainscapes in the west, south and east and two seasonal rivers, the Narus and the **Kidepo**, offer an impression of how wild Africa must have looked. **Mount Morongole** is the eastern peak and the 2,700-metre (8,856-foot) top of **Lotuke** marks the border with Sudan.

A flight from **Entebbe** airport reveals the almost surrealist mountain ridges. The intricate interplay of clouds, rain and sun adds to their uniqueness. Sporadic *manyattas* of the nomadic Karamajong tribesmen can be spotted from the air and clouds of smoke mark the areas being burnt when the rainy season has finished.

Unlike other Ugandan parks KVNP resembles the physical lay-out of game parks in the drier neighbouring countries of Kenya and Tanzania. But unlike Kenya, what makes Kidepo so attractive is the lack of noisy convoys of tourist buses: a drive through Kidepo will be a very lonesome and isolated experience.

The park is characterised by four main types of vegetation which are closely bound up with the seasonal flow of the two rivers. The **Valley of the Narus**, which is well watered, has savannah woodland, later gradually merging into fire-climax grassland in the south. Trees include several species of acacia. Apart from the savannah woodland, large areas are covered with a mixture of shrub steppe, bushland and

Far right, lions spend much of the day sleeping.

dry thicket.

Naturally, the animal population is closely associated with the various types of vegetation. Giraffe, elephant, buffalo and rare (in Uganda) Burchell's zebra favour the savannah woodland around the two rivers. Oryx, greater and lesser kudu, Grant's and Bright's gazelle and mountain reedbuck, klipspringer and ostrich can be found in the slightly drier areas.

Like everywhere else animal populations have decreased considerably since the 1970s and recent counts have not been available. KVNP used to be a stronghold for elephant and rhino but this is no longer the park's claim to fame.

During the Amin era, however, KVNP was saved some of the torment that befell other parks, both because of its remoteness and isolation and thanks to the excellent work of game warden Paul Ssali. The laudable efforts of this "warrior for wildlife" have been imaginatively described in John Hemingway's book, *No Man's Land.*

The Kidepo Valley even blossomed during the days of Amin: it seems the former dictator visited the park and took a keen interest in animals. From the lodge's balconies looking out over the yellow and green Narus Valley, the view is spectacular. Elephant, giraffe, zebra, antelope and waterbuck are easy to spot, especially during the dry season (December to April).

During the 1960s and 1970s KVNP was often fully booked. It had a shop selling frozen and canned foods and a lorry frequently brought new supplies. Although those days are over, the lodge has benefitted from a very dedicated and honest workforce, and has not been looted like the lodges in Murchison Falls. The white linen tablecloths, cutlery and wine glasses have survived the years of civil strife, and the 16 double rooms can be used at any time, provided visitors bring their own food and drink.

The future for KVNP looks somewhat brighter, although work still needs to be done to bring the area and facilities up to standard.

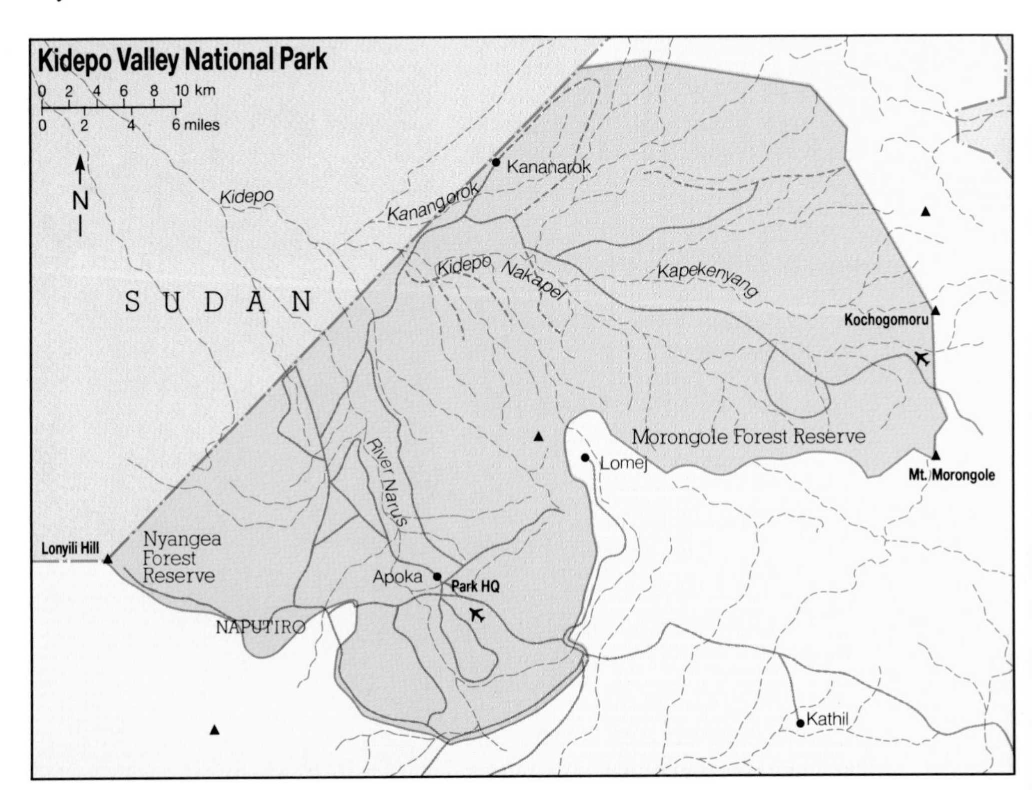

Lake Mburo

Lake Mburo National Park (LMNP) is the latest Ugandan national park—it was officially gazetted in 1982. However, the original 642 square kilometres (256 square miles) have turned into a political hotbed which has overshadowed its great potential as a holiday resort both for the local urban population and foreign tourists.

Earlier this century, the area was such an attractive place for wild animals that in 1936 the government was forced to declare it a controlled hunting area. The Kenyan born writer Brian Herne vividly describes those days—when the park boasted many lions—in his book, *Ugandan Safaris.* In 1964 the area was turned into a game reserve, having been cleared of tsetse fly in preceding years.

Conflicting interests: The clash of human and wildlife interests started in the mid-1970s. Cattle ranches blossomed at the edge of the reserve, and during the Amin years, ranch owners were able to claim huge tracts of land for themselves. At the same time the prevailing lawlessness opened up the reserve to *Bahima* herdsmen who immediately took over the good grassland and watering holes, and even poisoned the game which were competing for grazing pasture. When the reserve was declared a national park, no less than 6,000 people and 40,000 head of cattle had to be evicted. Lack of compensation and subsequent unrest in the country worsened the situation.

Tight negotiations between the ranchers, herdsmen, government, park authorities and conservationists finally resulted in an agreement to divide the park into two: 386 square kilometres (154 square miles) for the people evicted in 1982, and 256 square kilometres (102 square miles) for the wildlife. A task force is currently working to establish the exact boundaries.

Worth the effort: Despite its problems, LMNP remains worth visiting, if

Tawny eagle spreads its wings.

only for those animals that cannot be found elsewhere south of the Nile—impala, klipspringer, eland, roan antelope and oribi. Other species of game include numerous topi, waterbuck, bohor reedbuck, sitatunga, bushpig, bushbuck, porcupine, baboon, vervet monkey, greater galago, leopard, hyaena, hippo and buffalo. This is one of the few places in the country where numbers of animals increased between 1982 and 1985.

Already 258 species of birds have been identified. Future discoveries are likely to bring the actual number closer to 400.

Six bird species cannot be found in any other park. These include rufous bellied heron, black-throated barbet, red-faced barbet, tabora cisticola, green-capped eremomela and southern red bishop.

The slowly rolling hills and bright green plains have become dominated by *Acacia gerrardii* and *Acacia hockii* at the expense of grass. This alarming trend has been attributed to overgrazing by cattle. Grass becomes too short for regular burning, resulting in the unimpeded spread of acacia trees. Park authorities are considering reintroducing giraffe, which feed on acacia, in the hope of halting any further advance.

One advantage of LMNP is that the absence of lions makes it reasonably safe for visitors to walk through the park, although they should be cautious of the occasional lone buffalo.

Depending on the outcome of negotiations with the herdsmen, LMNP has the potential to become a short trip holiday resort, especially for the urban populations of **Kampala** and nearby **Masaka** and **Mbarara**. It is only a short drive to the park from all these towns. Roads and tracks have been badly damaged by the weather and lack of maintenance, but the current park staff are doing a commendable job in making the main circuits manageable again. Although facilities are somewhat lacking, this should not discourage anyone from visiting. There are excellent natural camping sites near the lake.

Helmeted guinea fowl—one of the many bird species found here.

Forests And Mountains

The ridges of the Ruwenzoris, Mount Elgon and the Virunga volcanoes display an unrivalled scenic splendour. But even more important will be Uganda's rich and unexplored forests which have the potential to revive the country's tourist industry. These include the Bwindi Impenetrable Forest, Kigezi Game Reserve and Kibale Forest Corridor Game Reserve.

The Ruwenzoris in the west, on the border with Zaire and the volcanoes in the south-west on the border with Rwanda, and Mount Elgon in the east next to Kenya offer excellent opportunities for mountain climbing, with long walks through unique flora where numerous animal species roam.

Plans are being discussed to turn part of the Ruwenzoris into a national park. The mountain ridge with the snow-capped peak of Margherita on Mount Stanley harbours elephant, buffalo, leopard, red forest duiker, giant forest hog, chimpanzee, blue monkey, Stuhlman's monkey and the Ruwenzori hyrax. It is the only place in Uganda where the extremely rare Ruwenzori colobus can be found.

Unfortunately, if the area is not soon gazetted into a game reserve or park, it may be damaged by the slow encroachment that is already taking place. Similarly, at Mount Elgon, there is great potential for tourism although cattle raiders have made this area relatively insecure.

The third mountain corridor is the **Kigezi Mountain Gorilla Reserve** in the south-west, dominated by the Virunga volcanoes. The 43-square-kilometre (17-square-mile) reserve lies between the towering peaks of mounts Muhabura and Gahinga.

The **Travellers' Rest Hotel** at **Kisoro** is the starting point for a trek to see the famous mountain gorillas (*Gorilla gorilla beringei*). Visitors need to check with the game warden in Kabale before

Scenic walk in the Ruwenzoris.

proceeding to Kisoro to register at the hotel, which will provide them with a guide. He only takes visitors up twice a week, preferably on Saturdays and Wednesdays.

A three- to four-hour climb takes you to the habitat of the mountain gorilla, though chances of actually seeing this huge primate are less than 100 percent. (A generous tip for the guide improves your chances!) But, the reserve offers other spectacular sights, including thick bamboo forest and the rare golden monkey (*Ceropithecus mitis kandti*), which make the climb worthwhile. Other animals to be seen include buffalo, bushbuck and giant forest hog.

The drive from Kabale to Kisoro is simply spectacular. This area is rightly referred to as the "Switzerland of Uganda", and is made up of green mountains, dangerous hair-pin bends, misty valleys, terraced fields, and idyllic lakes, such as the romantic **Lake Bunyonyi**, just a few kilometres from Kabale off the Kisoro road.

Jungle walks: In the south-west lies the 560-square-kilometre (224-square-mile) **Impenetrable Bwindi Forest/ GameSanctuary** which has one of the richest forest mammal faunas in Africa with 97 species. Most special is the mountain gorilla; other mammals include chimpanzee, elephant, and several species of duiker.

This well-watered area was gazetted as a game sanctuary in 1961, mainly because of the mountain gorilla whose numbers are currently estimated at 115. This figure has dwindled since the early 1960s and the government and conservationists are therefore extremely cautious about letting visitors into this area.

Much better suited for tourists is the **Kibale Forest Reserve**. Together with the adjacent **Kibale Forest Corridor and Game Reserve** (KFCGR) the area covers 900 square kilometres (360 square miles) close to the western town of **Fort Portal**, where accommodation is readily available.

The Kibale Forest Corridor and Game Reserve is quite well protected, especially around the research centre of the New York Zoological Society, only 16 kilometres (10 miles) from Fort Portal, which has been turned into a paradise for everybody who likes to walk in thick tropical rain forest. The reserve harbours the largest concentration of primates anywhere in the world: no less than 11 species can be found here in large numbers. Among them are: red colobus, black and white colobus, chimpanzee (the focus of American research), blue monkey, red-tailed monkey, L'Hoest monkey, mangabey, and olive baboon.

They can be spotted quite easily during a daytime walk through the dark pines and bright green jungle. The trek on foot has been facilitated by the 300 kilometres (186 miles) of paths and tracks which have been laid out by American researchers. A map and all the necessary information can be obtained from the research headquarters.

Other animals to be found in the forest reserve include two species of duikers, bushpig, bushbuck, warthog, giant forest hog, waterbuck, sitatunga, serval, golden, palm and civet cat.

However, KFCGR no longer harbours many large animals. The main purpose of gazetting the area was the now quite unbelievable fact that it provided a migratory route for elephants moving along a south-north axis, all the way from Zaire to Sudan, via Queen Elizabeth National Park, the Kibale Corridor, Katonga Game Reserve and Murchison Falls!

Today, massive encroachment in the KFCGR and elsewhere clearly prohibits this type of movement. Without doubt, forests are Uganda's most latent tourist attraction. The national parks, although spectacular, have to compete with neighbouring countries where facilities are much better.

The forests, however, are unique to Uganda, which is the only English-speaking country where mountain gorillas can be seen. The population of 600 to 800 chimpanzees in KFCGR cannot easily be found elsewhere. The animals are quite easy to spot because they have become used to human beings, although this does not detract from their natural behaviour.

Right, Lake Bunyonyi is near the Rwandan border.

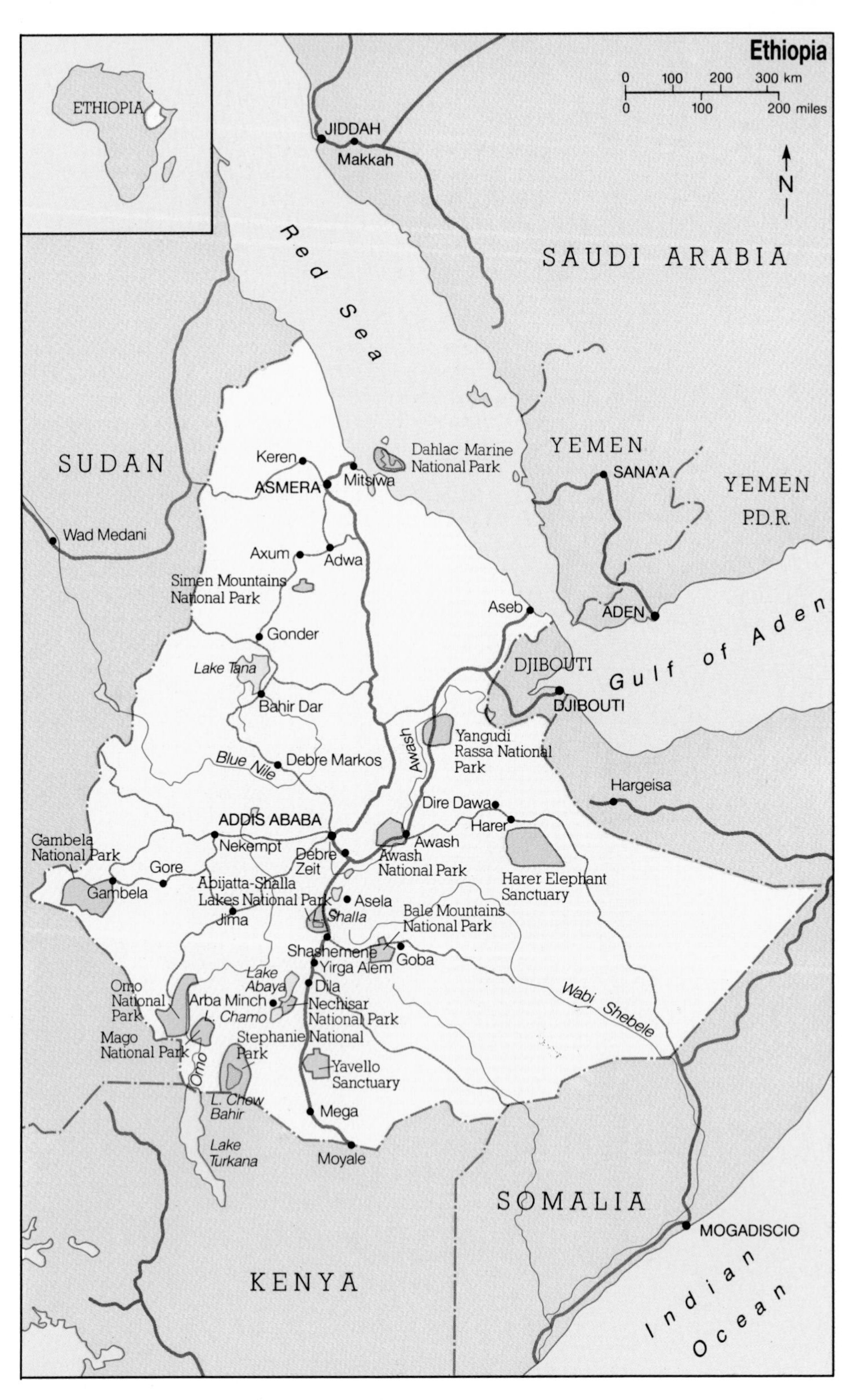
Ethiopia
0 100 200 300 km
0 100 200 miles
N
ETHIOPIA
JIDDAH
Makkah
Red Sea
SAUDI ARABIA
SUDAN
Keren
ASMERA
Mitsiwa
Dahlac Marine National Park
YEMEN
SANA'A
YEMEN P.D.R.
Wad Medani
Axum
Adwa
Simen Mountains National Park
Aseb
ADEN
Gonder
Lake Tana
DJIBOUTI
DJIBOUTI
Gulf of Aden
Bahir Dar
Awash
Yangudi Rassa National Park
Blue Nile
Debre Markos
Hargeisa
Dire Dawa
Harer
ADDIS ABABA
Gambela National Park
Nekempt
Awash
Debre Zeit
Awash National Park
Gore
Gambela
Abijatta-Shalla Lakes National Park
Asela
Harer Elephant Sanctuary
Jima
L. Shalla
Bale Mountains National Park
Shashemene
Goba
Yirga Alem
Lake Abaya
Dila
Omo National Park
Arba Minch
Nechisar National Park
Wabi Shebele
L. Chamo
Stephanie National Park
Mago National Park
Omo
Yavello Sanctuary
L. Chew Bahir
Mega
Lake Turkana
Moyale
SOMALIA
MOGADISCIO
KENYA
Indian Ocean

ETHIOPIA

Ethiopia is a vast wrinkled tableland, divided by two blocks of highlands with numerous peaks and valleys. The Ethiopian highlands were first formed 40 million years ago, when lava flowed over a huge area, both from volcanoes and from a gradual and insidious outpouring of molten rock. The Great Rift Valley subsided, splitting the highland block in two, to form the more extensive western highlands—on which the capital city of Addis Ababa is located—and the horseshoe-shaped eastern highland chain.

The country varies in altitude from some 100 metres (328 feet) below sea level in the Dalol Depression, to over 4,500 metres (14,760 feet) at the highest peak of Ras Dedjen in the Semien Mountains. In between these two extremes, 40 percent of Ethiopia is formed by the highland blocks lying 2,000 metres (6,560 feet) and more above sea level. The highlands have been cut off from similar high country in other parts of Africa for 20 million years by the arid lowlands surrounding them. Their size and isolation have ensured that many of the less mobile animal and plant species have evolved to cope with the unique conditions into forms not found elsewhere. Hence the very high numbers of endemic species found only in Ethiopia.

Eritrea, which covers the northern coastal strip and has its capital in Asmara, became a separate sovereign state in May 1993. The country's attractions include deep sea diving off the Red Sea and beautiful beaches such as Massawa.

While wildlife conservation is not a priority for the Ethiopian government, given the more pressing problems facing the country, tourists are welcome and facilities will continue to develop. Four of the national parks are already well developed and accessible to tourists. Most of the others can be visited, although it requires considerable organisation and expense on the part of the vistor, and a willingness to cope with the logistical problems involved in getting to little known areas.

Although this guide deals specifically with its wildlife parks, visitors should not confine themselves to an appreciation of Ethiopia's wildlife. Many elements have influenced the country over time, making Ethiopia a land of mystery and a land apart.

Preceding pages: **the eastern edge of Ethiopia's Sanetti Plateau.**

ABIJATTA-SHALLA LAKES

Situated in the Great Rift Valley, only 200 kilometres (124 miles) south of Addis Ababa, and in the **Lake Langano** recreational area, the **Abijatta-Shalla Lakes National Park** attracts numerous visitors. It was created primarily for its aquatic bird life, particularly those that feed and breed on lakes Abijatta and Shalla in large numbers.

The park comprises the two lakes, the isthmus between them and a thin strip of land along the shorelines of each. Developments have been limited to a number of tracks on land, and the construction of seven outposts. While attention is focused on the water birds, the land area does contain a reasonable amount of other wildlife.

Two different lakes: The two lakes are very different in character. Abijatta is shallow at about 14 metres (46 feet): Shalla has a depth of 260 metres (853 feet) and is calculated to hold a greater volume of water than all of the Ethiopian Rift Valley lakes put together. Abijatta is surrounded by gentle, grass-covered slopes and swathed in acacia woodlands. Shalla exudes a sense of mystery and foreboding, surrounded as it is by steep, black cliffs and peaks that reflect in its deep waters, which are liable to be whipped up by sudden storms and flurries of wind. It contains nine small, isolated islands, rarely visited since there are no boats on the lake. These islands provide an excellent breeding ground for many bird species.

The network of tracks in this park is always developing. At present you can enter at four different points, three of which are inter-connected. Approaching from Addis you first reach the **Horakello** entrance, where the small Horakello stream flows between lakes Langano and Abijatta. The stream mouth is a source of relatively fresh water, much frequented by water birds for drinking and bathing.

Abijatta itself is very alkaline but shallow, so flamingoes can be seen scattered over most of its surface, and especially along the windward edge where their algal food source concentrates. You can approach quite closely, but beware of treacherous deep mud if the lake is low. Large numbers of both greater and lesser flamingoes gather here, together with great white pelicans and a host of other water birds.

A track which runs for 20 kilometres (12 miles) along the treeline of the eastern shore of Lake Abijatta connects Horakello with the park headquarters further south at **Dole.** From here you can see other parts of Lake Abijatta and some mammal species, especially Grant's gazelle, warthog and occasionally the oribi.

Hot springs: The headquarters houses a small museum, currently being upgraded, which gives an excellent idea of the wealth of birdlife in the park. There are over 400 species recorded here, almost half the number recorded for the whole country. A further track leads on from Dole to the shores of Lake Shalla where hot steam, mud and water

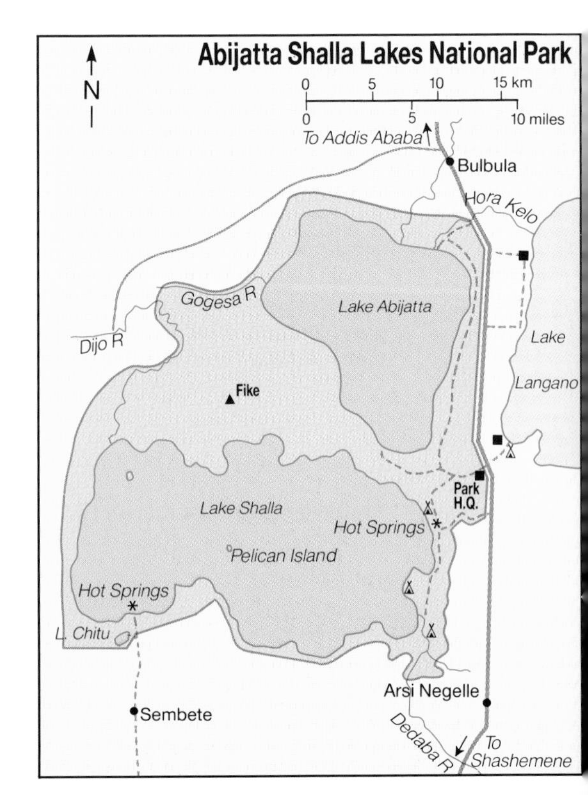

bubble to the earth's surface. Revered locally for their medicinal properties, the hot springs (*Filwoha*) have a sense of primaeval mystery about them, especially in the cooler early mornings. They are relics of the massive volcanic activity that has formed this amazing country and landscape.

A further entrance to this park exists in the south, where a rough track leads to another small hot spring area at **Ghike**. Here you can stay in a wooden self-help guest house, perched high on a cliff above the lake, with a view across the islands. There are plans to install a boat at the lake which will ferry small groups of people to the islands to observe the breeding colonies of thousands of great white pelicans and greater flamingoes. The great white pelican colony is estimated to be visited by up to 13,000 pairs annually, and is the most important breeding site for the species in the world.

There is no accommodation in the park but Lake Langano, which lies just over the main road marking the park boundary, has two reasonable hotels on its shores, the **Wabe Shebelle** and the **Bekelle Mola**, from which all parts of the park are easily reached. It is possible to camp at the hot springs and further south on the track east of Shalla, leading to the **Dedaba River** and outpost.

In association with the Abijatta-Shalla Lakes National Park is **Senkello Swayne's Hartebeest Sanctuary**, some 70 kilometres (43 miles) from the town of **Shashemene**, and close to the **Chike** entrance of the park. The sanctuary was established for this endemic subspecies of the hartebeest (*Alcelaphus buselaphus swaynei*) which once roamed the plains of Somalia and Ethiopia in thousands, but is now restricted to four small localities in Ethiopia. The sanctuary is small but well worth a visit. Set beneath a small rounded hill, over 2,000 of these rich, chocolate-coloured hartebeest are packed into this area of wooded grassland, along with bohor reedbuck (*Redunca redunca*), oribi and many different species of birds.

Unspoilt Lake Shalla.

AWASH

Situated only 225 kilometres (140 miles) south east of Addis, down in the Rift Valley, **Awash National Park** was created for its concentrations of arid grassland and acacia woodland fauna. Dominated by the volcano **Mount Fantale**, the valleys, plains and gorges contain an abundance of mammals and birds which are easily seen from the excellent network of tracks. A good tarmac road connects Awash with Addis, and there is accommodation at the **Kereyou Lodge**, which comprises large caravans set in a beautiful location overlooking the junction of the Arba and Awash river gorges. The nearby park headquarters, beside the **Awash River**, has a small informative museum and campsite, and is close to the **Awash Falls.**

Various circuits exist south of the main road, enabling you to travel through the **Illala Sala plains** where large numbers of beisa oryx, Soemmering's gazelle, ostrich and kori bustard (*Ardeatis kori*) can be seen. Look at these animals carefully, for besides the usual red-billed oxpeckers (*Buphogus erythorhynchus*) on their backs you will also see jewelled carmine bee-eaters (*Merops nubicus*) hitching rides, and using their mobile perches as lookout points to watch for the insects disturbed by the animals' feet. To see one of these birds nestled down between the wings of an ostrich, or riding a stern kori bustard is an unforgettable sight.

A small group of Swayne's hartebeest lives in the park where thousands used to roam in the past. They were reintroduced about 20 years ago, and their numbers are slowly increasing.

Simply gorgeous: Other circuits lead you along the edge of the spectacular **Awash River gorge** and through areas of dense wait-a-bit thorn scrub where careful observation will reveal both greater and lesser kudu, as well as scores of Salt's dikdik

Beisa oryx and passenger.

(*Madogua Saltiana*) at the cooler times of day. A journey along the upper reaches of the river above the falls will give you sightings of defassa waterbuck, warthog, vervet monkey and olive baboon. Observe these baboons carefully: Awash is the site ofa crossbreeding exercise between the olive and the hamadryas baboon (*Papio hamadryas*) and several of the troops in this southern area exhibit hybrids and individuals of both species.

Travel north of the main road into the bulk of the park with a wildlife scout, and take the circuit that leads to the **hot springs**. There you will see the hamadryas baboon in its pure form, the males with their spectacular silvery capes blowing in the wind as they come to drink from the warm streams. This species is absent from the rest of East Africa, but its presence in the mountains of Yemen across the Red Sea indicates that there were ancient links between these two areas.

The hot springs gush from the ground at over 35 C (95 F), collecting in deep, clear, blue pools—inviting but enervating for those who dare to take the plunge.

This circuit continues around Mount Fantale through attractive rolling country with faults, ridges, lava flows and eroded features. This needs a full morning or afternoon and takes you from the main gate back to the main road at the **Sabober gate.** Greater kudu are common here, as are herds of oryx at different times of the year. Don't expect giraffe and zebra since both species are curiously absent, although Grevy's zebra are present nearby and can sometimes be seen wandering within the park boundary.

An interesting track leads you up the steep slopes of Fantale, through the changing vegetation, to a spectacular view from the crater's rim. Early in the morning, steam can be seen rising from the vents in the centre; vultures glide lazily beneath you, and you may be able to catch the occasional glimpse of the bushy-tailed Chanler's mountain reedbuck (*Redunca fulvorufula chanleri*).

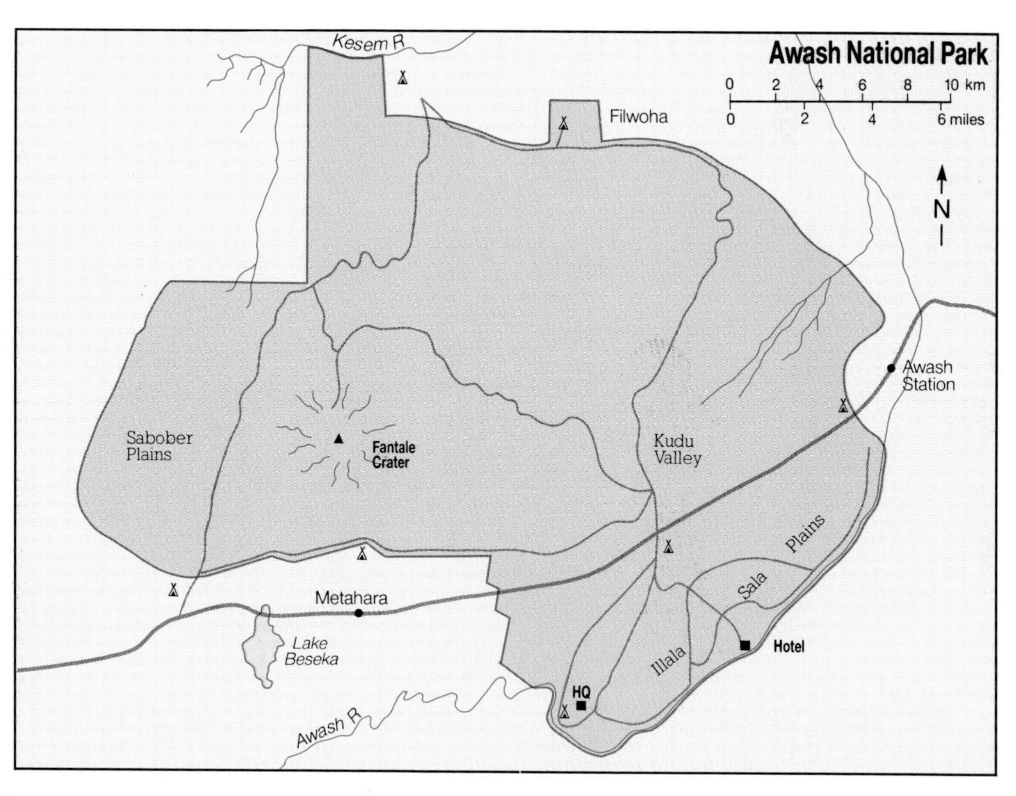

THE BALE MOUNTAINS

The road to the **Bale Mountains National Park** takes you 250 kilometres (155 miles) on tarmac south down the Rift Valley to Shashamene, and then 150 kilometres (93 miles) due east on a good gravel road across the high wheat-growing plateau and up into the mountain massif. Alternatively you can take the Nazareth road towards Awash, turn south to **Asella**, then take the gravel road south to the mountains. It is worth doing a circuit, going one way, returning the other, and taking in Awash and Abijatta-Shalla Lakes national parks in the process.

You first encounter the park as you ascend the mountains, travelling along part of its northern boundary. A steep juniper-filled valley, rising through giant heather (*Erica arborea*) moorlands and topped by volcanic plugs and dykes, forms the gateway to the mountains. After climbing to 3,600 metres (11,800 feet), visitors drop into the **Gaysay valley** and cross the small area of montane grasslands that form the northern extension of the park. The village of **Dinsho** nestles against the mountains here, and houses the park's headquarters, campsite, self-help lodge and museum.

Dinsho is located right in the north of the park with the bulk of the mountains lying to the south. To reach these, you can either go by vehicle 20 kilometres (12 miles) south into the **Web valley** along the **Simbirro** track, and continue by foot or on horseback to the central peaks area, or you can continue east by vehicle to **Goba**, the capital of the Bale region. Here the excellent **Goba Ras Hotel** provides a good base for exploration of the main part of the park.

Mountain high, valley low: A good gravel road climbs into the mountains south of Dinsho, rising through juniper, hagenia and hypericum and forest zones to heather moorlands, and eventually on to the flat expanse of the

A dip in the hot springs.

Sanetti Plateau. This is the highest all-weather road in Africa, at over 4,000 metres (13,120 feet) above sea level. The road travels for 30 kilometres (18 miles) along the plateau to the southern edge, and then performs a spectacular zigzag descent of the **Harenna Escarpment**. In 10 kilometres (six miles) it falls over 1,000 metres (3,280 feet) into the top of the **Harenna Forest**, and then descends gradually over the next 60 kilometres (37 miles) to the southern edge of the forest and park boundary.

The park is spread over 2,200 square kilometres (880 square miles). Approximately half of this comprises the high plateau and mountain area; the other half is the dense Harenna Forest. Altitude varies from 1,500 metres (4,920 feet) at the southern edge of the Harenna Forest, to 4,377 metres (14,357 feet) at the top of **Tullu Deemtu**, the highest peak in southern Ethiopia and outside the Semien Mountains.

The Gaysay Valley is the best place to see larger wildlife, either by vehicle or on horseback. Large numbers of the endemic mountain nyala (*Tragelaphus buxtoni*) occur here, forming the best concentration of the species in the world. In the wet season the flower-filled valley is popular grazing land for nyala, bohor reedbuck, grey duiker (*Cephalophus grimmia*), warthog, and Menelik's bushbuck (*Tragelaphus scriptus meneliki*). Serval and Egyptian mongoose (*Herpestes ichneumon*) are particularly common, and caracal are occasionally seen.

The mountain block comprises the Sanetti Plateau to the east, separated from the wide **Upper Web valley** by lava flows and skirted by steep heather moorlands. When allowed to grow unmolested the tree-like heather reaches heights of five metres (17 feet) and is difficult to penetrate, either on foot or on horseback.

High-altitude desert: The open Afro-alpine moorlands above the treeline are home to the simien fox (*Canis simensis*), a long-legged, long-snouted endemic member of the dog family, whose

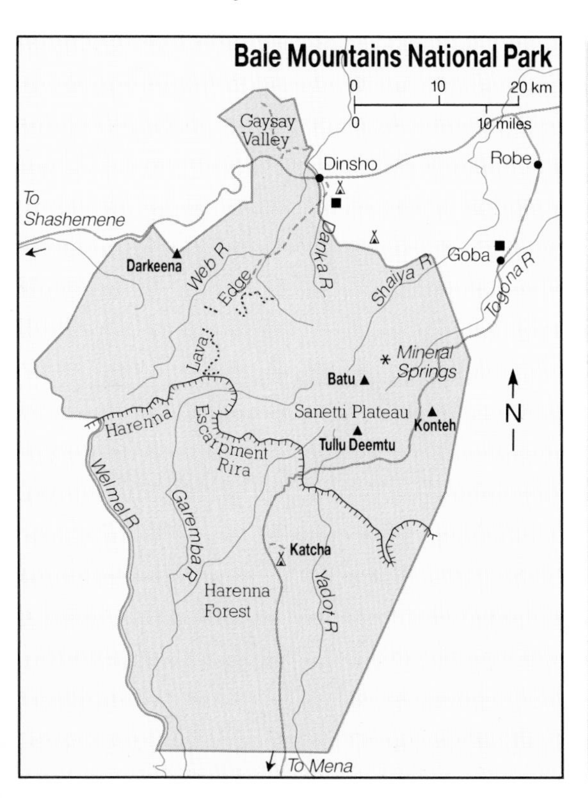

Far right, the attractive Bale Mountains.

overall population of less than 1,000, mainly here in Bale, gives cause for concern.

Other animals include low numbers of mountain nyala and at least seven rodent species, most endemic to Ethiopia. These occur in large numbers, scurrying away as you pass through and squeaking in indignation at the intrusion. Although not normally considered wildlife by most people, it is worth taking the time to watch these rodents as they try to detect the numerous predators which gather to feast on them.

Perhaps the most interesting animal is the giant molerat which is endemic to the Bale mountains. It has modified the habitat considerably since glaciation over 10,000 years ago. The molerats' constant burrowing, collection of surface vegetation and heaps of discarded rubbish near their nests have created a ploughed landscape dotted with mounds covered in different vegetation. Giant molerats weigh almost one kilogram (2.2 pounds) and are preyed upon by the simien fox which will wait patiently for hours over an open burrow.

In the high altitude desert there are no trees but several interesting plant species have adapted to the extreme conditions. Look in particular for a plant that has opted for gigantism—the endemic giant lobelia which is over five metres (17 feet) tall when it flowers.

Temperatures on the plateau in the dry season (November to February) can fall as low as -20 C (-4 F), but within 24 hours will rise to 26 C (79 F) at noon. This amazing change explains many of the adaptations and behaviour of the animals and plants in this area as they seek to avoid being alternatively baked, frozen and desiccated.

The best way to see the mountains is to hire horses from the park office in Dinsho. There are a number of routes starting there which can be designed to suit the time you have available. Simple circular routes returning to Dinsho can take one to three days; longer routes go through the Web valley, lava flows and across the plateau where your vehicle can meet you. Take a wildlife scout or

Mountain nyala in the Gaysay Valley.

guide from Dinsho to help you find your way and to lead you to the mineral springs, alpine tarns and cave shelters you would otherwise probably miss.

The Harenna Forest is well worth a visit, if only to add to a day trip from Goba. The escarpment is cloud-covered in the wet season but the forest below is often sunny and bright. It is worth diverging to the natural grass clearing of **Katcha**, 10 kilometres (six miles) below the village of **Rira**, and even camping there, to experience the forest with its rich, abundant bird and mammal life. Leopard, black serval (*Leptailurus serval*), lion, all three of Africa's pig species, Menelik's bushbuck, grey duiker, black and white colobus—better known in Ethiopia as "Guereza"—olive baboon, grivet (*Cercopithecus aethiops*) and Sykes monkey are commonly seen. Mountain nyala also occur here in low numbers.

Bale is a bird haven with many habitats at different altitudes which attract a broad variety of species. Over 200 different birds have been recorded and once the forest is properly studied, considerably more will probably be discovered. Fifteen of Ethiopia's 27 endemic bird species have been recorded there and a week spent bird watching in the various habitats should enable you to see a wide variety.

Bale is Ethiopia's flagship conservation area, containing so many different habitats and species typical of the Ethiopian Highlands that no visit to the country is complete without spending a reasonable amount of time there.

A day trip from Goba will take you to the **Sof Omar limestone caves**, 110 kilometres (68 miles) east of the town of **Robe**. The **Web River** flows from the mountains through a total of 15 kilometres (nine miles) of tunnel eroded over aeons by the water and its organic acid content. The journey there and back takes you through very different country to that of the mountains, and well illustrates the extremes of this fascinating nation: from frost and simien fox down to the lowlands and camel in a mere 150 kilometres (93 miles).

Giant lobelia can be found at high altitudes.

Nechisar and Other Parks

Set on the isthmus between two more of Ethiopia's Rift Valley Lakes—**Chamo** and **Abbaya**—**Nechisar National Park** is a jewel of magnificent scenery and plentiful wildlife. Park headquarters is located near the town of **Arba Minch** (Forty Springs), 500 kilometres (310 miles) south of Addis, passing through the town of Shashemene after 250 kilometres (155 miles) of tarmac. A good gravel road leads along the western edge of the rift to Arba Minch.

Good roads in the park lead through the **Kulfo River ground water forest** in the floor of the rift below Arba Minch. Across the river, the main route into the park passes across the isthmus between the lakes via a series of spectacular and rugged climbs across faults in the rift floor, and up an escarpment to the **Nechisar** (white grass) **Plains** for which the park is named. Four-wheel drive vehicles are essential.

Greater kudu are commonly seen crossing the isthmus, and this area is said to be one of the best locations for the species in Africa. Guenther's dikdik abound at every turn and a short track leads you to the **Crocodile Market** at the northern edge of Lake Chamo where monster crocodiles lie out on the lake shore at close quarters.

Once up on the plains, several different circuits take you through a mass of Burchell's zebra, Grant's gazelle, large troops of olive baboons and some Swayne's hartebeest, all against the backdrop of the highlands to the east, the jewelled lakes far below and the soaring mountains of the western highlands above Arba Minch.

It is well worth spending time in Kulfo River ground water forest if birds are your main interest. There is an excellent campsite set beneath giant fig trees beside the river in the forest. You may also camp at the hot springs on the far side of the Nechisar Plains, a good starting point for early morning obser-

Looking out on Lake Chamo.

vation of the plains' animals.

Yangudi Rassa: North of the Awash and still in the Rift Valley, **Yangudi Rassa National Park** was established for its population of wild ass (*Asinus africanus*). It is possible to stay in the little hotel in **Gewane** and follow park staff on regular trips for fleeting glimpses of these interesting animals in their semi-desert habitat.

Harer Elephant Sanctuary: This is worth a trip as an extension of a visit to the historic walled city of **Harer**. The elephants here are an endemic subspecies, but difficult to see because of limited viewing tracks.

Far south: At the southern end of Ethiopia, the **Omo and Mago national parks** are situated close together on each side of the **Omo River**. Good dry season roads exist, or you can fly with charter aircraft, having arranged with the official travel agency, National Tour Organisation (NTO) for ground transport to meet you. Good campsites have been made but with no facilities. Fuel is also difficult to procure in this area and you are advised to take all you need with you. Large numbers of larger wildlife species such as elephant, buffalo, eland and Lelwel hartebeest (*Alcelaphus buselaphus lelwel*) occur, but there are few usable tracks.

Yavello Sanctuary: This sanctuary is close to the main tarmac road south from Addis to the Kenya border. There are small hotels in **Yavello** town, and a few tracks from which to explore the area. Reasonable numbers of both lesser and greater kudu and gerenuk, Grevy's zebra, beisa oryx and dikdik can be seen, as well as ostrich and giraffe. The area is most significant, however, as a centre for a group of endemic birds: Stresemann's bush crow (*Zavattariornis stresemanni*), the degodi and Sidamo lark, the white-tailed swallow (*Hirundo megaensis*) and—further east—Prince Ruspoli's turaco (*Tauraco ruspoli*).

Other Areas: Developments are limited in most cases. Some, such as **Dahiak**, **Semien**, and **Cambella** are inaccessible to visitors.

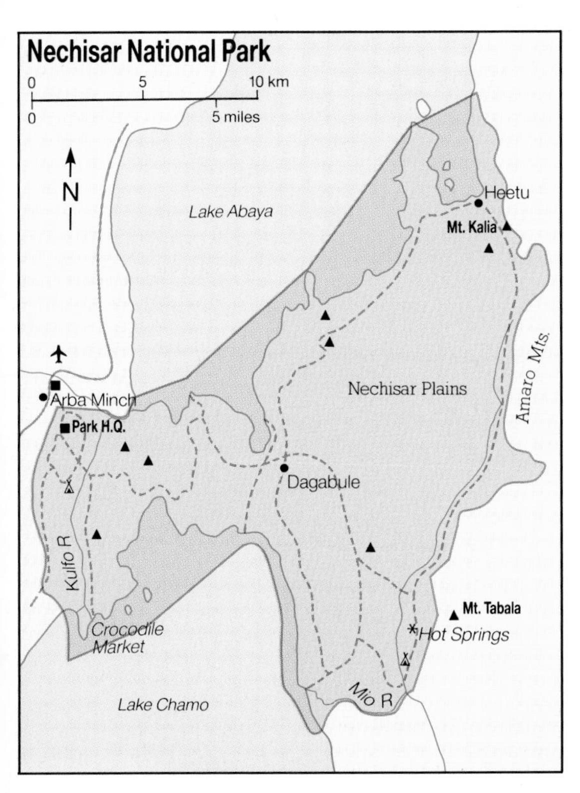

Far right, the Kulfo River ground water forest.

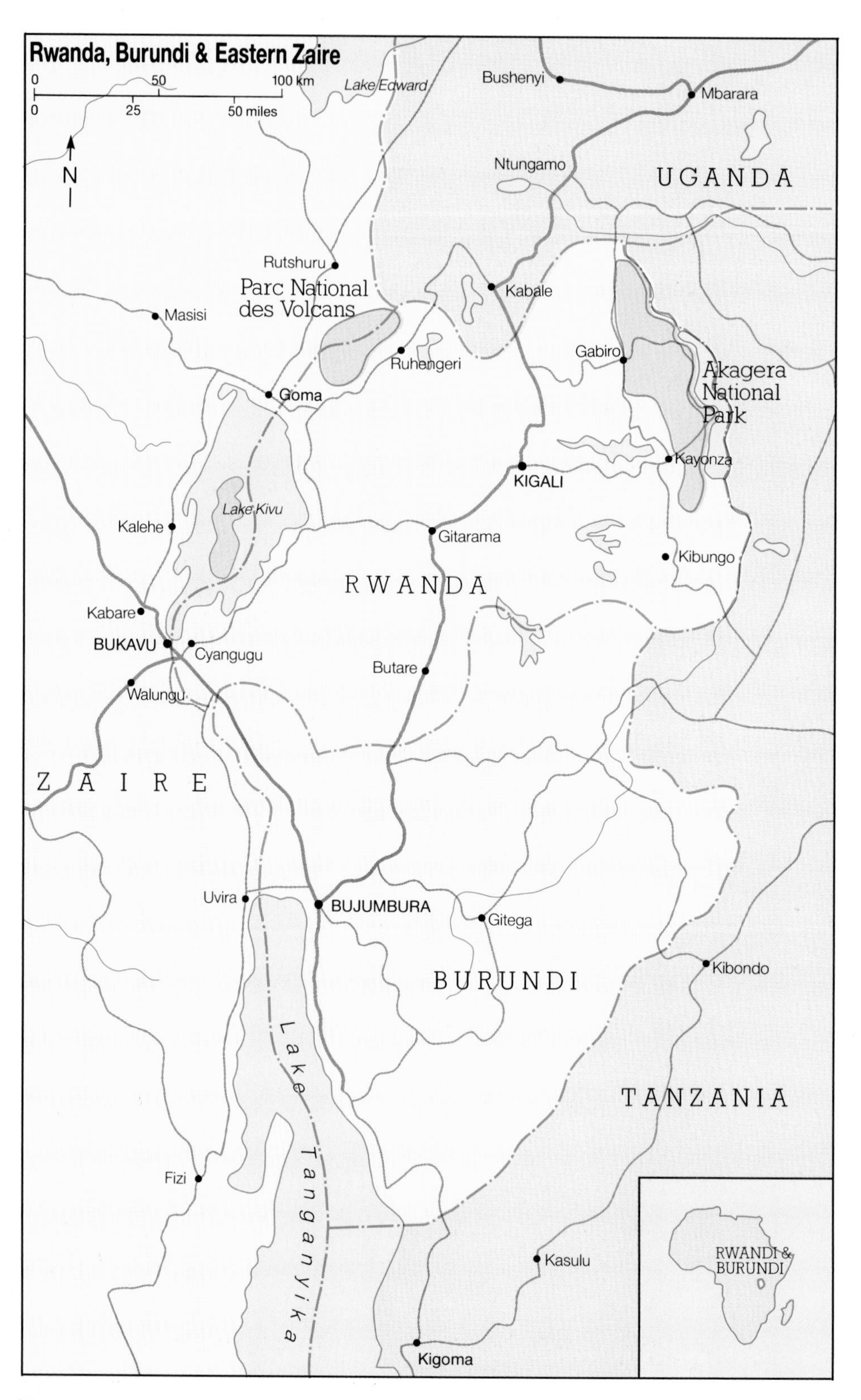
Rwanda, Burundi & Eastern Zaire
0
50
100 km
0
25
50 miles
N
Lake Edward
Bushenyi
Mbarara
Ntungamo
UGANDA
Rutshuru
Parc National des Volcans
Kabale
Masisi
Ruhengeri
Gabiro
Akagera National Park
Goma
Kayonza
KIGALI
Lake Kivu
Kalehe
Gitarama
Kibungo
RWANDA
Kabare
BUKAVU
Cyangugu
Butare
Walungu
ZAIRE
Uvira
BUJUMBURA
Gitega
Kibondo
BURUNDI
Lake Tanganyika
TANZANIA
Fizi
Kasulu
RWANDI & BURUNDI
Kigoma

RWANDA

Rwanda, a tiny country in the mountains of east central Africa just south of the equator, has been riven by a merciless civil war, and it will be some time before the country returns to normal. Nevertheless, it remains an important constituent in the East African Wildlife scene. The country borders Tanzania on the east, Uganda on the north, Zaire and Lake Kivu on the west, and Burundi on the south. Rwanda's land area of 26,340 square kilometres (10,290 square miles) is only about the size of some of East Africa's larger national parks, but before the civil war it had a population of more than six million, the highest population density on the African continent.

The country is mountainous in the extreme, varying from 1,200 metres (3,960 feet) to over 4,500 metres (14,760 feet). Much of the land is spectacularly beautiful. Steep mountains are cleft by deep precipitous valleys, many of them jewelled by lakes. There is an excellent system of well engineered roads, two major national parks and some good hotels and lodges.

A German explorer, Lieutenant Von Goetzen, was the first European to reach what is now Rwanda when he crossed the Akagera River (one of the headwaters of the Nile) near Rusumo, in the south-east of the country, in 1884. Here he found a feudal hierarchy headed by Tutsi nobles, members of one of the country's three indigenous tribes.

In 1899, after many centuries as a feudal society with social classes similar to those of medieval Europe, Rwandans were made to recognise German sovereignty. This lasted until World War I, when Belgium took over Rwanda (and neighbouring Burundi). Colonial ties were severed on July 1, 1962, with the formation of the Republic of Rwanda.

Rwanda's best known attraction is the gorilla tracking in the famous Parc National des Volcans, but while it is unsafe for tourists to visit the country, gorilla trekking can be undertaken in Uganda. There is a variety of wildlife at Akagera National Park and the scenery through Rwanda is a delight waiting to be seen again.

Preceding pages: the deep green valleys on the way to Ruhengeri are partially obscured by cloud cover.

AKAGERA

First created in 1934, this is Rwanda's largest national park, with an area of 2,500 square kilometres (1,000 square miles). It is undulating, even hilly, country with a mixture of bush and grassland savanna studded by an extensive chain of lakes that drain into the **Akagera River**, which forms the park's eastern boundary, before eventually flowing into Lake Victoria and becoming one of the sources of the Nile.

The park is 120 kilometres (75 miles) east of **Kigali** and **Hotel Akagera**, near the southern end of the park, can easily be reached in a morning from the capital, most of the journey being on smooth tarmac roads. A similar 150-kilometre (90-mile) stretch of tarmac links Kigali with the **Gabiro Guest House** in the north of the park.

Birds and more birds: At a fishing village on the shores of **Lake Ihema**, one can take a boat out to the nearby islands, alive with cormorants, darters, herons, storks and other water birds. Up to 12 species can be seen nesting here. Birds are the main strength of Akagera and it is a wonderful place for ornithologists; not only is bird life prolific but the park is far enough west across the African continent to have different species from the major countries of East Africa. Bennett's woodpecker, the golden-backed weaver, the short-tailed pipit, Souza's shrike, and the violet-crested turaco can all be seen in Rwanda more easily than in Kenya or Tanzania. More than 520 species have been recorded in Akagera National Park and competent ornithologists can hope to see as many as 100 on a single day.

A strong metal boat with a sun canopy has been launched on **Lake Mahindi** in the northern part of the park by the World Wildlife Fund who are anxious to encourage Rwanda's conservation programmes. From this boat visitors can view water birds and ease along the channels connecting Lake Mahindi to the Akagera River and other lakes. Here it is possible to see some very unusual birds skulking in the papyrus. The forest gallery on either side of the water channels is unique.

Game drives: There are several circuits in the park for game viewing. From Hotel Akagera one can take a half-day drive along the rolling hills of the **Ridge Circuit** or spend a full day going north to Gabiro Guest House. This is a journey of about 130 kilometres (80 miles) on a road reminiscent of a farm track in England. The undulating, lush countryside, studded with lakes, resembles a deserted English Lake District in the summer, were it not for the wildlife. There are very large numbers of impala, plenty of topi, waterbuck and many African Cape buffalo. Some of the old male buffalo have spectacular horns; spans of 1.1 metres (44 inches) are not uncommon. There is a very small number of elephants which were reared as orphans and now run free. Lion, leopard, serval cat, African hunting dogs, hyaena and jackal also exist here though none of them will be

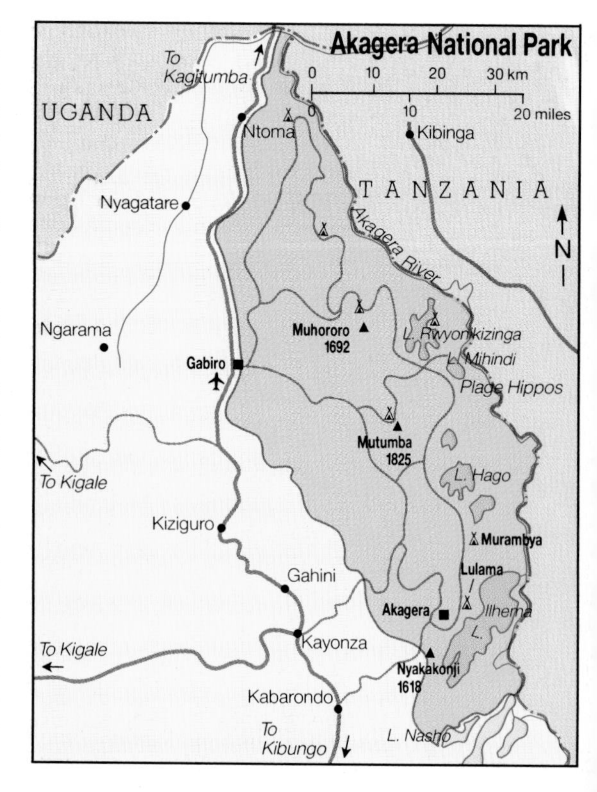

as easy to find as in the best of East Africa's parks.

Night driving is allowed in Akagera; aardvark, civet and genet are among the animals one might see after dark.

The road from Hotel Akagera to Gabiro passes, in turn, lakes Ihema, **Hago** and Mahindi. All these give lovely views and chances to observe hippos and other water animals. At Lake Mahindi there is a fine picnic site where hippos are much in evidence, often standing out of the water near clumps of papyrus.

Ten kilometres (six miles) further north, on the shores of **Lake Rwanyakizinga**, there is a wooden observation tower specially built to view sitatunga antelopes who live in the marsh. These animals are adapted to swamp life with long hooves that lie forwards and support them on soft ground. With binoculars it is often possible to enjoy good views of these comparatively rare creatures grazing in open glades.

Spending a full day over the drive from Hotel Akagera to Gabiro Guest House is an excellent way of touring the park and seeing its major attractions, and the guest house is a pleasant place to stay. Located on a large spread out site with many individual touches in its layout and design, Gabiro's individual stone cottages seem appropriate in a national park.

Though it is easy to return to Kigali from Gabiro many visitors will drive on to **Ruhengeri** and the Parc National des Volcans through scenery which becomes steadily more interesting. Deep, heavily cultivated valleys separate huge hills which march with increasing drama towards the Virunga volcanoes, Rwanda's frontier to the north-west. This is the home of the famous mountain gorillas, the country's most valuable wildlife asset. Before reaching Ruhengeri it is worth making a side trip to see beautiful **Lake Ruhondo** (about 10 kilometres before you reach Ruhengeri town). Nestled in a valley and studded with islands, this is one of the finest views in a country of memorable panoramas.

Beautiful Lake Ruhondo.

MOUNTAIN GORILLAS

Gentle giants: In the make-believe world of Hollywood the gorilla has been invested with all the aggression of *Homo sapiens* to create a parody which has no relation to the original, gentle creature whose form he takes.

Almost entirely vegetarians, mountain gorillas (*gorilla gorila beringei*) eat over 100 different plants (and some insects). They live in stable groups which vary in size between two and 30 individuals. Each group is dominated by a senior male, normally more than 15 years old (ages of up to 35 years are known in captivity but many workers think gorillas may live even longer). Such males stand about 1.75 metres (5 feet 9 inches) tall, weigh around 180 kilos (396 pounds) and are the undisputed leaders of their groups. Easily recognised by the saddle of silver hairs which form across their backs with increasing age, these 'silverbacks' weigh about twice as much as a mature female and are immensely strong without being aggressive. Such a silverback will normally have several sexual partners who will remain with him for life. There may be another, immature, male in the group between eight and 13 years old, along with other juveniles and infants. Because the gorillas have permanent relationships they form strong ties; but as in other species both males and females leave their natal groups when they reach sexual maturity so there is little or no inbreeding in normal populations. Males are driven out by the silverbacks and may spend several years alone before they can attract females to form a group of their own. Females usually leave their natal groups when two families meet at the overlapping edges of their territories.

Gorilla babies are born after a gestation period of nine months—give or take three weeks depending on the individual—and suckle until the age of 18 months, though they begin eating vege-

Looks fierce, but this silverback is only yawning.

tation at about 10 weeks. Once she has reached the age of eight and is sexually mature, a female will have a baby roughly every four years, but 30 percent will be lost through infant mortality. Because of their bulk and weight gorillas move and climb slowly being careful not to break branches. Strictly diurnal, at night they sleep in crude nests made of leaves and branches, usually on the ground but sometimes on platforms which they make a few feet up in trees. Typical behaviour involves waking at dawn, feeding for two hours, resting for a few hours and then feeding again. At night the whole group will sleep within 50 metres (165 feet) of each other. But they are continuously on the move, normally sleeping at a new site each night.

Gorillas communicate via a variety of grunts, screams and roars; they also beat their chests with their hands, producing an intimidating slapping sound which travels a long way in the forest.

Gorilla tracking: To go gorilla tracking in the Parc National des Volcans you must have a permit to visit one of the four groups which are habituated to humans. In 1989 these were known as Group 9, Group 11, Group 13 and the *Suza* Group. Permits for Group 9 are sometimes available for purchase at the park office at Kinigi, 11 kilometres (6.5 miles) from Ruhengeri, where you meet up with your guide. All other permits are sold in advance to travel companies specialising in African safaris, so it is much easier to work with one of these in order to avoid disappointment. In 1989 a permit cost US$172. Only six tourists per day are allowed with each gorilla group and this rule is strictly enforced. People with colds, flu or stomach problems are not allowed because of the danger of infecting the gorillas.

Each party of visitors is allotted a guide—usually English speaking—whose charges are included in the cost of the permit. Porters are available for hire at park headquarters and are highly recommended as gorilla tracking can be extremely strenuous. To locate your allotted group, the guide will lead you to where the gorillas were sighted on the previous day and then literally track them through the forest. The terrain is very rough, often over steep slopes covered with matted vegetation. Be prepared for rain, mud and giant stinging nettles. Long trousers, leather gloves and a waterproof jacket will help.

With any group you might be out in the forest for many hours; usually a trek of one to three hours will bring you to the gorillas. After reaching the animals each party spends one hour with them before returning to park headquarters. This is adequate time to enjoy very good, close-up sightings.

Appropriate behaviour is essential when gorilla watching and your guide will advise you on this. Hard-eyed stares are as aggressive to gorillas as they are to ourselves; and you should not stand higher up than the gorillas in an open space. No doubt because these rules are observed, accidents with tourists are unknown; however, silverback males often make a false charge at some point during the day, giving visitors the thrill of a lifetime.

Babysitting in the jungle.

Parc National des Volcans

Most visitors to the **Parc National des Volcans**—and indeed to the country of Rwanda—come to see gorillas. Thanks to a very successful conservation project it is now possible to virtually guarantee close-up sightings of these massive apes and many hundreds of people enjoy this privilege each year.

The Virunga Volcanoes: Rwanda, Uganda and Zaire come together at the summit of 4,127-metre (13,540-foot) **Muhabura**—one of an extensive range of volcanic peaks in central Africa which form the boundaries between these three countries. **Karisimbi** ("white shell"), at 4,507 metres (14,786 feet), is the highest in the chain—tall enough to create the frequent showers of snow which justify its name. The border between Rwanda and Zaire runs from Karisimbi over the crests of **Visoke**, **Musinde**, **Sabynyo** (at 500,000 years the oldest of the volcanoes) and **Gahinga** to Muhabura.

These mountains bring an incredible 10 percent of Rwanda's rain to the 0.5 percent of the country which is made up by the Parc National des Volcans. This high rainfall (183 cm/72 inches annually) sustains the vegetation typical of high mountains in east and central Africa; on the lower slopes there is mixed woodland—though much of the original woodland has been felled to make way for agriculture. Then comes a zone of dense bamboo which extends from about 2,286 metres (7,500 feet) up to between 2,591 metres (8,500 feet) and 2,988 metres (9,800 feet) depending on the local area. Above the bamboo is an ancient forest of gnarled *Hagenia abyssinica* trees, many of them draped with mosses and decorated with orchids and ferns. Among the dominant *hagenia*, St. John's Wort (*Hypericum lanceolatum*) is another common tree which often supports *Unsea* lichen or 'Old Man's Beard' with its long trailing strands so reminiscent of Spanish moss.

The volcanic peak of Muhabura.

Above the *hagenia* and *hypericum* forest, at about 3,354 metres (11,000 feet) is the so called sub-alpine zone, an open moor-like area with a fascinating array of plants found only at high altitude in the tropics. Here giant senecios and lobelias grow to five metres (15 feet) and fields of everlasting *helichrysum* flowers decorate the ground. But few visitors venture so far up these volcanoes.

Early roots: The oldest national park in Africa, the Parc National des Volcans was originally created as part of the Parc National Albert in 1925, largely due to the influence of an unusual American, Carl Ethan Akeley (1864-1926). The son of a farmer from upstate New York, Akeley became an expert at mounting museum specimens and he made several expeditions to Africa. Akeley's Hall of African Mammals remains one of the major exhibits at the American Museum.

In 1921 he visited the Belgian Congo, partially in order to collect gorilla specimens for the museum, and became fascinated by the great apes. Although he did take four gorillas for the museum Akeley became alarmed at the extent of hunting and persuaded the Belgian government to create the Parc National Albert—named after the then King of Belgium. (Akeley died on a subsequent visit to gorilla country and was buried at Kabara, on the Ugandan side of the Virunga volcanoes, in 1926.)

In those days both modern Rwanda and Zaire came under the aegis of the Belgium Government so the Parc of 1925 included most of the mountains of the Virunga Volcanoes in Rwanda and Zaire. In 1960 the Parc was divided into two new parks: in Zaire the Parc National des Virunga and in Rwanda the Parc National des Volcans.

Owing to the population pressure for new agricultural land the Parc National des Volcans has been reduced in size since 1960; it now has an area of 125 square kilometres (50 square miles). But Zaire's Parc National des Virunga and the Kigezi Gorilla Sanctuary in Uganda are both contiguous with the

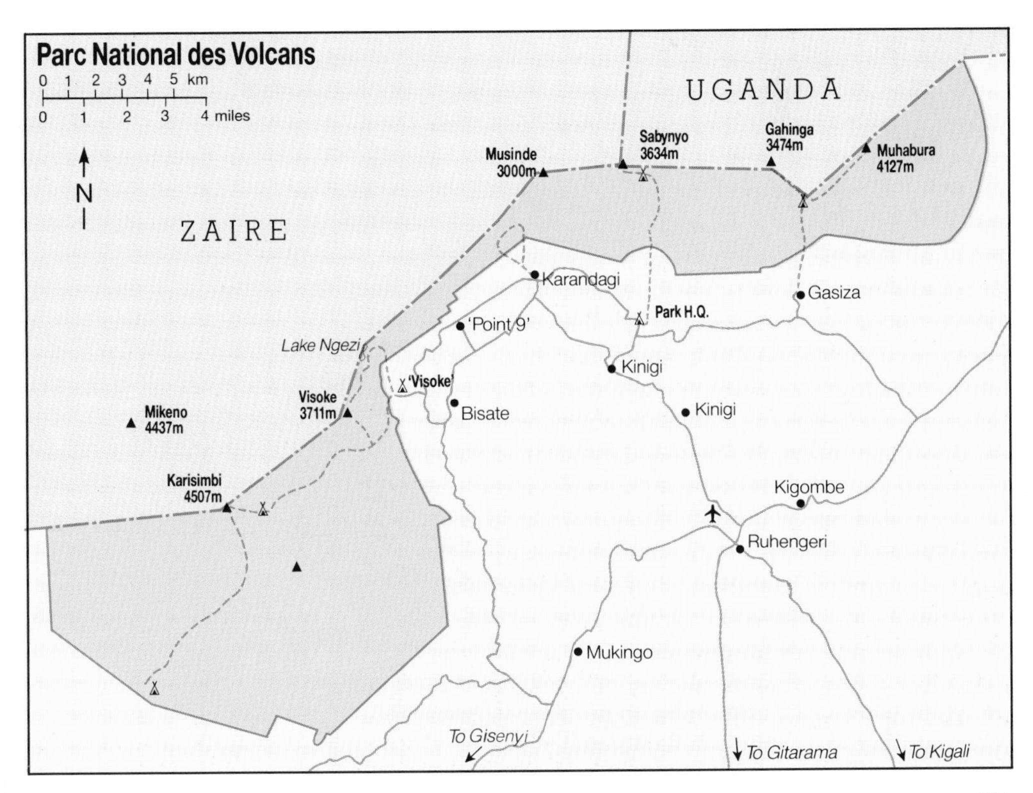

Rwandan park so the total area of this protected mountain habitat is still 375 square kilometres (150 square miles).

Gorilla Conservation: Many dedicated individuals have contributed to the conservation of the mountain gorillas of the Virunga volcanoes. After Akeley's influence in setting up the original Parc National Albert, the next notable conservationist was Walter Baumgartel, who established the Travellers' Rest Hotel in Kisoro, on the Ugandan side of the volcanoes, used for many years by tourists and research scientists working on gorillas. It was here that George Schaller, the eminent wildlife scientist, stayed in 1960 when he conducted a study involving over 450 hours of observing of wild gorillas.

Since then there has been enormous world-wide interest in gorilla research, much of it stimulated by the fact that gorillas are such near relatives to ourselves. Only four species of mammals—all primates—share the characteristics of being without tails, having two nipples and having five digits on both hands and feet, one of which is an opposable thumb. These are the three great apes—chimpanzees, orangutans, gorillas—and *Homo sapiens*. Many workers, notably Drs Louis and Mary Leakey (famous for their work on the fossils of early man in Africa) believed it was important to study gorillas in their natural habitat in order to find out more about how early man may have behaved in prehistoric times. Dr Leakey was among those who encouraged Dian Fossey to begin her research on the gorillas, which she started with Walter Baumgartel in 1963. By late 1967 she was establishing a new camp and the **Karisoke Research Centre.** Her book *Gorillas in the Mist* has done much to publicise the plight of these great apes. Dian Fossey was murdered in the park in 1985, probably by poachers, but her work, made famous through many documentary and feature films, still continues through the Mountain Gorilla Project.

Many dedicated scientists have worked here on gorilla research and on

Eyeball-to-eyeball with a mountain gorilla.

habituating them to humans so that it is possible for visitors to enjoy them. There are many encouraging signs that these aims are being fulfilled. The tourism side of conservation is making a substantial contribution to Rwanda's foreign exchange and there have been no poaching losses of gorillas in those groups which are visited by tourists. Indeed some scientists believe that in the 1980s the numbers of gorillas in the Parc National des Volcans may have increased slightly. In a 1986 census a total of 279 individuals were counted.

Nyungwe Forest: In south-west Rwanda, mid-way between the towns of **Cyangugu** (at the southern end of Lake Kivu) and **Butare**, is the **Foret Naturelle de Nyungwe**, an extensive area roughly 60 kilometres (37 miles) long and 30 kilometres (19 miles) wide.

This beautiful tropical forest is rich in plants, butterflies, birds and primates. There are nine different species of monkeys; notable are the Angolan black and white colobus (*Colobus angolensis)* which can sometimes be seen in troops of up to 300. Very different from the Abyssinian black and white colobus (*Colobus abyssinicus*), frequently seen in the mountain forests of East Africa, the Angolan has no white cape around the lower part of the body. Like all colobus they spend most of their time high in the trees, seldom coming down to the ground.

Another monkey found here is L'Hoest's monkey (*Cercopithecus l'hoesti*). This is a beautiful animal with a red back and a prominent ruff of white whiskers around its dark face; unique in its class of monkeys L'Hoest's has a prehensile tail with a hooked tip.

The area is also good for ornithologists: Ruwenzori and great blue turacos, Western green tinkerbirds, purple-breasted and regal sunbirds are some of the notable species to be found here. There are tracks through the forest which make it easy to walk; but it is wise to take a local guide from Cyangugu or one of the villages. The **Ibis Hotel** in Butare could also be used as a base when visiting Nyungwe.

Lake Kivu and the Hotel Meridien at Gisenyi.

EASTERN ZAIRE

Parc National de Kahuzi Biega is named after the two volcanic peaks of **Kahuzi** (3,308 metres/10,850 feet) and **Biega** (2,790 metres/9,120 feet), the highest peaks in the western wall of the western branch of the Great Rift Valley. The park offers excellent gorilla tracking. Here one sees the Eastern Lowland gorilla (*Gorilla gorilla graueri*) which has minor differences from the *Gorilla gorilla berengei* of the Parc National des Volcans in Rwanda. The Kahuzi Biega gorillas have lower foreheads, shorter hair and longer arms, but they are every bit as impressive as their Rwandan cousins.

Unless your journey is planned by a travel company which is making arrangements for you, gorilla trackers should book in advance through the IZCN (Institut Zairois Pour la Conservation de la Nature) office, 185 Avenue President Mobutu, Bukavu.

Fees must be paid at the park entrance office at **Tshivanga**, 31 kilometres (20 miles) from Bukavu on the actual day of the excursion. In 1989 the fees were US$100 per person per day. Eight people are allowed in each party and gorilla tracking begins at 9 a.m. each day. In 1989 only three groups of gorillas were habituated to human visitors.

The Kahuzi Biega forest is very beautiful; the trees, plants, butterflies and birds all add great interest to a visit. Occasionally visitors are unable to reach the gorillas because of elephants and buffalo; but the guides and trackers are experts at avoiding problems.

Mount Kahuzi and Mount Bugulumiza can both be climbed and give fine views of Lake Kivu. Modest fees must be paid at the park office at Tshivanga.

Mount Kahuzi is a five-hour excursion for strong, well-clad mountain walkers. There is a good track and a guide can be found at the **Kahuzi Patrol Post** at the base of the mountain on the western boundary of the park, reached via the tarmac road which leads

Far left, sitting in a gorilla nest. Below, the Virunga volcanoes.

from Tshivanga to Bunyakiri, Irangi and Walikale.

Mount Bugulumiza has a well-kept track which rises gently for six kilometres (four miles) to an open plateau with spacious views. Four-wheel drive vehicles can negotiate the track and there is a hut near the summit, a good spot for a picnic lunch.

A different world: Reached in 90 minutes from **Bukavu** (or an hour from Kahuzi Biega), the village of **Irangi** is in a world by itself. Here at only 800 metres (2,625 feet) the climate is warmer and more humid. The village is at the edge of a tropical rain forest on the banks of the fast flowing and beautiful **Lwana River**. More luxuriant than the montane forests of the gorillas, Irangi has fantastic bird life and incredible butterflies. King Leopold III's **hunting cottage**, built on the river bank in 1957, is still intact at modest charges. Not far from the cottage a wobbly but safe footbridge, woven from rattan and bamboo, gives access to the forest paths.

Parc National des Virunga: Originally part of the Parc National Albert, this huge park has an area of 8,000 square kilometres (3,200 square miles). The park contains the Zairean side of the Virunga volcanoes which form the watershed between the largest of the continent's rivers, the Nile and the Zaire (formerly the Congo). One of these, the **Nyiragongo**, at 3,170 metres (10,400 feet), is still active and can be climbed with a guide from the town of **Goma** at the northern end of Lake Kivu. From time to time there are obvious signs of volcanic activity with steam jets and even red-hot lava being visible inside the crater.

Some 160 kilometres (100 miles) north of Goma, gorilla safaris are available at **Djomba**. From a scenic spot in the shadow of Mount Sabynyo, guests at the comfortable log cabins of the **Djomba Intrepids Club** can make daily treks to visit two mountain gorilla families. Nearby Hutu villages also provide some colourful insight into tribal life.

Market day at Djomba.

INSIGHT GUIDES
TRAVEL TIPS

Kenya

Tanzania

Uganda

Ethiopia

Further Reading

KENYA

Getting Acquainted

The Land

Area: 224,900 sq miles/582,600 sq km.
Capital: Nairobi.
Population: 21.4 million.
International dialling code: 254 (Nairobi: 2).
Time zone: Three hours ahead of GMT.
The place: One can find the entire spectrum of geographical features in Kenya, from sea level to 5,200 metres (17,000 feet). Beaches, desert, forest, savannah and snow are all present, with accompanying fauna.

Climate

Temperatures vary according to altitude and situation. The coastal and lake areas are humid; average annual high and low daytime temperatures range from 22°C to 30°C (72°F to 87°F). Usually a fresh breeze blows from the ocean. The plateau and Nairobi, at an altitude of 1,675 metres (5,500 feet), are cool in the mornings and evenings with constant winds in the dry seasons (January to March and June to October). Average annual high and low temperatures are 10°C to 26°C (50°F to 79°F).

Culture & Customs

The population of Kenya includes Africans, Indians, Europeans and Arabs.

Common sense, tact and politeness, regardless of situation, should be applied across the board. A few guidelines: show respect for authority, primarily the president, members of parliament, judges, the police force and army.

Do not take photos of military installations or government figures or indeed of any person without their consent. Usually a fee will be demanded, between Ksh 100 and Ksh 150 is standard.

Never destroy or deface money, the flag or any picture of the president – these are offences punishable by jail.

Black market vultures lurk on street corners, approaching the innocent tourist with fantastic exchange rates. Dealings with these people may well lead to jail, or deportation, or both.

Prostitution is rife, and primary concern should be the risk of disease. AIDS is a growing problem in Kenya as in the rest of the world.

The authorities take a hard line with drugs, and can make life very uncomfortable for anybody caught taking them into or out of the country.

Nudity is illegal. Skimpy clothing may be frowned upon by the locals in rural and Muslim coastal areas. To call a waiter, use the word *bwana* which means "mister".

Government & Economy

Executive power is held by the President Daniel arap Moi, who is aided by his vice president and cabinet chosen from the National Assembly (parliament). This legislature consists of 158 publicly elected members, 12 presidential nominees, the Attorney General and the Speaker. The process of Government in Kenya is democratic. The first multi-party elections took place in December 1992. There are now 11 registered political parties.

Kenya is one of the more prosperous countries in black Africa, due to a good record of sound governing and stable politics. Kenya's primary exports are coffee and tea. Other exports include pineapples, string beans, pyrethrum and other cash crops.

Tourism is another primary source of foreign exchange. Kenya is also regarded as the regional (East Africa) centre for financial and commercial activities.

Planning The Trip

What To Bring

Kenya uses a 240 volts, 50 hertz supply. If you want to use appliances that run on 110 volts, bring a converter. Plugs are three-pin, square peg (as in Britain).

What To Wear

As a rule, clothing should be loose, comfortable and casual. In Nairobi and further upcountry it can get cold at night so a sweater or jacket is necessary. Safari clothing can be easily and economically bought in Nairobi. Drip-dry clothes are convenient and dry fast. *Khangas* and *kikoys* (brightly coloured, patterned sarongs) are useful as beachwear, pyjamas or wraparound skirts. *Bata* safari boots are cheap and durable footwear.

If your visit falls in the rainy season, buy an umbrella in Nairobi.

Entry Regulations

Visas & Passport

All visitors must have a valid passport. Visas are required by everyone, except for the citizens of most Commonwealth countries, and some other countries with which Kenya has reciprocal waiver agreements. The Commonwealth exceptions are Australia, Nigeria and Sri Lanka. Indian, Bangladeshi and Pakistani citizens also require visas. At present, visas can be obtained at the port of entry, for a fee of about US$20, or the equivalent in convertible currency. A "Visitors Pass" is stamped in your passport upon arrival. This is usually valid for one month but can be extended to three months at Department of Immigration offices in Nairobi, Kisumu and Mombasa. Officials will want to see a return ticket.

Customs

A reasonable amount of personal effects, including cameras and film can be imported duty-free. A deposit (foreign currency, refundable) may be required for radios, tape recorders, video or movie equipment and computers. Duty-free allowances are 200 cigarettes or 50 cigars, one litre of spirits and a quarter of a litre of perfume. Banned imports include agricultural or horticultural produce, pets, game trophies, pornographic literature and illegal drugs.

Health

Malaria is endemic in Kenya below altitudes of 1,830 metres (6,000 feet). You should start taking malaria prophylactics two weeks before you enter Kenya; continue taking them during your stay and for one month after you return home.

Inoculations against yellow fever and cholera are not mandatory unless you are arriving from the Far East, Central America, South, Central and West Africa.

AIDS is a problem in East Africa and the rule of thumb on prostitutes is look and dance, but don't touch. Blood is screened for AIDS at Nairobi Hospital. In an emergency where blood donations are needed, contact your embassy or high commission.

Swimming in slow-flowing or still rivers or lakes can result in bilharzia (*Schistosomiasis* – an unpleasant disease that resides in river snails). In the sea, swimmers are protected from large marine predators by a barrier reef. There are the usual coastal hazards, however, including sea urchins, fire coral, and the occasional poisonous fish or ray. Wear tennis shoes if walking around rock pools.

The sun is directly overhead, and a silent danger to all unwary tourists. Tropical sunburn is painful and dangerous, so avoid it; exposure must be gradual. Use sunblock creams, and make children wear T-shirts.

Above 1,525 metres (5,000 feet) the visitor may feel lightheaded or tired but one will soon get used to it. When climbing above 3,463 metres (13,000 feet), watch out for symptoms of altitude sickness (*pulmonary oedema*) which can be fatal if the afflicted person is not removed immediately to a lower altitude.

In rural areas boil all water before drinking. Water in Nairobi is safe to drink but most hotels and restaurants now serve bottled water.

Medical insurance is available locally. There are plenty of doctors in town, and the *Flying Doctor* will rescue stomped or mauled tourists from the wildest areas.

Money

The Kenya shilling is divided into 100 cents. Notes are issued in denominations of Kshs 10, 20, 50, 100, 200 and 500. Coins are in Kshs 1 and 5 and cents 5, 10 and 50. The Kshs 20 note is sometimes referred to as a "Kenya Pound" (K£) and the slang for the shilling is "Bob".

Exchange rates are not fixed and there could be fluctuations or substantial change at any time, so it's important to check current rates before departure. However, an indicative rate in late 1994 was Kshs 45 to the US dollar.

On departure, Kenyan shillings held by visitors can be changed back into foreign currency. No Kenyan money may be exported from the country so it is important to get rid of all of it. Destruction of any Kenyan money is an offence.

Major credit cards are accepted in some establishments throughout the country.

Getting There

By Air

Visitors travelling to Kenya by air will arrive at either Jomo Kenyatta International Airport (JKA), Nairobi, or the slightly smaller Moi International Airport, Mombasa. Carriers from Europe, America, Asia and within Africa all offer scheduled services to Nairobi. Check with your travel agent for details.

Standard amenities are available in JKA, including a 24-hour bank, a post office, shops, cafés, restaurants and bars. Porters and handcarts are also available. Taxis are the main form of transport from the airport to the city centre. Pay about Ksh 800. If you are part of a group tour, they should provide a minibus with driver; alternatively one can always hire a self-drive vehicle from the major car hire companies at one of the desks in the arrival hall.

By Rail

Mombasa is mainly a cargo and military port and there is an outside possibility that one could obtain a berth on a cargo ship from a European port. A delightful option is a stopover in Mombasa as part of a luxury liner trip (*Cunard/Ellerman*). These normally include short safaris around the country if desired by passengers.

By Road

Overland trips are usually arduous and time consuming, especially if you're undertaking one on an individual basis. Some tour companies offer exotic trips – check the advertisements in international newspapers. One company, S*iafu*, has offices in Nairobi and London. To do it yourself, you would need four-wheel drive vehicles which are equipped for extensive travelling across desert terrain. Also, consult with the embassies of countries through which you intend to travel. Minimum required documentation includes International Driving Licence, International Certificate of Insurance and International Touring documents – Carnet de Passage and Triptique.

The Automobile Association (AA) in London has up-to-date information on self-drive trips throughout Africa.

Practical Tips

Emergencies

Security & Crime

Kenya is quite a secure place, despite the numerous legends about tourists getting eaten (either by wildlife or locals), knifed or generally assaulted. Bag snatching and mugging occasionally take place, but shout "thief" (mwizi) and if there are any people around they will descend upon the hapless miscreant, and mete out swift and severe punishment – mob justice. Violent crime rarely involves tourists, but in a threatening situation do what you are told, and remain quiet and calm. Do not walk about at night through dark areas of town, or on deserted beaches. Always lock your car, and hide valuables in the boot, or better still, in the hotel safe. Dial 999 for police.

Pharmacies

There are plenty of chemists in Kenya, with well stocked drug stores. If you cannot get your exact prescription filled, there is usually an alternative drug available. The best policy is to bring enough special medication to cover the duration of your visit. Chemists are closed on Saturday afternoons, Sundays and public holidays. During these times the name of the duty chemist is usually posted on drug store doors, or can usually be found in the newspapers.

Weights & Measures

Metric is used throughout the country – weights are measured in kilogrammes (1 kg = 2.205 pounds); distances are in kilometres (1 km = 0.6214 miles); and liquids are measured in litres (1 litre = 1.76 pints).

Business Hours

These vary, according to the store. Generally stores are open from 9am to 5.30pm, with a break for lunch between 1pm and 2pm. Some stores are open till evening, and also over the weekends. In Mombasa, trade begins even earlier, from 7am to 12.30pm, and then a siesta till 3pm or 4pm to avoid the heat. They re-open till about 10pm. Bargaining takes place in markets and small stores, especially in Mombasa.

Most banks offer a currency exchange facility, and they are open on weekdays from 9am to 2pm. Some are open in the morning on the first and last Saturdays of each month. At JKA there are 24-hour branches for visitors. Banking can be a lengthy process, so be patient.

Tipping

The rules you use at home when tipping should apply here. For instance, add 10 percent to a restaurant bill unless it's there already. In that case, slip the waiter Ksh 20 – 30. The biggest tip will probably go to your tour driver if he has turned out to be helpful and responsive on the trip. Maybe as much as US$5 per person per day of safari. Porters expect Ksh. 20 – 30 per bag.

Religious Services

Whatever your religious inclination, Kenya will probably have a representative church, chapel, mosque, synagogue or temple. Freedom of worship is taken seriously, and missionaries are everywhere. Check the daily papers for services.

Media

Radio & Television

The English speaking radio station and one television station are operated by the government. There is one other privately owned television station, Kenya Television Network, which offers a good variety of programmes together with CNN.

Newspaper

There are 3 English-language daily newspapers: the *Standard*, the *Nation*, and the *Kenya Times*. There is one weekly newspaper, the *East African Chronicle*, which contains news from all around East Africa.

Postal Services

There is a post office in most major shopping centres in Nairobi, and the system is efficient. Mail can also be sent from major hotels. International and local speedpost and parcel services are offered by several independent operators.

Poste Restante is free: the main pick up point is the Central Post Office, Haile Selassie Avenue. Telegrams can be sent by phone; call the operator (900) to ask for help.

Telephone & Telex

International and local calls can be made with ease (direct dial or operator assisted). Payphones are plentiful, and cardphones (for international calls) are starting to come into vogue.

Internal and external direct dial telex and fax facilities are also available. Check with your hotel for the nearest location.

Tourist Offices

Official Information Bureaux are situated in Kimathi Street (tel: 23285) in Nairobi, and near the "tusks" on Moi Avenue in Mombasa. Otherwise, local tour companies are usually glad to help. Publications are plentiful – maps, guides and *What's On* pamphlets are available at most bookshops.

Embassies & Consulates

A complete list of all embassies, consulates and High Commissions in Kenya can be found in the complimentary monthly magazine *What's On*, available at hotels and tour companies throughout Nairobi.

Special Information

African Fund for Endangered Wildlife (AFEW), P.O. Box 15004, Nairobi, tel: 891658. Established in 1978 specifically to save the endangered Rothschild's giraffe, AFEW runs a Nature Education Centre on the property of the founders, Jock and Betty Leslie-Melville. At the centre, students feed giraffes, see wildlife films and take nature walks through the adjacent indigenous forest. Although it is a small family foundation, AFEW makes a commendable contribution to Kenyan wildlife conservation.

African Wildlife Foundation (AWF), P.O. Box 48177, Nairobi, tel: 223235. AWF was established in 1961 in Washington D.C., USA. It opened its African field office in Nairobi in 1963 to help develop and supervise worthy projects. The foundation supports a number of conservation and environmental education programmes and the development of protected area management, planning and operations. It helped launch, and continues to support, two wildlife colleges for African wildlife managers, in Tanzania and Cameroon.

Elsa Wild Animal Appeal (Elsa), P.O. Box 30092, Nairobi, tel: 742121. Formed by the late Joy Adamson, Elsa is funded by the generous royalties from the films and works originating from Joy Adamson's books. The main policies of the trustees are conservation education, particularly among indigenous people in Africa, and assistance to the Kenya Government in retaining wildlife habitats. Elsa has developed a baby animal wildlife clinic at Nairobi National Park and a Conservation Centre at Elsamare on Lake Naivasha, where conservationists can stay at a reasonable cost.

The David Sheldrick Memorial Appeal, P.O. Box 15555, Nairobi. This appeal was established in 1977 following the death of David Sheldrick, founder warden of Kenya's Tsavo National Park. David's closest friends form the advisory committee which governs policy and direction. The appeal has raised funds to support anti-poaching efforts, education and translocation of animals and has compiled a field manual, *The Wilderness Guardian*, which covers park development and wildlife management and conservation.

The East African Wild Life Society (EAWLS), P.O. Box 20110, Nairobi, tel: 748170/1/2/3. This is the oldest local organisation, founded in 1956 with a worldwide membership of 15,000. EAWLS publishes a popular bi-monthly magazine, *Swara*, and a scientific journal for its members.

The society is active in conservation education, research, anti-poaching, animal capture and relocation and wildlife policy formulation and facilitation. It also runs a shop and welcomes both members and donations from the public.

Gallmann Memorial Foundation (GMF), P.O. Box 45593, Nairobi. GMF was set up by Kuki Gallmann in memory of her husband Paolo and son Emmanuel who were keen conservationists. It supports ecological, environmental and educational programmes on Ol Ari Nyiro Ranch in northern Kenya in addition to several other projects in East Africa.

Wildlife Conservation International (WCI), P.O. Box 45593, Nairobi, tel: 221699, 224569. A division of the New York Zoological Society, it designs, funds and directs a wide range of projects.

Wildlife Clubs of Kenya (WCK), P.O. Box 40658, Nairobi, tel: 740811, 742564. Founded in 1969 by a group of boys in a boarding school, WCK has since spread to schools countrywide and is the most grassroots movement. Members organise trips, seminars, rallies and workshops. Its main objectives are to spread interest, awareness and knowledge about wildlife and the environment. Due to tremendous growth, operations are now being decentralised.

Worldwide Fund for Nature (WWF), P.O. Box 40658, Nairobi, tel: 740811, 742564. Founded in 1961, the regional office of WWF in Nairobi manages and develops projects throughout Africa. P.O. Box 40075, Nairobi, tel: 332963.

Photography

Film is easy to get, but expensive. Processing is fast, with good results. If there is no hurry, take your unprocessed films home, taking the usual precautions through airport X-ray security machines. Bring a telephoto lens for long distance wildlife shots. Panoramic views look best on wide angle lenses. 100/200 ASA film should cover most eventualities, but 1000 ASA is good for firelight or water hole scenes where a flash would not work. Mornings (until 10am) and afternoons (after 3pm) are the best time to take pictures; the lack of shadows at midday flattens out pictures. Keep your camera protected against heat and dust, and do not leave it lying about or it will get stolen.

Getting Around

Domestic Travel

By Air

Kenya Airways, the international airline, provides domestic services to and from Nairobi, Mombasa, Malindi and Kisumu. A number of smaller private companies (including Air Kenya) are based at Wilson Airport, Nairobi. These offer regular charter services to Maasai Mara, Samburu and Lamu. Ring Africair at 501210 for further information and to rent planes.

By Rail

Kenya Railways operates sleeper trains from Nairobi to Mombasa and Kisumu, tel: 221211 ext. 2700-2 for information.

Public Transport

Privately-owned *matatus* (the alternative to buses) can be anything from a Peugeot saloon to an Isuzu bus. Minibuses are very popular. *Matatus* are usually more frequent, faster, fuller and more expensive than the buses. The national bus company, Kenya Bus Service (KBS), has cheap rates and covers an extensive network. KBS headquarters is located around the junction between Accra and River roads. Their buses are always packed with commuters morning and evening. Private long haul bus companies serve most towns in Kenya.

Private Transport

One can rent self- or chauffeur-driven cars from a number of local and international car hire firms (Hertz, Avis etc.)

Drive on the left. Get your national drivers' licence endorsed at Nyayo House. Some rural roads are bad (muddy when wet), and one might meet the occasional reckless driver, but locals are willing to help lost tourists. When travelling out of town, always take a spare tyre, some tools, extra petrol and water. If one is going a long way into the bush, serious planning is required. A local guide (check tour companies), at least two vehicles and numerous extra supplies are strongly recommended.

On Departure

Arrive at the airport two hours before your flight leaves. On domestic flights, Ksh 100 must be paid before boarding. After checking in, identify your bags for customs. Prior to boarding the aircraft there are normal passport, ticket and security checks.

Where To Stay

Hotels & Lodgings

Nairobi

5-STAR (CLASS A)

The Grand Regency, P.O. Box 40511, Nairobi, tel: 211199; fax: 21720. City Centre – Lots of brass and formal atmosphere.

Hilton International Hotel, P.O. Box 30364, Nairobi, tel: 334000; fax: 339462. City Centre – all you would expect from this multinational chain. Mostly businessmen use it.

Hotel Inter-Continental, P.O. Box 30353, Nairobi, tel: 335550; fax: 210675. Within the central city grid. Similar to the Hilton in that it is a businessman's hotel.

Nairobi Safari Club, P.O. Box 43564, Nairobi, tel: 330621; fax: 331201. Upmarket, suites only hotel in the centre of town.

Nairobi Serena Hotel, P.O. Box 46302, Nairobi, tel: 725111; fax: 725184. Stylish enterprise of the Aga Khan; set in a park of the edge of the city grid. Excellent food and service.

Norfolk Hotel, P.O. Box 40064, Nairobi, tel: 335422/33; fax: 336742. Historic favourite of visitors; fully modernised; a short walk to the city centre.

Safari Park Hotel & Casino, P.O. Box 45038, Nairobi, tel: 802493-6; fax: 802477. Out of town on the Thika Road. Large hotel with all the facilities. Very well-appointed.

Windsor Golf & Country Club, P.O. Box 45887, Nairobi, tel: 802259/802210; fax: 802322. Fifteen minutes' drive from the centre of Nairobi. Very attractive hotel with the main feature being the 18-hole championship golf course.

4-STAR (CLASS B)

Panafric Hotel, P.O. Box 30486, Nairobi, tel: 720822-8; fax: 726356. On a hill overlooking the central grid; full amenities and services.

Six-Eighty Hotel, P.O. Box 43436, Nairobi, tel: 332680; fax: 332908. Modern town hotel that is reasonably priced.

New Stanley Hotel, P.O. Box 30680, Nairobi, tel: 333248-51; fax: 229388. City Centre; safari epicenter is the hotel's pavement cafe.

Mayfair Court Hotel, P.O. Box 74957, Nairobi, tel: 748288/748278; fax: 746826. Newly re-furbished hotel in the suburb of Westlands.

3-STAR (CLASS C)

Boulevard Hotel, P.O. Box 42831, Nairobi, tel: 227567/8/9; fax: 334071. Just beyond the Norfolk; smallish garden hotel; excellent value for money.

Fairview Hotel, P.O. Box 40842, Nairobi, tel: 723211; fax: 721320. Long-established, set in lawns; reasonable accommodation for families.

Hotel Ambassadeur, P.O. Box 30399, Nairobi, tel: 336803/09; fax: 211472. Close to the Hilton, a good Indian restaurant, the Safeer, is incorporated.

Jacaranda Hotel, P.O. Box 14287, Nairobi, tel: 448713/7; fax: 448977. Family style in the suburb of Westlands. Recently taken over by a major local hotel chain.

Hostels

YMCA Hostel, P.O. Box 30330, Nairobi, tel: 724116/7; fax: 728875.

Youth Hostel, Ralph Bunche Road, Nairobi, tel: 723012.

YWCA Hostel, P.O. Box 40710, Nairobi, tel: 724699.

The Coast

5-Star (Class A)

SOUTH COAST

Diani Reef Grand Hotel, P.O. Box 35, Ukunda, tel: Diani 2175/2723; fax: 2196.

Golden Beach Hotel, P.O. Box 31, Ukunda, tel: Diani 2625/2054; fax: 2321.

Indian Ocean Beach Club, P.O. Box 40075, Nairobi, tel: Diani 3730; fax: 3557.

Kaskazi Beach Hotel, P.O. Box 135, Ukunda, tel: Diani 3170.

Leisure Lodge Club, P.O. Box 84383, Mombasa, tel: Diani 2011/2; fax: 2159.

NORTH COAST

Mombasa Inter-Continental Hotel, P.O. Box 83492, Mombasa, tel: 485811; fax: 485431.

Mombasa Beach Hotel, P.O. Box 90414, Mombasa, tel: 471861.

Nyali Beach Hotel, P.O. Box 90581, Mombasa, tel: 471551.

Reef Hotel, P.O. Box 82234, Mombasa, tel: 471771/2; fax: 471349.

Serena Beach Hotel, P.O. Box 90352, Mombasa, tel: 485721; fax: 485453.

Whitesands Hotel, P.O. Box 90173, Mombasa, tel: 485763; fax: 485652.

4-Star (Class B)

SOUTH COAST

Africana Sea Lodge, P.O. Box 84616, Mombasa, tel: Diani 2624; fax: 2145.

Jadini Beach Hotel, P.O. Box 84616, Mombasa, tel: Diani 2622; fax: 2269.

Safari Beach Hotel, P.O. Box 84616, Mombasa, tel: Diani 2726; fax: 2357.

NORTH COAST

Bamburi Beach Hotel, P.O. Box 83966, Mombasa, tel: 485611; fax: 485900.

Plaza Hotel, P.O. Box 88299, Mombasa, tel: 485321.

Severin Sea Lodge, P.O. Box 82169, Mombasa, tel: 485001.

Traveller's Beach Hotel, P.O. Box 87649, Mombasa, tel: 485121; fax: 485678.

3-Star (Class C)

SOUTH COAST

Beachcomber Club, P.O. Box 54, Msambweni, tel: Diani 52033; fax: 3112.

Trade Winds Hotel, P.O. Box 8, Ukunda, tel: Diani 2016.

Two Fishes Hotel, P.O. Box 23, Ukunda, tel: Diani 2101-4; fax: 2106.

NORTH COAST

Kenya Beach Hotel, P.O. Box 95748, Mombasa, tel: 485821.

Neptune Beach Hotel, P.O. Box 83125, Mombasa, tel: 485701.

Sun 'N' Sand Beach Hotel, P.O. Box 2, Kikambala, tel: 32621/32008; fax: 32133.

Whispering Palms Hotel, P.O. Box 5, Kikambala, tel: 320045.

Mombasa Town

The Manor Hotel, P.O. Box 84851, Mombasa, tel: 314643; fax: 311952.
Oceanic Hotel, P.O. Box 90371, Mombasa, tel: 311191/312838.
The Outrigger Hotel, P.O. Box 82345, Mombasa, tel: 220822/3.

Watamu, Malindi & Lamu

NORTH COAST

5-STAR (CLASS A)
Hemingways, P.O. Box 267, Watamu, tel: 32624; fax: 32256.

4-STAR (CLASS B)
Peponi Hotel, P.O. Box 24, Lamu, tel: 33421/2/3; fax: 33029.
Turtle Bay Beach Hotel, P.O. Box 457, Malindi, tel: 32226/32080; fax: 32268.

3-STAR (CLASS C)
Ocean Sports, P.O. Box 100, Watamu, tel: 32008; fax: 32266.
The Driftwood Beach Club, P.O. Box 63, Malindi, tel: 20155.
Lawfords Hotel, P.O. Box 20, Malindi, tel: 20440.
Watamu Beach Hotel, P.O. Box 300, Malindi, tel: 32001/32010.

Up-Country Hotels

5-STAR (CLASS A)
Mt Kenya Safari Club, P.O. Box 35, Nanyuki, tel: 22960/1.

4-STAR (CLASS B)
Aberdare Country Club, P.O. Box 58181, Nairobi, tel: 216920/40; fax: 216796.
The Outspan Hotel, P.O. Box 47557, Nairobi, tel: 335807; fax: 340541.
Lake Baringo Club, P.O. Box 47557, Nairobi, tel: 335807; fax: 340541.
Lake Naivasha Country Club, P.O. Box 47557, Nairobi, tel: 335807; fax: 340541.
Naro Moru River Lodge, P.O. Box 18, Naro Moru, tel: 0176-62622/62212; fax: 0176-62211.

3-STAR (CLASS C)
Golf Hotel, P.O. Box 42013, Nairobi, tel: 229751 or Kagamega, tel: 0331-30150.
Safariland Lodge, P.O. Box 72, Naivasha, tel: 0311-21034/47; fax: 0311-21216.
Sirikwa Hotel, P.O. Box 3361, Eldoret, tel: 31655.
Sunset Hotel, P.O. Box 215, Kisumu, tel: 41100-4.
Tea Hotel, P.O. Box 75, Kericho, tel: 20280.

Game Lodges

5-STAR (CLASS A)
Amboseli Serena Lodge – Amboseli National Park, P.O. Box 48690, Nairobi, tel: 711077; fax: 718103.
Borana Lodge – Timau, P.O. Box 24397, Nairobi, tel: 568804; fax: 564945.
Keekorok Lodge – Maasai Mara National Reserve, P.O. Box 47557, Nairobi, tel: 335807; fax: 340541.
Mara Serena Lodge – Maasai Mara National Reserve, P.O. Box 48690, Nairobi, tel: 711077; fax: 718103.
Rusinga Island Fishing Lodge – Lake Victoria, P.O. Box 58581, Nairobi, tel: 216920/40; fax: 216796.
Taita Hills Lodge, P.O. Box 30624, Nairobi, tel: 334000; fax: 339462.
Salt Lick Lodge, P.O. Box 30624, Nairobi, tel: 334000; fax: 339462.
Sarova Shaba Lodge – Shaba National Reserve, P.O. Box 30680, Nairobi, tel: 333233; fax: 211472.
Samburu Lodge – Samburu National Reserve, P.O. Box 47557, Nairobi, tel: 335807; fax: 340541.
Samburu Serena Lodge – Samburu National Reserve, P.O. Box 48690, Nairobi, tel: 711077; fax: 718103.

4-STAR (CLASS B)
Amboseli Lodge & Kilimanjaro Safari Lodge – Amboseli National Park, P.O. Box 30139, Nairobi, tel: 227136/337510; fax: 219982.

The Ark – Aberdare National Park, P.O. Box 58581, Nairobi, tel: 216920/40; fax: 216796.
Kilimanjaro Buffalo Lodge – Amboseli National Park, P.O. Box 30139, Nairobi, tel: 227136/337510; fax: 219982.
Kilaguni Lodge – Tsavo West National Park, P.O. Box 30471, Nairobi, tel: 336858; fax: 218109.
Lion Hill Lodge, P.O. Box 30680, Nairobi, tel: 333233; fax: 211472.
Mara Sopa Lodge – Maasai Mara National Reserve, P.O. Box 72630, Nairobi, tel: 220182/336088; fax: 331876.
Ngulia Lodge – Tsavo West National Park, P.O. Box 30471, Nairobi, tel: 336858; fax: 218109.
Treetops Lodge – Aberdare National Park, P.O. Box 47557, Nairobi, tel: 335807; fax: 340541.
Voi Safari Lodge – Tsavo East National Park, P.O. Box 30471, Nairobi, tel: 336858; fax: 218109.

3-STAR (CLASS C)
Buffalo Springs Lodge, P.O. Box 30471, Nairobi, tel: 336858; fax: 218109.
Mountain Lodge, P.O. Box 30471, Nairobi, tel: 336858; fax: 218109.
Lake Nakuru Lodge, P.O. Box 70559, Nairobi, tel: 212405/226778; fax: 230962.
Maralal Safari Lodge, P.O. Box 45155, Nairobi, tel: 211124/334177; fax: 214261.
Olkurruk Mara Lodge, P.O. Box 30471, Nairobi, tel: 336858; fax: 218109.

Tented Camps

5-STAR (CLASS A)
Africa Expeditions Private Tented Camp – Loita Hills, P.O. Box 24598, Nairobi, tel: 561882/561959; fax: 561457/561054.
Africa Expeditions Private Tented Camp – Maasai Mara National Reserve, P.O. Box 24598, Nairobi, tel: 561882/561959; fax: 561457/561054.
Finch Hattons – Tsavo West National Park, P.O. Box 24423, Nairobi, tel: 604321/2; fax: 604323.
Governor's and Little Governor's Camps – Maasai Mara National Reserve, P.O. Box 48217, Nairobi, tel: 331871/2; fax: 726427.
Kichwa Tembo Camp – Maasai Mara National Reserve, P.O. Box 74957, Nairobi, tel: 746707/750780; fax: 746826.
Larsens Camp – Samburu National Reserve, P.O. Box 47557, Nairobi, tel: 335807; fax: 340541.
Mara Intrepids Club – Maasai Mara National Reserve, P.O. Box 74888, Nairobi, tel: 338084/335208; fax: 217278.
Mara Safari Club – Maasai Mara National Reserve, P.O. Box 58581, Nairobi, tel: 216920/40; fax: 216796.
Samburu Intrepids Club – Samburu National Reserve, P.O. Box 74888,

Nairobi, tel: 338084/335208; fax: 217278.
Siana Springs Camp – Maasai Mara National Reserve, P.O. Box 74957, Nairobi, tel: 746707/750780; fax: 746826.

4-STAR (CLASS B)

Delamere Camp – Lake Elmenteita, P.O. Box 48019, Nairobi, tel: 335935/331191; fax: 216528.
Fig Tree Camp – Maasai Mara, P.O. Box 40683, Nairobi, tel: 221439/218321; fax: 332170.
Island Camp – Lake Baringo, P.O. Box 58581, Nairobi, tel: 216920/40; fax: 216796.
Kindani Camp – Meru National Park, P.O. Box 56118, Nairobi, tel: 445795/444494; fax: 440749.
Mara River Camp – Maasai Mara National Reserve, P.O. Box 48019, Nairobi, tel: 335935/331191; fax: 216528.
Tsavo Safari Camp – Tsavo East National Park, P.O. Box 30139, Nairobi, tel: 227136/337510; fax: 219982.
Ziwani Tented Camp – Tsavo West National Park, P.O. Box 74888, Nairobi, tel: 338084/335208; fax: 217278.

Central Booking Offices

Some out-of-Nairobi hotels and lodges have a central booking system based in Nairobi. The management groups and appropriate numbers are listed below:

Alliance Hotels, P.O. Box 49839, Nairobi, tel: 332825/337508/220149; fax: 219212.
Africana Sea Lodge – Diani Beach
Jadini Beach Hotel – Diani Beach
Naro Moru River Lodge – Naro Moru
Safari Beach Hotel – Diani Beach
African Tours & Hotels, P.O. Box 30471, Nairobi, tel: 336858; fax: 218109.
Buffalo Springs Lodge – Buffalo Springs National Reserve
Kabarnet Hotel – Kabarnet
Kilaguni Lodge – Tsavo West National Park
Milimani Hotel – Nairobi
Mombasa Beach Hotel – North Coast
Mountain Lodge – Mount Kenya
Ngulia Safari Lodge – Tsavo West National Park
Olkurruk Mara Lodge – Maasai Mara National Reserve
Sunset Hotel – Kisumu
Trade Winds Hotel – Diani Beach
Voi Safari Lodge – Tsavo East National Park
Block Hotels, P.O. Box 47557, Nairobi, tel: 335807; fax: 340541.
Keekorok Lodge – Maasai Mara National Reserve
Indian Ocean Beach Club – Diani Beach
Lake Baringo Club – Lake Baringo
Lake Naivasha Country Club – Lake Naivasha
Larsens Camp – Samburu National Reserve
Nyali Beach Hotel – North Coast
Outspan Hotel – Nyeri
Samburu Lodge – Samburu National Reserve
Shimba Lodge – Shimba Hills National Park
Treetops – Aberdare National Park
Hilton Lodges & Hotels, P.O. Box 30624, Nairobi, tel: 334000; fax: 339462.
Nairobi Hilton Hotel – Nairobi
Taita Hills Lodge – Taita Hills
Salt Lick Lodge – Taita Hills
Lonrho Hotels, P.O. Box 58581, Nairobi, tel: 216920/40; fax: 216796.
The Ark – Aberdare National Park
Aberdare Country Club – Mweiga
Island Camp – Lake Baringo
Mara Safari Club – Maasai Mara National Reserve
Mount Kenya Safari Club – Nanyuki
Norfolk Hotel – Nairobi
Ol Pejeta – Nanyuki
Rusinga Island Fishing Club – Lake Victoria
Sweetwaters Tented Camp – Nanyuki
Serena Lodges & Hotels, P.O. Box 48690, Nairobi, tel: 711077; fax: 718103.
Nairobi Serena Hotel – Nairobi
Mara Serena Lodge – Maasai Mara National Reserve
Samburu Serena – Samburu National Reserve
Serena Beach Hotel – North Coast
Sarova Hotels, P.O. Box 30680, Nairobi, tel: 333233; fax: 211472.
New Stanley Hotel – Nairobi
Ambassadeur Hotel – Nairobi
Panafric Hotel – Nairobi
Lion Hill Lodge – Lake Nakuru National Park
Sarova Shaba Lodge – Shaba National Reserve
Whitesands Hotel – North Coast
Mara Sarova Lodge – Maasai Mara National Reserve
The Conservation Corporation, P.O. Box 74957, Nairobi, tel: 746707/750298/750780; fax: 746826.
The Mayfair Court Hotel – Nairobi
The Windsor Golf & Country Club – Nairobi
Kichwa Tembo Camp – Maasai Mara National Reserve
Siana Springs Tented Camp – Maasai Mara National Reserve

Eating Out

Where To Eat

Throughout Kenya, one can find an international variety of restaurants. Dining out guides are available at bookstores. In general, food is of excellent quality and very reasonably priced.

Note that wines and spirits are imported and therefore cost their European equivalent in Kenya shillings.

Following is a list of restaurants and eateries:

African Heritage Cafe, Banda St, tel: 337507. Local specialities.
Akasaka Restaurant, Muindi Mbingu St, tel: 220299. Japanese cuisine.
Alan Bobbe's Bistro, Koinange St, tel: 224945. French and expensive.
Carnivore Restaurant, Langata Road, tel: 501775. Barbecued meats and great atmosphere.
Daas Ethiopian Restaurant, Ralph Bunche Road, tel: 727353.
Dawat Restaurant, Shimmers Plaza, Westlands, tel: 749337. Indian.
Gringo's Restaurant, Limuru Road, tel: 521231/2. Tex-Mex.
Haandi Restaurant, The Mall, Westlands, tel: 448294. Excellent Indian but not cheap.
Horseman Restaurant, Ngong Road, tel: 882033/882133. International cuisine and Karaoke bar.
Minar Restaurant, Sarit Centre, Westlands, tel: 748340. Indian.
Nawab Tandoori, Muthaiga Shopping Centre, tel: 740209. Indian.
Pagoda, Shankardass House, Moi Avenue, tel: 227036/230230. Chinese.
Siam Thai Restaurant, Unga House, Westlands, tel: 751727/751728.
Tamarind Restaurant, Harambee Ave,

tel: 338959/217990. Seafood specialities on offer.
Toona Tree, International Casino, tel: 744477. Italian/International.

Attractions

Culture

Museums

Visit the National Museum of Kenya: it has some excellent displays of stuffed animals and tribal artifacts. Next to it is the snake park where the inhabitants are very much alive, from cobras to crocodiles.

Art Galleries

Indigenous art is flourishing in Kenya, and there is a different exhibition every week. Try Gallery Watatu, French Cultural Centre, African Heritage Ltd, and Paa-ya-Paa gallery for up and coming African art. There is also an art gallery at the National Museum which is wholly reserved for work of East African artists.

Theatres

The National Theatre, Phoenix Players, Braeburn Theatre and French Cultural Centre all offer classical, traditional and contemporary stage productions of excellent standard.

Movies

There are about 10 movie theatres in Nairobi: the Nairobi, Kenya and 20th Century offer the best facilities and latest movies. Check newspapers for listings.

Music

Check the entertainment page of the daily newspapers. Otherwise the Carnivore offers live band entertainment, and the International Casino complex is a good place to eat, drink, dance and gamble.

Diary Of Events

Hotels generally provide traditional dancing and other activities. Otherwise, visit the Bomas of Kenya – an interesting experience close to Nairobi. Check the entertainment page of daily newspapers or the *What's On* guide for information.

Tours

Travel Packages

There are many tour operators in Kenya that offer safaris ranging from day trips to two-week excursions. Some of the more extensive tours can be booked through travel agents abroad, and most (except for the extremely exclusive) can be booked in Nairobi. Minibus tours are most common, economical, and they cover most of Kenya. However, for more remote areas four-wheel drive trucks are relied on and light aircraft can be chartered to anywhere. Tour information can be found at any travel agent or hotel, but shop around for a specific preference because the selection is wide.

Tour Operators

Further information about safaris throughout Kenya can be found at the following specialist tour companies:
Abercrombie & Kent Ltd, P.O. Box 59749, Nairobi, tel: 334955; fax: 215752.
Africa Expeditions Ltd, P.O. Box 24598, Nairobi, tel/fax: 561882/561959/561054.
Ker & Downey Safaris Ltd, P.O. Box 41822, Nairobi, tel: 553212/556466; fax: 552378.
Let's Go Travel Ltd, P.O. Box 60342, Nairobi, tel: 340331; fax: 336890.
Rafiki Africa, P.O. Box 76400, Nairobi, tel /fax: 561562.

Shopping

Most towns offer the tourist something in the way of curios. Mostly, these consist of baskets, bracelets and other jewellery, wood and stone carvings, batiks and other fabrics. Argue, haggle, beg for discounts, threaten to walk away and use every other trick in the book to beat prices down. Tourists will be approached with "elephant hair" or "ivory" curios – usually fake, and definitely illegal. Do not buy ivory; Kenya's elephant population is being decimated by poachers.

There are several clothing shops near the large hotels; other items from batteries to brandy are available.

Sports

Check the clubs directly (Impala, Nairobi, Parklands, Muthaiga) for most field and court activities. Horse racing takes place most Sundays at the Ngong Road racecourse. Deep sea fishing and other watersports can all be arranged through hotels on the coast. The Safari Rally takes place every Easter, and is internationally acknowledged as the most gruelling race of its kind.

Sports Clubs

Temporary memberships are offered to visitors in most sports clubs around Kenya. To avoid a wasted trip it's advisable to telephone the clubs concerned to confirm the availability of squash or tennis courts, golf course, etc.
Impala Club, Ngong Road, P.O. Box 41516, Nairobi, tel: 568573. Tennis, Squash, Rugby, Football, Hockey and Cricket.
Nairobi Club, Ngong Road, P.O. Box 30171, Nairobi, tel: 336996. Squash, Tennis, Cricket, Hockey, Bowls and Basketball.
Parklands Sports Club, Ojijo Road, P.O. Box 40116, Nairobi, tel: 742829. Tennis, Squash, Rugby, Hockey, Cricket and Snooker.

Golf Clubs

Karen Country Club, tel: 882801/2.
Muthaiga Golf Club, tel: 767754/5.
Royal Nairobi Golf Club, tel: 724215.
Sigona Club, tel: (0154) 32431.
Windsor Golf & Country Club, tel: 803231.

Sailing

Nairobi Sailing Club, Nairobi Dam, P.O. Box 49973, tel: 501250.

Mountaineering

Mountain Club of Kenya, P.O. Box 45741, Nairobi, tel: 501747.

Horse Racing

Jockey Club of Kenya, P.O. Box 40373, Nairobi, tel: 566109. Racing most Sundays (check the newspapers). First race at 2.15pm. Admission: Adults Shs. 100, Children Shs. 20. Silver Ring free.

Lions Clubs

Mombasa (Central), P.O. Box 82569, Mombasa, tel: 25061.
Mombasa (Pwani), P.O. Box 81871, Mombasa, tel: 20731.
Nairobi (Host), P.O. Box 47447, Nairobi, tel: 742266.
Nairobi (Central), P.O. Box 44867, Nairobi, tel: 338901.
Nairobi (City), P.O. Box 30693, Nairobi, tel: 27354.
Nairobi (North), P.O. Box 42093, Nairobi, tel: 21251.
Nairobi (Westlands), P.O. Box 42539, Nairobi, tel: 556020.
Nairobi (Kikuyu), P.O. Box 47301, Nairobi, tel: 24023.
Lions International (District 411), P.O. Box 45652, Nairobi, tel: 331709.

Rotary Clubs

Eldoret, P.O. Box 220, Eldoret, tel: 2936.
Kilindini, P.O. Box 99067, Mombasa, tel: 25157.
Mombasa, P.O. Box 90570, Mombasa, tel: 25924.
Nairobi, tel: 742269.
Nairobi North, c/o P.O. Box 30751, Nairobi, tel: 21624.
Nairobi South, c/o P.O. Box 46611, tel: 337041.
Rotary International (District 920), c/o P.O. Box 41910, tel: D Gov 24128.

Other Clubs & Associations

Nairobi Photographic Society, P.O. Box 49879, Nairobi, tel: 337129. Meetings are held at the St John's Ambulance Headquarters (behind Donovan Manle Theatre) at 8.30pm on the first and third Thursdays of each month.
Mombasa Rowing Club, P.O. Box 82037, Mombasa, Club House at Ras Liwatoni next door to the Outrigger Hotel. Daily membership for tourists – Shs 10.
Kenya Divers Association, P.O. Box 95705, Mombasa, tel: 471347.
The Aquarist Club of Kenya, P.O. Box 49931, Nairobi, tel: 25975 (day), tel: 559281 or 746636 (evening). Meets at 10am the second Sunday of every month in the Fairview Hotel. Visitors and prospective members welcome.
The Caledonian Society of Kenya, P.O. Box 4075, Nairobi, tel: Sec 520400 (evenings).
American Women's Association, P.O. Box 47806, Nairobi, tel: 65342. Membership Chairman.
Nairobi Branch of the Royal Society of St George P.O. Box 48360, Nairobi, tel: 891262.
Geological Club of Kenya, P.O. Box 44749, Nairobi.
Geographical Society of Kenya, P.O. Box 41887, Nairobi.
African Cultural Society, P.O. Box 69484, Nairobi, tel: 335581. Cultural Festivals, Lectures and Theatre.

Language

English is understood by many people in up-country Kenya, but not so much at the coast which is predominantly Muslim, speaking the Afro-Arab-Indian mix *Swahili.* It's not a difficult language and it's worth learning even a few words.

English is taught in schools all over the country, so there is always someone who will understand what you're talking about, even in the remote bush. At the coast, more locals are responding to the European continental tourist invasion and speak German, French and Italian. The following Swahili words follows may come in handy:

General

Mr	*Bwana*
Mrs	*Bibi*
Miss	*Bi*
I	*Mimi*
You	*Wewe*
He, She	*Yeye*
We	*Sisi*
They	*Wao*
What?	*Nini?*
Who?	*Nani?*
Where? (Place)	*Mahali gani?*
Where? (Direction)	*Wapi (Upande gani?)*
When?	*Hini?*
How	*Vipi*
Why?	*Kwanini?*
Which?	*Ipi? (gani)*
To eat	*Kukula*
To drink	*Kukunywa*
To sleep	*Kulala*
To bathe	*Kuoga*
To come	*Ijayo*
To go	*Ku-enda*
To stop	*Kusimama*
To buy	*Kununua*
To sell	*Kuuza*
Street/road	*Barabara*
Airport	*Uwanja wa Ndege*
Shop	*Duka*
Money	*Fedha*
Cent	*Senti*
Hotel	*Hoteli*
Room	*Chumba*
Bed	*Kitanda*
Food	*Chakula*
Coffee	*Kahawa*
Beer	*Tembo (or Pombe)*
Cold	*Baridi*
Hot	*Moto*
Tea	*Chai*
Meat	*Nyama*
Fish	*Samaki*
Bread	*Mkate*
Butter	*Siagi*
Sugar	*Sukari*
Salt	*Chumvi*
Bad	*Mbaya*
Today	*Leo*
Tomorrow	*Kesho*
Now	*Sasa*
Quickly	*Haraka*
Slowly	*Pole-pole (pol-i...pol-i)*
Hospital	*Hospitali*
Police	*Polici*

Numbers

One	*Moja*
Two	*Mbili*
Three	*Tatu*
Four	*Ine*
Five	*Tano*
Six	*Sita*
Seven	*Saba*
Eight	*Nane*
Nine	*Tisa*
Ten	*Kumi*
Eleven	*Kumi na moja*
Twelve	*Kumi na mbili*
Thirteen	*Kumi na tatu*
Twenty	*Ishirini*
Twenty-one	*Ishirini na moja*
Twenty-two	*Ishirini na mbili*
Twenty-three	*Ishirini na tatu*
Thirty	*Thelathini*
Forty	*Arobaini*
Fifty	*Hamsini*
One hundred	*Mia moja*
One thousand	*Elfu moja*

More Words

Hello	*Jambo*
How are you?	*Habari?*
I am well	*Mzuri* (good, fine, etc.)
Thank you	*Asante (sana)* (very much)
Goodbye	*Kwaheri*
Where is the Hotel?	*Hoteli iko wapi?*

Where does this road lead to?	*Njia hii ina-endawapi?*
Please help me push this car*	*Tafadhali nisaidie kusukuma gari*
Please change this wheel	*Tafadhali badilisha* gurudumu hili
Good morning	*Habari ya asubuhi*
Good afternoon	*Habari ya mehana*
Good evening	*Habari ya jioni*
Please come in	*Karibu ndani* tafadhali
Please sit down	*Keti tafadhali*
You're welcome	*Una karibishwa*
Where do you come from?	*Ume kuja kutoka wapi?*
I come from	*Nime toka*
What is your name?	*Jina lako nani?*
My name is	*Jina langu ni...*
Can you speak Swahili?	*Waweza kuongea kiswahili?*
Yes	*Ndiyo*
No	*Hapana*
Only a little	*Kidogo tu*
I want to learn more	*Nataka kujifunza zaidi*
How do you find Kenya?	*Waonaje Kenya?*
I like it here	*Hapa napenda*
The weather is hot, isn't it?	*Hewa hapa in joto sivyo?*
Yes, a little	*Naiyo kidogo*
Where are you going?	*Una kwenda wapi?*
I am going to	*Nak wenda*
Turn	*Geuka kulia*
Turn left	*Geuka kushoto*
Go straight	*Enda moja kwa* moja
Please stop here	*Simama hapa* tafadhali
How much?	*Ngapi?*
Wait a minute	*Ngoja kidogo*
I have to get change	*Ni badilishe pesa kwanza*
Excuse me	*Samahani*
Where is the toilet?	*Wapi choo?*
In the back	*Upande wa nyuma*
Where may I get something to drink?	*Naweza kupata wapi kinywaji?*
One cup of coffee	*Kikombe kimoja* cha kahawa
How much does this cost?	*Inagharimu pesa ngapi?*
That's quite expensive	*Waweza kupunguz*
I will buy it	*Nita nunua*

TANZANIA

Getting Acquainted

The Land

Area: 362,700 sq miles/939,800 sq km.
Capital: Dodoma.
Population: 27 million.
International Dialling Code: 255 (Dadoma: 61).
Time Zone: three hours later than GMT.
The Place: Lying mostly above 3,000 ftt (4,000 metres). Tanzania borders the Indian Ocean and contains the continent's largest lakes: Nyasa, Tanganyika and Victoria, and its highest mountain, Kilimanjaro (19,340 ft/ 5,895 metres).

Climate

The rainy season is from November to May when the days are warm and it can be humid, especially on the coast.

Rains bring changes in vegetation and migratory patterns of wildlife, and can affect the roads too. The short rains in November and December help to settle the dust. In places facing the ocean rains slow or cease during January and February and it's a good time for climbing mountains. The heavier period of rains is from March to May when some low-lying or poor roads become impassable.

The high plateau country means cool nights and pleasant-to-warm days. The coastal strip is hot and humid, especially from December to March, while breezes make May to October rather more pleasant.

The higher mountains can be cold in June and July, when there is no rain and the grass is long. In August and September strong winds and fires make the air hazy.

At all times of the year, be prepared for dust. The fine dust around volcanoes is especially insidious and can affect binoculars, cameras and contact lenses. Bring plenty of plastic bags and fluids for protection.

Language

Swahili is spoken throughout Tanzania (see page 338). It is a *Bantu* language with a lartge component of Arabic words. Many English words have been assimilated, such as *sharti* meaning "shirt" or *Kekei* which is "cake".

Planning The Trip

What To Wear

Light cotton clothing is best for all conditions; trousers and shorts, skirts and open-necked shirts. A sweater or jacket is useful, especially in the highlands and during the cold season from June to August. A waterproof jacket offers good protection from rain, wind and sun.

On the coast and in other Muslim areas "revealing" outfits such as swimsuits or short skirts and shorts should not be worn in public places. However, locally made cotton cloths are very useful to buy to wear as a wraparound, for a shawl, extra towel or seat cover etc: *Kangas* are patterned clothes with a rectangular border, usually sold in pairs; continuously patterned cloth without a border is called *kitenge*: the heavier, tasselled *kikoi* is usually worn by men. Hats, sunglasses and sun block creams are essential. Insect repellents are necessary against mosquitoes, tsetse flies, nuisance flies, midges, fleas or ticks. Repellents with 100 percent "deet" (diethyltoluamide) are the most effective (*Jungle Formula*).

What To Bring

Bring a torch with spare batteries and a water bottle with a cup top. Water sterilisation tablets are also useful.

Entry Regulations

Visas & Passports

Visas must be obtained in advance by all visitors except citizens of Britain and the Commonwealth and the

Scandinavian countries. Entry is prohibited to residents and nationals of South Africa and to anyone with a South African stamp in their passport.

Although it is theoretically possible to change unspent Tanzanian shillings back to your own currency on departure, in practice there is seldom the opportunity to do so. Unless entering the country at an airport with a bank on the premises, it is usually easier to change money at hotels with exchange facilities than at one of the crowded banks.

A departure tax is levied on all passengers embarking on any international or domestic flight at any airport in Tanzania.

Health

Visit Tanzania in good health. Hospitals and doctors are few and far between. Check with your doctor for any vaccinations you might need. Yellow fever and cholera vaccinations are officially required: many visitors also get vaccinated against tetanus, polio and typhoid. A gamma-globulin shot gives some protection against infectious hepatitis. Bring malarial prophylaxis and any medications that you need or habitually use.

Many travellers who come to Africa with a safari company will have taken out temporary memberships with the Flying Doctors. Check with your travel agent or write directly for your own individual membership: Flying Doctor Service, c/o Kilimanjaro Christian Medical Centre (KCMC), P.O. Box 3010, Moshi, Tanzania.

Getting There

By Air

The main international airports at Dar es Salaam, Kilimanjaro and Zanzibar are served by several international airlines. The national airline, *Air Tanzania*, has domestic flights to all major cities. Very few domestic air charter companies fly small groups to more remote locations.

By Rail

There are three main routes: the Central Line, from Dar es Salaam via Dodoma and Tabora to Kigoma, with a side branch from Tabora to Mwanza; the Tanga line, from Tanga to Moshi and Arusha; and the Tanzam or Uhuru line, from Dar to the Zambian border. Sleeping accommodation is available but should be booked well in advance. All services are generally slow and crowded.

By Road

Very few roads are surfaced although a determined effort is being made to upgrade the major routes. You can expect to experience a wide range of conditions, even on a short journey.

Most visitors arrange their travel with one of the many tour operators based in Arusha or Dar es Salaam, who offer safaris to suit every pocket. Public transport is cheap but normally very overcrowded. There are taxi services in most large towns but it is wise to agree upon a price before the trip. It is virtually impossible to hire a self-drive car, but chauffeur-driven vehicles are available.

If you bring your own car, you should note that driving is on the left, traffic signs are international but very sparse, and fuel is subject to occasional shortages so an ample reserve should be carried, especially if you are visiting remote areas.

Where To Stay

Lodges & Hotels

Lodges and hotels offering services of a high standard or expressly for visitors are available in many wildlife areas. There are also many less expensive hotels and inns. Prices fluctuate greatly and there is often an extra charge for service (usually 5 percent), a government hotel levy (12 percent to 17 percent) and a sales tax on all bills.

Dar es Salaam

Kilimanjaro Hotel, P.O. Box 9574, Dar es Salaam, tel: 21281. Large with a swimming pool and overlooks the harbour.

Motel Agip, P.O. Box 529, Dar es Salaam, tel: 23511. Located near the centre of town.

Oysterbay Hotel, P.O. Box 2261, Dar es Salaam, tel: 68631. Located to the north of the city on the beach front. It has a pleasant and relaxed atmosphere.

Beach Hotels near Dar es Salaam

Bahari Beach Hotel, P.O. Box 9312, Dar es Salaam, tel: 47101.

Kunduchi Beach Hotel, P.O. Box 9331, Dar es Salaam, tel: 41061.

Arusha Town

Mount Meru Hotel, P.O. Box 877, Arusha, tel: 2711/2728/2717.

Mount Meru Game Lodge, P.O. Box 427, Arusha, tel: Usa River 43 or contact Abercrombie & Kent office: Arusha 7803. Small with a charming atmosphere and is adjacent to the game sanctuary.

Ngare Sero Lodge, P.O. Box 425, Arusha, tel: Arusha 3629. Small in exceptionally lovely grounds with natural forest and springs.

Ol Donyo Orok Lodge, P.O. Box 735, Arusha, tel: 7020.

Momela Lodge, P.O. Box 418, Arusha, tel: Arusha 3038. Bungalows or rondavels and is just outside the park.

Gombe National Park

Campsites and a rest house; arrange through TANAPA.

Mahale National Park

Campsites can be arranged through TANAPA. Before or after visiting these parks you may have to stay in Kigoma.

Mahale Mountains Tented Camp, Greystoke Safaris, P.O. Box 1373, Kigoma.

Kigoma

New Kigoma Railway Hotel, Private Bag, Kigoma, tel: Kigoma 64.

Kilimanjaro National Park

Kibo Hotel, P.O. Box 102, Marangu, tel: 4. Located on the slopes below the park gate and very spacious. Climbs can be arranged through reception.

Marangu Hotel, P.O. Box 40, Marangu, tel: 11. Lovely gardens and climbs can be arranged through reception.

Climbing Kilimanjaro requires booking huts either through the hotels above (or in Moshi town), or through TANAPA which also has two comfortable hostels at the Marangu Entry Gate.

Lake Manyara

Gibb's Farm, P.O. Box 6084, Arusha, tel: 6702; fax: 7877.

Lake Manyara Hotel, P.O. Box 3100, Arusha, tel: 3842.

Maji Moto Tented Camp, P.O. Box 40097, Nairobi, tel: 331825; fax: 212656.

Mikumi National Park

Mikumi Wildlife Lodge, P.O. Box 2485, Dar es Salaam, tel: 23491. Located in the centre of the park and is built around a water hole.

Mikumi Wildlife Camp, P.O. Box 1097, Dar es Salaam, tel: 68631. Informal but comfortable.

Ngorongoro Conservation Area

Ngorongoro Crater Lodge, P.O. Box 751, Arusha, tel: 3530. Rustic, comfortable and perched on the crater rim.

Ngorongoro Sopa Lodge, P.O. Box 1823, Arusha, tel: 6886.

Ngorongoro Wildlife Lodge, P.O. Box 3100, Arusha, tel: 3173.

Ndutu Safari Lodge, P.O. Box 6084, Arusha, tel: 6702; fax: 7877.

Selous Game Reserve

Mbuyu Safari Camp, P.O. Box 5350, Dar es Salaam, tel: 31597. Located on the Rufiji River and provides luxurious tents. It was named after the giant baobabs all round.

Stiegler's Gorge Camp, P.O. Box 9320, Dar es Salaam, tel: 48221. Overlooks the gorge, and has charming chalets.

Behobeho Safari Camp, P.O. Box 2261, Dar es Salaam, tel: 68631. Has *bandas* to rent which are situated on hill slopes with lovely views.

Rufiji River Camp, P.O. Box 20058, Dar es Salaam, tel: 63546. Tents perched above the river, great views.

Serengeti National Park

Grumeti River Camp, P.O. Box 40097, Nairobi, tel: 331825; fax: 212656.

Lobo Wildlife Lodge, P.O. Box 3100, Arusha, tel: 3173.

Seronera Wildlife Lodge, P.O. Box 3100, Arusha, tel: 3173.

Tarangire National Park

Oliver's Camp, P.O. Box 425, Arusha, tel/fax: 057 8548/3108.

Tarangire Safari Lodge, P.O. Box 2703, Arusha, tel/fax: 7182.

Campgrounds

Camping is only permitted at authorised sites which you should try to book beforehand to avoid disappointment. Most campsites have minimal or no facilities despite the high price paid for their use, so you should be prepared to dig your own toilet and bury or burn litter and you should travel with sufficient water containers for your basic needs. Most campsites are unprotected, so camping is at your own risk.

Direct questions concerning facilities, campsites, roads, etc. should be addressed to the appropriate park or reserve at its head office:

Tanzania National Parks TANAPA (all parks), P.O. Box 3134, Arusha, Tranzania, tel: Arusha 3181 ext: 1386.

Wildlife Division (all game reserves and controlled areas), P.O. Box 1994, Dar es Salaam, Tanzania, tel: Dar es Salaam 27271.

Ngorongoro Conservation Area Authority (NCAA), P.O. Box 776, Arusha, Tanzania, tel: 3339.

Booking & Fees

Bookings and fees will be handled by the tour operator, but if you organise your own safari you should note these points: an entry fee is payable for every 24-hour period spent in the park, measured from your time of entry. Non-residents must pay this in foreign exchange and it is advisable to carry sufficient cash or travellers' cheques for this purpose. If you are camping, you must pay an additional camping fee per night.

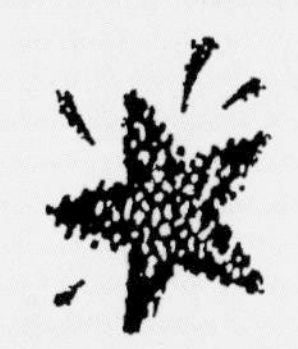

Attractions

Shopping

What to buy

Tanzania is the home of Makonde onde art-carvings in African ebony of intertwined people or openwork "spirit" forms. Other types of carvings, masks, necklaces and beadwork, baskets and batiks, etc., can be found in all curio shops in the larger towns and at villages along the routes to the national parks.

Books of any kind are hard to find so bring field guides. However, Tanzania National Parks sells very useful guides to the parks, available at the park headquarters in Arusha in the Conference Centre.

Most hotels have gift shops with their own selection of goods.

Markets

Notable markets are at Mwanza, Morogoro and Dar es Salaam; the small market at Mto-wa-Mbu near Manyara has a wide variety of curios and a tradition of exchange and barter for clothing, hats, sunglasses, etc. Ask your driver or guide for help if you want to bargain. Locally made k*anga* and *kitenge* cloth also make excellent gifts.

UGANDA

Getting Acquainted

The Land

Area: 91,000 sq miles/237,000 sq km.
Capital: Kampala
Population: 18.7 million
International dialling code: 256 (Kampala: 41)
Time zone: three hours ahead of GMT.
The place. Astride the equator, it is divided into swampy lowlands, a wooded plateau and desert. Lake Victoria, the world's third-largest lake, is on the southern border.

Climate

The climate is extremely enjoyable throughout the year. The hottest months are January and February when temperatures can reach over 30°C (86°F). The relatively coolest months are July and August.

Higher areas in the southwest, west and east, are colder, so bringing a sweater to wear in the evenings is recommended. The highest town is Kabale. The forests and mountains experience a lot of rain so some kind of waterproof protection is advisable.

Most of the country experiences rain throughout the year, especially during the late afternoon and at night. The dry season starts in December and lasts until early March. The longer rains fall from March to May and during early September. Only the northeast part of Karamoja has a comparatively dry climate.

Culture & Customs

Ugandans are generally considered extremely generous, hospitable and polite. Following the British tradition, men are addressed as "Sir" and women as "Madam".

A handshake starts and ends almost every conversation. Women, especially in the southern and western rural areas, often kneel when greeting an elder person. Kissing is almost never done in public.

Food which is offered when visiting someone's house should be eaten. It is considered an insult if it is not.

Ugandans are good talkers and love a friendly chat or argument, so don't be suprised if people greet you on the streets or simply start talking to you.

Government & Economy

Uganda has had eight presidents since the country gained independence in 1962 and has been plagued by tribal and religious differences. Economic and political decline started with the takeover by Idi Amin in 1971. He was ousted by Tanzanian and liberation forces in 1979.

A year later elections were held and officially won by former president Milton Obote. But according to many, they were rigged, and those dissatisfied with the results went into the bush and started a guerilla war. In January 1986 Yoweri Kaguta Museveni, leader of the NRA, brought a coalition government to power. A process of democratisation followed the elections.

"Resistance Committees" play a vital role in this process. These committees are elected by the Ugandan people and eventually are represented in the Ugandan parliament by the National Resistance Council. There have been three constitutions since independence.

Political instability remains a problem, especially in the north and east which have been hit by rebel activities. A peace agreement in 1988 brought 35,000 rebels out of the bush, and security has dramatically improved.

The economic situation is closely related to the political turmoil. A recovery programme supported by IMF and World Bank put great emphasis on restoration of roads, diversification of exports and rehabilitation of the basic industrial sector. Agriculture dominates the economy.

Language

English, the official language, is understood by most people in urban areas. *Luganda* is the most common local language. *Swahili* is widely spoken in the northern part of the country (*see page 338*).

Planning The Trip

What To Wear

Casual wear is most appropriate, although shorts, miniskirts and very open blouses should be avoided in general.

The sun can be very hot so good sun-block cream is essential.

Entry Regulations

Visas & Passports

Valid passports are required for all visitors to Uganda.

Countries exempted from visas include: members of the Economic Union, Commonwealth countries, United Arab Emirates, all PTA countries, Saudi Arabia, Japan, South Korea and the United States of America. Visas are valid for 3 months and are extendable on request. Uganda missions abroad issue visas for a fee. Multiple visas are charged at US$20, single visas US$10 and transit visas at US$5. Foreign journalists wishing to undertake coverage within the country have to give the Ministry of Information 14 days' notice.

Customs

You can bring in one bottle of spirits, 225 grams (half a pound) of tobacco, one bottle of perfume, and cats and dogs if you can prove they have had the necessary vaccinations.

Tax is payable on expensive personal items such as video cameras.

Health

Certificates of inoculation against cholera and yellow fever are necessary. Immigration officials at Entebbe Airport check for these.

Malaria is widespread, so bring prophylactic tablets (chloroquine and paludrine).

Water should only be drunk in the best hotels, otherwise it should be boiled. Typhoid is still common.

Medical facilities in Uganda are under review. The main hospital in Kampala is Mulago Hospital, but Nsambya Hospital is generally considered to be the best.

It is advisable to bring along sterile needles as a precaution against AIDS.

Money

You are allowed to bring in any amount of foreign exchange. The highest denomination in Ugandan currency is the 5000 shilling note; the other notes being 50, 100, 200, 500 and 1000. Foreign exchange bureaux and banks have similar rates and the black market for dollars is virtually non-existent. Traveller's cheques in US$ or pounds sterling are the safest bet.

Public Holidays

1 January	New Years Day
26 January	NRM Government Anniversary Day
8 March	Women's Day
Good Friday	
Easter Monday	
1 May	Labour Day
3 June	Martyr's Day
9 June	Heroes Day
9 October	Independence Day
25 December	Christmas Day
26 December	Boxing Day

Getting There

By Air

Currently Sabena and British Airways are the only European air companies connecting Western Europe with Uganda. Sabena has three flights a week between Brussels and Uganda's Entebbe International Airport some 40 km (25 miles) from Kampala. British Airways has flights from London to Entebbe and vice versa twice a week. Gulf Air, Ethiopian Airlines, Egypt Air also have stopover flights between Entebbe and London.

Another possibility is to go through Kenya by Kenya Airways which has flights from London to Entebbe via Nairobi four times a week. Uganda Airlines has connecting flights for British Airways at Nairobi Airport four times a week. Uganda Airlines and Kenya Airways together have 21 flights a week from Entebbe to Nairobi and 19 flights per week from Nairobi to Entebbe.

Frequency of flights to Entebbe may increase in the near future. Air France intends to start a regular service between Paris and Entebbe.

On leaving the country by plane, every passenger has to pay US$20 airport tax. This can also be paid in local currency.

Airlines

Sabena, Sheraton Hotel Arcade, P.O. Box 3966, Kampala, tel: 259880/234200.

Kenya Airways, Kimathi Avenue, P.O. Box 6969, Kampala, tel: 256506/259472.

Uganda Airlines, Airways House, Colville Street, P.O. Box 5740, Kampala, tel: 232990/241996/241987.

British Airways, 23 Kampala Road, P.O. Box 3464, Kampala, tel: 256695/257414.

Egypt Air, 8/10 Entebbe Road, Kampala, tel: 241276/233960.

Ethiopian Airlines, 1 Kimathi Avenue, Kampala, tel: 254796/7.

Gulf Air, Pioneer Mall, Kampala Road, P.O. Box 7507, Kampala, tel: 230524.

By Road

Uganda can easily be reached by road from Kenya via Busia and Malaba. Access by road is also possible via Rwanda, Zaire and Tanzania. The road to Sudan is temporarily closed because of rebel activities both in northern Uganda and southern Sudan.

Practical Tips

Electricity

Most towns have an irregular power supply of 240 volts. The main hotels use a generator in case of power cuts. Protection for expensive equipment against power surges is advisable and a small flashlight is very useful.

Business Hours

Most shops open at 8.30am and close at 5.30pm. The small shops in the various suburbs remain open until about 11.30pm. On Sundays only these shops are open.

Banks open at 8.30am and close at 2pm Monday to Friday. Main banks are Uganda Commercial Bank, Barclays Bank, Nile Bank, Bank of Baroda, The Cooperative Bank, Standard Chartered Bank, Stanbic Bank of Uganda, The Libyan Arab Bank and the new Greenland Bank located opposite the Bank of Uganda, offering computerised services up to 10pm Monday to Sunday.

Uganda Commercial Bank, Kampala Road, P.O. Box 973, Kampala, tel: 234710/23.

Barclays Bank, Kampala Road, P.O. Box 2971, Kampala, tel: 232597.

Nile Bank, 22 Jinja Road, P.O. Box 2834, Kampala, tel: 231904.

Grindlay's Bank, Kampala Road, P.O. Box 7131, Kampala, tel: 230811/2.

The Post Office opens at 8am and closes at 2pm from Monday to Friday. On Saturday it closes at 12pm.

Tipping

A tip in a restaurant should usually be between 5 percent and 10 percent. Given their very basic salary, waiters are usually very happy to accept a tip. Boys who help you with your shopping expect a 500 to 1000 shilling tip. Taxi drivers usually calculate their tip in the bill. Bargaining is very common in Kampala, in markets, with street-sellers and in most of the informal shops. Actual prices to be paid will often be two-thirds of what is asked. Outside Kampala, and even outside the narrow city centre, people usually ask the right prices.

Religious Services

Most Ugandans are either Protestant, Catholic or Muslim. There are churches and mosques for all these beliefs. Kampala has the central Bahai temple for Africa.

Media

Uganda enjoys considerable freedom of press with a varying number of newspapers and magazines both in English and the local Luganda. The most popular daily is the *New Vision*, a government-owned daily whose editorials manage to maintain some degree of independence and has improved in content and quality over the years. There is also the *Monitor*, an independent thrice-weekly, and the weekly *Uganda Confidential*.

There are three television stations; Uganda Television, Sanyu Television and CNN International. Radio Stations include; Radio Sanyu, Capital Radio and Radio Uganda. Broadcasts are in English and other local languages at different times of the day.

Telephone & Telex

The Central Post Office in Kampala has counters for international telephone calls, fax and telex. The offices for overseas telegrams is in Entebbe Road in Kampala. It also has a fax machine and is open 24 hours daily.

Uganda's telephone communications have been upgraded, and it should now be possible to make direct calls outside East Africa. Otherwise, calls must be made through the operator by dialling 0900 or 0905.

Useful Numbers

990: Telegrams and telegram enquiries
901: Directory enquiries
902: Telephone call enquiries
903: Time announcement in English

It should be noted that many telephones are commonly out of order.

Getting Around

From The Airport

There are "special taxis" from Entebbe Airport to Kampala which are quite expensive at around 30,000 shillings (negotiable). An irregular bus service to town is available and minibuses, also referred to as "taxis", take people to the capital.

Domestic Travel

By Air

Several local companies offer air services to various parts of the country. The most reliable include:

Bel-Air, Jinja Road, Kampala, tel: 257716.
Mission Aviation Fellowship, Kampala, tel: 54186.
Speedbird Aviation Services, Kampala, tel: 231290.

Travel Packages

There are a few companies who organise tourist trips to several parts of the country. These will include camping, either luxury or budget, gorilla and chimpanzee trekking and lodge safaris.

Africa Expeditions Ltd, P.O. Box 24598, Nairobi, tel/fax: 561959/561882/561457/561054.
African Pearl Safaris, Embassy House, Kampala, tel: 233566/7; fax: 235770.
Blacklines Tours & Safaris, P.O. Box 6968, Kampala, tel/fax: 255288/255267.
Hot Ice Ltd, P.O. Box 151, Kampala, Uganda, tel: 236777/267441; fax: 242733.
Rafiki Africa, P.O. Box 76400, Nairobi, tel/fax: 561562.

Public Transport

Kampala has just re-established its bus services and has a "taxi park" where private mini-buses follow regular bus routes.

Services from Kampala to most other major towns exist and buses leave from the area close to the "taxi park".

"Special taxis" can be found in town, especially near the Nakasero Market and at the main hotels. There are no particular signs for these taxis and the price is negotiable.

Private Transport

Hertz car hire services operate at the airport. But it should be noted that hiring a car in Uganda is very expensive, never less than US$100 per day. To reach most parts of the country you will need a four-wheel drive vehicle. Please note that there are no breakdown services outside Kampala. If your car breaks down you will have to stop another passing car and go to the nearest village or town to get help. Petrol may be a problem on long trips. Always carry spare jerrycans.

Hertz car hire services, Spear Touring Safaris Co. Ltd., Spear House, 22 Jinja Road, P.O. Box 5914, Kampala, tel: 59950 or 32395 (Airport tel: 26518).

Where To Stay

Hotels

Kampala has a number of good albeit expensive hotels. The following hotels are considered the best with self-contained accommodation and televisions in all rooms:

Nile Hotel Complex, tel: 235904.
Kampala Sheraton Hotel, tel: 244590.
Hotel Equatorial, tel: 250780.

Other hotels include the Fairway Hotel, Hotel Diplomate, Silver Springs, Speke Hotel which is slightly cheaper, Reste Corner Hotel, Lion Hotel and Summer Hotel. The best restaurants with continental food can be found in the Nile Hotel Complex, Kampala Sheraton Hotel, Hotel Equatorial and the Fairway Hotel.

There are many other places to stay in Kampala which can satisfy even the lowest budgets. Up country now, most towns have at least one hotel offering accommodation and good food. Service can be a little slow and electricity and water may be a problem, but comfortable hotels may be found in:

Lake Victoria Hotel, P.O. Box 15, Entebbe, tel: 042-20891.
Sunset Hotel, P.O. Box 156, Jinja, tel: 043-20115.
Crested Crane Hotel, P.O. Box 444, Jinja, tel: 043-21513/4/5.
White Horse Inn, P.O. Box 11, Kabale, tel: 0486-22020.
Masindi Hotel, P.O. Box 11, Masindi.
Mountains of the Moon, P.O. Box 38, Fort Portal, tel: 0493-2513.
Marquerita Hotel, P.O. Box 90, Kasese, tel: 0493-4013.
Mount Elgon Hotel, P.O. Box 670, Mbale, tel: 045-3611/2.
Lake View Hotel, P.O. Box 165, Mbarara, tel: 0485-21392.
Mweya Safari Lodge, P.O. Box 22, Kasese, Queen Elizabeth National Park, tel: 0493-4266.

African Pearl Safaris Homestead, Kampala, Bwindi National Park, tel: 233566/7.

Eating Out

What To Eat & Drink

The local food is matoke (mashed green banana) with ground nut sauce, or *luwumbo* (chicken boiled in banana leaf). The local alcoholic beverage is *Uganda waragi*, a distilled spirit not unlike gin. Several locally distilled, much stronger, variations exist in the rural areas. Uganda has three readily available beers that can be differentiated by blue, black or white caps. *Creps* is the locally produced soft drink, made out of pineapples.

Where To Eat

The big hotels offer international food, mainly chips with fish or steak. To sample Ugandan food, go to the outdoor **UNICFE** restaurant, or to **Campus View** in Wandeguya, not too far from Makerere University.

Another restaurant well worth trying is **Nile Grill**, Kampala Road, P.O. Box 1295, Kampala, tel: 233522.

Attractions

Culture

Kampala has one **National Museum** on Kitante Road, with exhibits explaining the history and culture of the country. There is also a music gallery with a display of traditional instruments.

The capital has two art galleries: **Nummo Gallery** is just behind the Sheraton Hotel and exhibits contemporary art by Ugandan artists. At Makerere University there is a small art gallery near the Faculty of Fine Arts, which is well worth a visit.

The **National Theatre** in the city centre has regular plays, but usually in Luganda. It also has musical events and a cinema. Big banners in town announce the programme.

National Theatre, Siad Barre Avenue, P.O. Box 3187, Kampala, tel: 254567.

The **Alliance Française** housed in the National Theatre building offers French films with English sub-titles every Wednesday. Other cinemas no longer exist; they have been transformed into video parlours.

Nightlife

One sign of improved security is the abundance of discos and nightclubs which can be found all over the city. Pulsations and Silk Club are very popular as is the Ange Noir. The Half London Pub and the Telex Bar have live entertainment and are popular with expatriates and locals alike.

Shopping

The UNICFE shop on Bombo Road, a few kilometres outside the city centre, has an excellent variety of Ugandan crafts. The country is famous for its batiks, mats and baskets.

A string of shops in the city centre close to the Imperial Hotel (currently under renovation) sells a similar range of souvenirs and crafts. Here prices are negotiable.

Sports

Soccer is the most popular sport, played every Saturday at 4pm in the Nakivubo Stadium. Lugogo Stadium on Jinja Road has seven tennis courts, open to everybody. The American Club in Makindye has another two, but here you need an introduction from a member. The Club also has a clean swimming pool. Sheraton Hotel has a lovely swimming pool for hotel guests only. A well-kept golf course can be found along Natete Road, just outside the city centre.

ETHIOPIA

Getting Acquainted

The Land

Area: 395,000 sq miles/1,023,000 sq km.
Capial: Addis Ababa.
Population: 53 million
International Dialling Code: 251 (Addis Ababa: 1)
Time zone: three hours ahead of GMT. The popular advertisment for the country boasts about "the land with 13 months of sunshine!" This follows from the continued use of the Julian calendar. There are 12 equal length months of 30 days each, and one month of five days – six in a Leap Year. New Year's Day is on September 11 (September 12 in a Leap Year).
The place: On the Red Sea, Ethiopia has a central plateau wiith mountains rising to 15,000 feet (4,000 metres). The main river is the Blue Nile

Climate

Ethiopia has a varying climate depending upon the altitude and orientation of any location. Generally speaking, there are two wet periods and two dry periods, but in many higher locations the wet periods run into each other, giving one long wet and one dry season. The months of October to March are usually dry and warm, but with very cold nights at high altitude. From May to June it may be dry. Wet periods are characterised by warmer nights and cooler days.

Culture & Customs

Ancient cultures that have evolved in isolation in the vast Ethiopian highlands have left their stamp on the countryside, especially in the form of buildings unique to the continent. The monuments at Lalibela, Gondar, Harer, and Axum are some of the most famous. Good hotels exist near most of these monuments.

Government & Economy

In May 1993, after three ruinous decades of civil war, Eritrea, with its capital in Asmera in the north, became a separate sovereign state.

Language

The official language is Amharic, which has a beautiful ancient script. English is widely understood, as is Italian.

Planning The Trip

What To Wear

Respectable western attire is acceptable. Be prepared for all climatic situations if you intend to travel widely. High altitudes are cool but at lower altitudes you may wish to wear shorts. Be prepared for cooler nights. If you intend to go into the high mountains (Bale and Semien) you will need warm, wind and waterproof clothing.

Entry Regulations

Visas & Passports

All nationalities require a visa to enter Ethiopia, except Kenyan citizens. Tourist visas valid for up to 30 days can be obtained from Ethiopian embassies abroad. They can also be obtained on arrival at Bole airport, provided your airline allows you to fly without a visa. Tourist visas can be extended in Addis. Transit visas are also available for brief stopovers, your passport being held at the airport in exchange for a transit visa card. A two-year multiple entry visa is available at US$70.

Eritrea: Visas can be obtained from Ethiopian embassies abroad, or at the airport on production of a letter of introduction from your local embassy.

Customs

Duty-free allowances include 200 cigarettes, 50 cigars or a pound of tobacco; one litre of alcoholic beverage; one litre or two bottles of perfume.

Special permission is required to import and use *professional* video cameras. Handycams and other video cameras normally used by tourists are permitted. The serial numbers of these video cameras are recorded on the currency form. Special import permission is no longer required for binoculars and they are usually not recorded on the currency form.

If you buy some of the old Ethiopian jewellery or artifacts available in most souvenir shops in Addis Ababa and other cities, it is wise to have these approved for export by the National Museums of Ethiopia, either by the shop owner or by yourself. This avoids the possibility of customs officials confiscating some of your souvenirs as "protected antiquities". No need to worry about obviously new items.

Health

A valid inoculation certificate against yellow fever is required upon entry to Ethiopia. A cholera inoculation certificate is needed if you have come within six days from an area where the disease is prevalent.

Diseases are no more common in Ethiopia than in other parts of Africa and, provided simple precautions are taken, you should have no major problems. Some people will find the spicy local food strong for their digestions at first. Malaria occurs in some areas of the country and prophylactic medication should be taken. Medicines are readily available and well-stocked pharmacies are a feature of most towns. Addis has a good selection of hospitals which can cope with most situations and problems, but, to avoid complications, you are advised to bring your own stock of prescription medicines and sterile needles.

Money

The official exchange rate is fixed to the US dollar. It is advisable to change your currency in Addis before proceeding on your travels within the country. Tourists are advised to change their money at banks rather than at the hotels because of the exchange rate.

You are only allowed to take Et birr 10 out of the country. Amounts in excess of this, but less than Et birr 100 may be changed back at the airport before departure, provided you can prove from your currency form that you imported it as foreign currency when you arrived, and can show exchange receipts. Amounts in excess of Et birr 100 must be changed back at the main bank in Addis. American Express cards are acceptable at a few large government hotels.

Getting There

By Air

Addis Ababa is sometimes known as the "diplomatic capital of Africa" due to the presence of both the Organisation of African Unity (OAU) and the UN Economic Commission for Africa (ECA). As a result, the city is well-served by international air routes, particularly by the national carrier Ethiopian Airlines (EAL), deservedly known as the best airline on the African continent. Bole International Airport is only six kilometres (four miles) from Addis city centre, and receives flights direct from Europe (including London, Frankfurt, Rome, Moscow, Berlin, Athens), Asia (including Jeddah, Abu Dhabi, Bombay, Peking) and many African cities (including Nairobi, Entebbe, Dar es Salaam, Djibouti, Lusaka, Harare, Khartoum, Cairo and many West African cities). Additional airlines serving Addis Ababa include Kenya Airways, Lufthansa, Aeroflot, Alitalia, Air Djibouti, Yemenia.

By Road

It is possible to drive to Ethiopia from Kenya but the road is not good and there have been reports of bandit attacks. This method of travelling to Ethiopia should therefore be avoided.

Useful Addresses

Government Agencies:

Ethiopian Tourism Commission (ETC), P.O. Box 2183, Addis Ababa, Ethiopia, tel: 15-98-79; 44-74-70; telex: 21067 ETC. In overall charge of tourism, planning, promotion, and the production of most promotional material.

National Tour Organisation (NTO), P.O. Box 5709, Addis Ababa, Ethiopia, tel: 15-29-55; 15-91-86; telex: 21370. The national travel agent, concerned with the operation of tourism, vehicles, guides, itineraries, permits etc.

Ethiopian Tourism & Trading Corporation (ETTC), P.O. Box 5640, Addis Ababa, Ethiopia, tel: 18-06-41; telex:

21411. Concerned with production of some brochures, booklets, postcards, and souvenir materials and management of duty-free shops.

Ethiopian Hotels Corporation (EHC), P.O. Box 1263, Addis Ababa, Ethiopia, tel: 15-27-00; telex: 21067. Manage and operate all government hotels where tourists stay, including Kerayu Lodge in Awash National Park.

Ethiopian Wildlife Conservation Organisation (EWCO), P.O. Box 386, Addis Ababa, Ethiopia, tel: 44-59-70; 15-44-36; telegram: WILDGAME. Management of all Conservation Areas (National Parks, Sanctuaries, Wildlife Reserves, Controlled Hunting Areas); control of hunting; control of wildlife utilisation.

Ethiopian Airlines (EAL), Main City Ticket Office, Churchill Road, P.O. Box 1755, Addis Ababa, tel: 517000/512222. Overseas offices, 4th Floor, Foxglove House, 166 Piccadily, London W1V 9DE, tel: 071-4919119.

Practical Tips

Electricity

The mains supply is 220 volts and 50 cycles AC. Rural hotels and lodges may have their own generators, usually of the same voltage as the mains.

Media

International communications include a good postal system, telephone and telex. Internally there is a widespread postal system, a telephone system which is improving daily as micro-wave links are constructed, and some radio communication.

There are two daily newspapers in English, *The Ethiopian Herald* and *The Monitor* in addition to daily Amharic papers.

Getting Around

By Air

Ethiopian Airlines' domestic service reaches over 40 destinations within Ethiopia, served with Twin Otter, ATR 42, Boeing 737 and Boeing 727 aircraft. There is also a charter branch of the airline called Admas which goes to off-route destinations.

Taxis

The NTO locate their cream-coloured Mercedes taxis at all major hotels in the capital, and at Bole International Airport. Fares are fixed and receipts given. In addition, NTO is able to provide chauffeur-driven cars for destinations away from Addis. These vary from saloon cars to comfortable Toyota Landcruisers, and 60-seater air-conditioned Mercedes buses, capable of reaching surprisingly remote and rugged parts of the country. NTO also has a few self-drive cars.

On Foot & Horseback

One of the joys of Ethiopia is to travel on foot or horseback in the mountainous areas along routes still rarely travelled and uncrowded, where you could not take a vehicle if you wanted to. Horses and mules are still an important form of transport in rural areas and can usually be hired by the day for an agreed price.

On Departure

There is now a departure tax of US$10. Check-in is two hours before for international flights. It is very important to confirm your onward flight at least three days before departure.

Where To Stay

Hotels

You are no longer obliged to stay in one of the Government hotels in Addis Ababa, although they are still some of the better hotels in town. There is also the Airport Hotel near the international airport, and several other small, newly built hotels that offer reasonable accommodation.

Eating Out

Where To Eat

Where you eat depends upon your constitution and sense of adventure and discovery. Addis can offer many cuisines, including European, Italian, Chinese, Armenian and Indian, but you will do the country an injustice if you don't try its own dishes. There are infinite possibilities, but most hotels will offer national food of some variety. Special national food restaurants exist at the Ghion and National Hotel, and at the Filwoha Hotel, where you eat in an attractive old Ethiopian building. Further detailed information is available in the excellent guide to Addis, produced by the Ethiopian Tourist Trading Corporation.

What To Eat

Ethiopia offers a cuisine unlike any other in Africa and you either like it or you don't! Most do. An endemic grass (*tef*) provides the staple carbohydrate source for most of the country. From it a large thin circular pancake (*injera)* is made, which is the vehicle for all foods. A large variety of sauces and stews, generically known as *wat*, is available. These vary from mild (*alichas*) from the north to fiery red (*ky wat*) made with vegetables (*atkilt*), meat (*siga*), pulses (*shiro*), chicken (*doro*), eggs (*inkulal*), liver fillet and tripe (*dulet*), and so on. *Wat* is traditionally placed on the *injera*, which is laid out flat on a circular, colourful, woven basketware table tray and cover combined (*massob*). You then proceed to tear off pieces of the *injera* from the edges, wrap it around a morsel from the centre, and deftly pop it into your mouth with one hand. At the end of the meal, your hostess will, on occasion, scoop up tasty morsels (*gursha*) and put them into your mouth to ensure you really are satisfied and her cuisine has been appreciated!

Fear not – even in remote areas western cuisine is available if the above description should worry you! So is Italian cuisine, especially various pastas, cooked with Ethiopian herbs and spices.

Drinking Notes

Ethiopia produces a remarkable array of drinks, including standard western "fizzies", four varieties of bottled beer, a wonderful mineral water named after its source *Ambo*, over 10 varieties of wine, a local champagne, brandy, ouzo and gin, and local *arakis* – clear liqueurs that are not sweet and sticky, made from coffee, milk, honey, plum, orange, lemon, and even from the *kosso* tree (*Hagenia abyssinica*) which is prized as a cure for tapeworm for those who overindulge in raw meat! A favourite!

Attractions

Tours

Most people visit Ethiopia for its rich culture. The best time to visit is during one of the religious festivals. Following is the list:

January 7	Ethiopian Christmas
January 19	Epiphany
March 2	Id Alfater/Ramadan
April 21	Ethiopian Good Friday
April 23	Ethiopian Easter
May 9	Id Al Adaha/Arafat
August 8	Birthday of the Prophet Mohammed
September 11	New Year
September 27	Meskel

Private Tour Operators

Africa Expeditions, P.O. Box 24598, Nairobi, Kenya, tel/fax: 561882/561959/561054/561457.

Universal Travel Agency, P.O. Box 821556, Addis Ababa, tel: 204824/204889/710058; fax: 710240.

Safeway Travel & Tours, P.O. Box 8449, Addis Ababa, tel: 511187.

Forship Travel Agency, P.O. Box 39754, Addis Ababa, tel: 552159/551493; fax: 553300.

Galaxy Express Services, P.O. Box 8309, Addis Ababa, tel: 510355; fax: 511236.

Wildlife Viewing

There are many endemic species in Ethiopia. Most remaining wildlife now exists in national parks and sanctuaries.

The tourist infrastructure is still very basic: you may have to work hard to see these animals but it makes the experience all the more rewarding. Bird trips are proving extremely popular as ornithologists seek to add Ethiopia's 27 endemic bird species to their life lists.

Trekking

The mountainous areas of Ethiopia are what make the country so very different. The abysses dictate that roads pass along ridge tops for the most part, so to see the spectacular scenery you must walk, or use horses and mules. Organising such trips in the Bale Mountains is relatively easy, but there are many other locations, where even a few hours walking or on horseback can prove very rewarding. Such places include the Kessem River gorge, the Chacha River gorge, the Arssi Mountains, Kuni Mountain, Mulu Valley, Debre Libanos and a host of others. Trekking on foot can also be very tough but rewarding in the arid lowlands.

Whitewater rafting trips of two weeks' duration are regularly run on the lower reaches of the Omo River. Whitewater rafting trips of shorter duration (1–2 days) are also sometimes available on the Awash River. Contact Africa Expeditions Ltd, P.O. Box 24598, Nairobi, tel/fax: 561959/561882/561457.

Shopping

What To Buy

The highlands of Ethiopia have influenced the art and clothing. Examples of both are readily available to purchase, and make excellent and, in many cases, practical presents. Choose from beautiful white cotton *shamma* ware, in the form of shawls, bags, stoles, and blankets; ancient (and more recent) silverware, as beads, famous Ethiopian crosses, bracelets and earrings; more expensive and usually modern gold articles; modern paintings on parchment; old church relics; parchment books; colourful Harer basketry; Ethiopian herbs and spices; delicious Ethiopian coffee or horse artifacts such as the colourful saddle cloths.

Where To Buy

All these and much more are available, in formal tourist shops near the main Post Office on Churchill Road, near the National Theatre, in hotel curio shops and, of course, at the famed *Mercato* – reputed to be Africa's biggest market. Country markets, often held on a Wednesday, also yield a host of "real" artifacts, practical and fascinating items of everyday use to their makers and purchasers alike.

If you buy some of the old Ethiopian jewellery or artifacts available in most souvenir shops in Addis Ababa and other cities, it is wise to have these approved for export by the National Museums of Ethiopia, either by the shop owner or by yourself. This avoids the possibility of customs officials confiscating some of your souvenirs as "protected antiquities". No need to worry about obviously new items.

Further Reading

Other Insight Guides

There are nearly 200 titles in Apa Publications' Insight Guides series, covering continents, countries, regions and cities all over the world.

Titles on East, West and Southern Africa include *Gambia and Senegal, South Africa* and *Namibia.* In North Africa, the series covers *Morocco, Tunisia, Egypt, Cairo* and *The Nile.*

Insight Guide: Indian Wildlife is another richly photographed and information-packed guide to one of the world's great nature regions. Other titles include *Amazon Wildlife, The Great Barrier Reef* and *US National Parks East* and *US National Parks West.*

Insight Guide: Kenya is an up-to-date guide to this dramatic country, giving you all the background of life there today and showing you how to make the most of a visit.

Insight Pocket Guides

Also available are Insight Pocket Guides, which have short, timed itineraries and a handy pull out map. Titles include *Insight Pocket Guide: Kenya, Morocco, Tunisia* and *Seychelles.*

Morocco, the threshold of Africa, is a heady mix and *Insight Pocket Guide: Morocco* captures all aspects. From the cities of Fez, Marrakesh and Tangier to snow-capped mountain ranges.

Insight Pocket Guide: Seychelles takes the visitor on a journey through deserted islands and on fascinating hikes through the land caressed by the Indian Ocean.

Art/Photo Credits

Photography by

Iain Allan 174, 175, 176, 177, 178, 179, 180, 181, 259, 297
Karl Ammann 2,/14/15, 20/21, 22, 43, 48/49, 54, 60, 62, 64, 66/67, 69, 70/71, 72, 73, 74/75, 76/77, 78, 79, 80, 82, 83, 84, 85, 86, 87, 88, 89, 90, 92, 93, 96, 97, 98, 100, 101, 102, 103, 104, 105, 108, 109, 110, 111, 114/115, 119, 123, 124, 125, 127, 130, 142/143, 149, 155, 163, 188/189, 200, 201, 205, 209, 211, 212, 215, 222, 223, 226, 227, 228, 229, 230, 233, 235, 239, 248, 249, 251, 252, 253, 255, 261R, 262, 268, 270, 273, 278, 279, 280, 282, 290, 292L, 293R, 295, 296, 320, 321, 324, 326R, 327, 328
James Ashe 120, 121
Alan Binks 264, 265
David Breed 166
Bill Campbell 151, 152
David Coulson 225, 234
Courtesy of Robin Hurt Safaris 167
Courtesy of Ulf Aschan 30, 31, 148
Peter Davey106/107, 112, 113, 129, 131, 158, 160, 162, 165, 214, 241, 244/245, 258L, 260, 267, 271, 272, 274, 276, 277, 284/285, 291
Fred de Vries 299
David Keith Jones 9, 16/17, 18/19, 25, 26, 36, 37, 38, 39, 41, 42, 46, 47, 50/51, 56/57, 58, 59, 61, 65, 68, 81, 94, 95, 98, 99, 116/117, 122, 126, 128, 144/145, 146/147, 153, 156, 157, 164, 168, 170/171, 172, 182/183, 184/185, 186/187, 194, 196, 197, 198, 202, 203, 204, 207, 208, 213, 216, 218, 219, 220, 224, 236, 237, 238, 240, 243, 250, 257, 258R, 263, 269, 275, 281, 283, 288, 314/315, 319, 322, 325, 326L
Chris Hillman 45, 91, 118, 161, 300/301, 305, 306, 308, 309, 310, 311, 312, 313
Nicky Martin 44, 169, 231
Mary Evans Picture Library 28, 32, 33
The Kobal Collection 132/133, 134, 135, 136, 137, 138, 139, 140, 141
Topham Picture Library 29, 34/35, 52/53, 150
Daniel Stiles 24, 27

Maps Berndtson & Berndtson

Visual Consultant V. Barl

Index

A

B

M

Z

A B C D E F G I J a b c d e f g h i j k

The Insight Approach

The book you are holding is part of the world's largest range of guidebooks. Its purpose is to help you have the most valuable travel experience possible, and we try to achieve this by providing not only information about countries, regions and cities but also genuine insight into their history, culture, institutions and people.

Since the first Insight Guide – to Bali – was published in 1970, the series has been dedicated to the proposition that, with insight into a country's people and culture, visitors can both enhance their own experience and be accepted more easily by their hosts. Now, in a world where ethnic hostilities and nationalist conflicts are all too common, such attempts to increase understanding between peoples are more important than ever.

Insight Guides:
Essentials for understanding

Because a nation's past holds the key to its present, each Insight Guide kicks off with lively history chapters. These are followed by magazine-style essays on culture and daily life. This essential background information gives readers the necessary context for using the main Places section, with its comprehensive run-down on things worth seeing and doing. Finally, a listings section contains all the information you'll need on travel, hotels, restaurants and opening times.

As far as possible, we rely on local writers and specialists to ensure that the information is authoritative. The pictures, for which Insight Guides have become so celebrated, are just as important. Our photojournalistic approach aims not only to illustrate a destination but also to communicate visually and directly to readers life as it is lived by the locals.

Compact Guides
The "great little guides"

As invaluable as such background information is, it isn't always fun to carry an Insight Guide through a crowded souk or up a church tower. Could we, readers asked, distil the key reference material into a slim volume for on-the-spot use?

Our response was to design Compact Guides as an entirely new series, with original text carefully cross-referenced to detailed maps and more than 200 photographs. In essence, they're miniature encyclopedias, concise and comprehensive, displaying reliable and up-to-date information in an accessible way.

Pocket Guides:
A local host in book form

However wide-ranging the information in a book, human beings still value the personal touch. Our editors are often asked the same questions. Where do *you* go to eat? What do *you* think is the best beach? What would you recommend if I have only three days? We invited our local correspondents to act as "substitute hosts" by revealing their preferred walks and trips, listing the restaurants they go to and structuring a visit into a series of timed itineraries.

The result is our Pocket Guides, complete with full-size fold-out maps. These 100-plus titles help readers plan a trip precisely, particularly if their time is short.

Exploring with Insight:
A valuable travel experience

In conjunction with co-publishers all over the world, we print in up to 10 languages, from German to Chinese, from Danish to Russian. But our aim remains simple: to enhance your travel experience by combining our expertise in guidebook publishing with the on-the-spot knowledge of our correspondents.